10 0183866 2
D0547986
WITHDRAWN
FROM THE LIBRARY

Molecular Evolution

A Phylogenetic Approach

Roderic D.M. Page
University of Glasgow

Edward C. Holmes
University of Oxford

b
Blackwell
Science

Editorial Offices:
Osney Mead, Oxford OX2 0EL
25 John Street, London WC1N 2BL
23 Ainslie Place, Edinburgh EH3 6AJ
350 Main Street, Malden
MA 02148 5018, USA
54 University Street, Carlton
Victoria 3053, Australia
10, rue Casimir Delavigne
75006 Paris, France

Other Editorial Offices:
Blackwell Wissenschafts-Verlag GmbH
Kurfürstendamm 57
10707 Berlin, Germany

Blackwell Science KK
MG Kodenmacho Building
7–10 Kodenmacho Nihombashi
Chuo-ku, Tokyo 104, Japan

First published 1998
Reprinted 1999

Set by Setrite Typesetters Ltd, Hong Kong
Printed and bound in the United Kingdom at the University Press, Cambridge

For further information on Blackwell Science, visit our website:
www.blackwell-science.com

DISTRIBUTORS

Marston Book Services Ltd
PO Box 269
Abingdon, Oxon OX14 4YN
(*Orders*: Tel: 01235 465500
Fax: 01235 465555)

USA
Blackwell Science, Inc.
Commerce Place
350 Main Street
Malden, MA 02148 5018
(*Orders*: Tel: 800 759 6102
781 388 8250
Fax: 781 388 8255)

Canada
Login Brothers Book Company
324 Saulteaux Cresent
Winnipeg, Manitoba R3J 3T2
(*Orders*: Tel: 204 837-2987)

Australia
Blackwell Science Pty Ltd
54 University Street
Carlton, Victoria 3053
(*Orders*: Tel: 3 9347 0300
Fax: 3 9347 5001)

A catalogue record for this title is available from the British Library

ISBN 0-86542-889-1

Library of Congress
Cataloging-in-publication Data

Page, Roderic D.M.
Molecular evolution: a phylogenetic approach/Roderic D.M. Page, Edward C. Holmes.
p. cm.
Includes bibliographical references and index.
ISBN 0-86542-889-1
1. Molecular evolution.
2. Evolutionary genetics.
I. Holmes, Edward C. II. Title.
QH390. P34 1998
572.8′38—dc21 98-4696
CIP

Contents

Acknowledgements

We thank Simon Rallison for commissioning the book, and for negotiating the contract at the same time that R.D.M.P.'s wife, Antje, was working in Blackwell's royalties department. This, of course, was merely a fortuitous coincidence. In any event, Antje's gentle prodding helped speed the completion of one half of the book. Ian Sherman shepherded the book to its conclusion with great patience in the face of our ludicrously optimistic assessments of when we would be finished. Several anonymous referees provided very helpful comments while Tim Anderson, John Brookfield, Mike Charleston, Nick Grassly, Rosalind Harding, Peter Holland, Mark Ridley and Vince Smith read chunks of the manuscript (some very large) for which we are extremely grateful. Jake Baum's input into part of Chapter 4 was also much appreciated. Finally we thank Paul Harvey, Wyl Lewis, Mark Ridley and Rachel Urwin for encouragement and inspiration.

Roderic D.M. Page
Glasgow

Edward C. Holmes
Oxford

Chapter 1
The Archaeology of the Genome

1.1 The nature of molecular evolution

Although a sometimes unpleasant occupant of our respiratory tracts, *Haemophilus influenzae*, a small Gram-negative bacterium, was an unlikely candidate to symbolise a revolution in molecular biology. But this is exactly what happened in July 1995 when the entire 1 830 137 DNA base pairs of its genome was published—the first of a free-living organism. A new era in biological science had begun. Soon after *Haemophilus influenzae* came the first complete genome from a eukaryote—that of the yeast *Saccharomyces cerevisiae*, followed by *Methanococcus jannaschii*, the first representative of the third domain of cellular life, the Archaea. In the next few years molecular biology will claim its biggest prize—the 3.3 billion bases that make up the genome of *Homo sapiens*.

DNA sequences are valuable because they provide the most detailed anatomy possible for any organism—the instructions for how each working part should be assembled and operate. Much of modern biology now relies on unravelling the information stored within gene sequences and this is true of evolutionary studies, where gene sequences are now recognised as an invaluable document of the history of life on earth. It is the aim of this book to show what evolutionary information is written into gene sequences and how this information might be recovered. This is the science of molecular evolution.

Take, for example, *Haemophilus influenzae*, *Saccharomyces cerevisiae* and *Methanococcus jannaschii*. Until recently, most textbooks divided cellular organisms into the eukaryotes, which possess a cell nucleus, and the prokaryotes, which do not. This tidy world was upturned in the 1970s when Carl Woese and colleagues, using the highly conserved 16S ribosomal RNA (rRNA) gene, showed that there were in fact *two* very different groups of prokaryotes—the Eubacteria like *Haemophilus influenzae*, now simply referred to as the Bacteria, and the Archaebacteria whose members include *Methanococcus jannaschii*, now known as the Archaea (Fig. 1.1). Until the rise of molecular biology in the 1970s, this third great branch of life was lost from us but with molecular phylogenies we have learned that the Archaea are in fact probably more

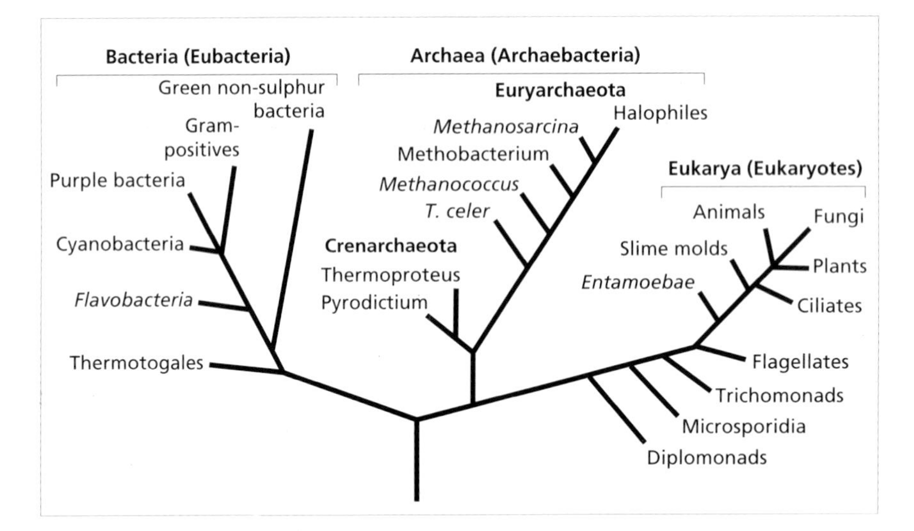

Fig. 1.1 Phylogenetic relationships between members of the three domains of cellular life—the Archaea, Bacteria and Eukarya, based on rRNA. The Archaea can be further divided into the Crenarchaeota (also known as the eocytes) and the Euryarchaeota (methanogens and halophiles). Although popular, this tree is by no means universally accepted. For example, there is still some debate as to whether the Crenarchaeota are in fact more closely related to the eukaryotes than they are to the other Archaea. From Morell (1996) and originally Olsen and Woese (1993), with permission.

closely related to the eukaryotes than they are to the Bacteria, even though they lack a cell nucleus and represent some of the most extreme forms of life on earth. *Methanococcus jannaschii*, for example, lives on deep-sea hydrothermal chimneys ('white smokers'), at pressures of 200 atmospheres and temperatures of 85°C! Gene sequences clearly contain a unique and important archaeological record of life's first tentative steps.

The importance of 16S rRNA for those interested in the early evolution of life lies in its slow evolution, which allows the historical record preserved in its gene sequences to be kept relatively intact. Other genes evolve a good deal more rapidly and so allow us to reconstruct historical pathways that have been trodden only recently. One such example of evolution in the fast lane are the genes which make up the human immunodeficiency virus (HIV), the cause of the disease AIDS. HIV evolves about a million times faster than human genes, which is why developing effective drugs and vaccines is such a problem. This super-fast rate of evolutionary change also means that sequences from this virus can be used to retrace its spread through populations. Studies of this sort have had some dramatic results. For example, in 1990 the Centers for Disease Control (CDC) in Atlanta received reports of AIDS in a young woman in Florida whose only risk of HIV infection was seemingly that she had previously been treated by a dentist suffering from AIDS. A subsequent investigation then uncovered a number of the dentist's other former patients who were also HIV infected. Could it be that these people were somehow infected by their dentist? A phylogenetic tree reconstructed on part of the envelope (*env*) gene of the virus revealed that those patients with no other risk factors for HIV infection had sequences closely related to those of the dentist, strongly suggesting that he had infected them, whilst the sequences from two patients who could have been infected in other ways were separated from the dentist on the tree (Fig. 1.2). The HIV genome had therefore stored evolutionary information, in the form of the mutations which had accumulated between transmission events, which could recount the very recent history of its spread.

Despite the attention given to it today, the ability to sequence entire genomes is just the latest in a series of milestones which mark the development of molecular evolution. The roots of this science were laid early in this century by George Nuttall, a Cambridge biologist, who (along with a contemporary, Uhlenhuth, working in Germany) mixed sera and antisera from different species in an attempt to discover the 'blood relationship' between them. The

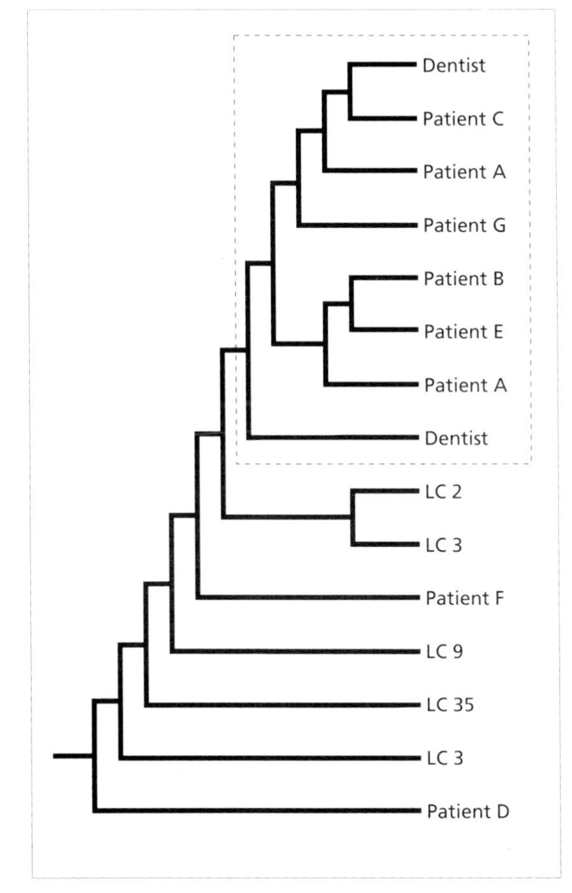

Fig. 1.2 The case of the Florida dentist. Each branch represents the sequence from part of the envelope (*env*) gene of HIV-1. Viral sequences were obtained from the dentist and seven of his former patients (labelled A to G), also infected with the virus. Five of these patients (A, B, C, E and G), have sequences very closely related to those of the dentist (boxed), suggesting that he infected them. Two of his other former patients (D and F) had other risk factors for HIV infection and their viruses are separated from the dentist by sequences taken from local controls (LC)—HIV-infected individuals living within a 90-mile radius of the dentist's surgery. Because HIV-1 is so variable, two different sequences are included for the dentist and patient A. Data taken from Ou *et al.* (1992).

idea was that the more closely related the species, the stronger the cross-reaction between sera and antisera. Today we know that this depends on the extent of genetic similarity between them. While Nuttall's techniques appear crude by today's standards, his work establishes the most important principle in molecular evolution—that the degree of similarity between genes reflects the strength of the evolutionary relationship between them.

In the 50 years that followed Nuttall's work, studies of evolution at the molecular level made little progress, largely because there was a paucity of data to work with. This sits in stark contrast to the revolution which took place in the theoretical wing of evolutionary biology resulting in the 'neo-Darwinian synthesis' of the 1930s. This synthesis, the basis of modern evolutionary thought, also forms a backdrop to many of the later debates that enveloped molecular data when it first became available. In this book we hope to describe an even more modern synthesis—that of molecular biology with phylogenetics—which enables us to explain the amazing diversity of genomes.

The 1950s witnessed a blossoming of molecular evolution. Two events were primarily responsible for this renaissance. The first was the discovery, by James Watson and Francis Crick in 1953, of the molecular structure of DNA—the double helix. This proved to be the key piece in the jigsaw that revealed DNA as the molecule responsible for carrying between generations the instructions for how organisms should be assembled and work correctly. At its most fundamental level, evolution can be thought of as changes in the structure of DNA. The second event took place in 1955 when Fred Sanger and colleagues, also working in Cambridge, published the first comparison of amino acid sequences (the product of DNA) from different species, in this case of the protein insulin from cattle, pigs and sheep. Although less famous than the breakthrough of Crick and Watson, this was the first study to reveal how species differed at the molecular level: cattle, pigs and sheep had three amino acid differences, indicating that their insulins had evolved along with their more obvious anatomical features.

By the early 1960s, the amino acid sequences of a variety of proteins had been determined. The next task was to accurately recover evolutionary information from these sequences. This required a simple mathematical description of the process of gene sequence change over time. In other words, it was necessary to build a *model* of molecular evolution. Models make it easier to reconstruct past events and make predictions about future changes, and are an important part of molecular evolution. Although the first models were developed almost 40 years ago, many of their basic elements are still relevant today and will be encountered many times in this book. The most basic was the assumption that evolution at the molecular level was a largely stochastic process, dominated by chance events. It was also realised that the number of sequence changes we observe between genes might not be the same as the number that has actually taken place: because there are only four DNA bases and 20 amino acids there is a chance that past mutations will be masked by those which have occurred more recently. As we shall see, recovering the true number of gene sequence changes which have accumulated over time is one of the central tasks of molecular evolution.

A final and highly controversial idea that came out of early models of molecular evolution was that genes might pick up mutations at fairly regular intervals, so that there is a 'molecular clock' of evolution. Although this is still hotly debated, it is one of the most important properties of gene sequence data as it enables us to place evolutionary history within a timeframe. If genes make good timekeepers, our view of the past will be much more complete.

Although the prospect of using protein sequences as evolutionary tools was generating considerable excitement among biochemists during the 1950s and 1960s, convincing evolutionary biologists of their worth was proving to be more difficult. This can be illustrated by the intense debate that surrounded attempts to resolve one of the biggest questions in evolutionary biology and one which we will encounter often in this book—the location of the human branch of the tree of life. The commonly held view was that humans were phylogenetically distinct from the great apes (chimpanzees, gorillas and orang-utans), being placed in different taxonomic families, and that this split occurred at least 15 million years ago (Fig. 1.3a). This date was based on fossils belonging to an extinct animal called *Ramapithecus* which was thought to be the first member of the human lineage, largely because of its human-like thick molar tooth enamel. As fossil remains of this species were placed at around 15 million years old, the split between humans and great apes must have occurred before this time. Or so it was thought.

This view of human origins was shattered in 1967 when Vince Sarich and Allan Wilson, using an update of the methods of Nuttall and Uhlenhuth, measured the extent of immunological cross-reaction in the protein serum albumin between various primates. The results were striking: humans, chimpanzees and gorillas were genetically equidistant and clearly distinct from the orang-utan. Furthermore, assuming that serum albumin was a good molecular clock which could be calibrated by the split between apes and Old World monkeys some 30 million years ago, Sarich and Wilson estimated that human, chimpanzee and gorilla separated at only 5 million years ago.

The work of Sarich and Wilson is important on a number of levels. It was one of the first examples of **molecular systematics**—the use of gene sequences to reconstruct phylogenetic relationships. More than that, it changed our perspective on human origins and opened the 'molecules versus morphology' debate about which data set are the best markers of evolutionary relationship. Although less contentious today, this debate has influenced many studies in molecular evolution. In the case of human origins, many palaeontologists assumed their interpretation of the fossil record was correct and that the molecular evidence was somehow flawed. In

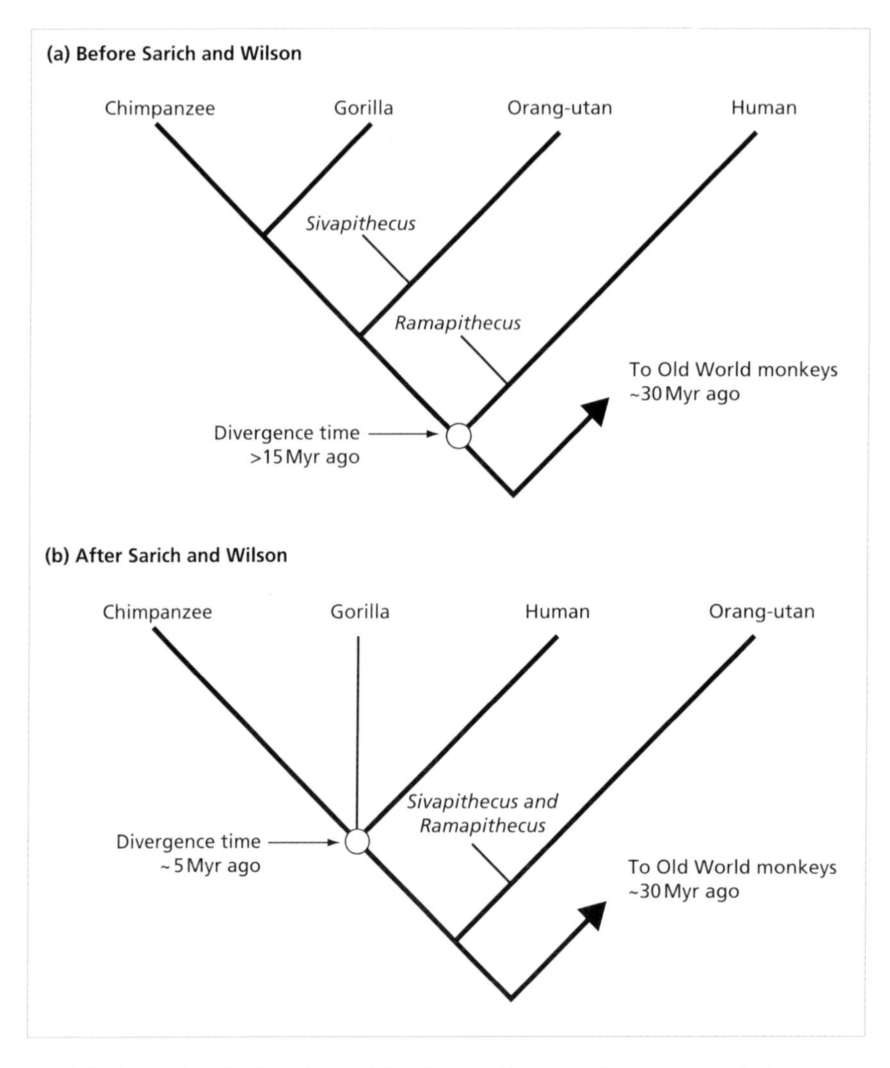

Fig. 1.3 How molecular data changed the picture of human origins. Tree (a) depicts the commonly held view of primate relationships during the 1960s, while tree (b) was that produced by Sarich and Wilson using immunological (serum albumin) distances. This tree, along with a divergence time of approximately 5 million years, has subsequently been confirmed using other molecular data, although human and chimpanzee now appear to be more closely related to each other than either is to the gorilla.

particular, it was claimed that the tick of the human molecular clock might have slowed in comparison with those of other primates so that divergence times would be underestimated. The argument was not resolved until the discovery, in 1980, of remains from another fossil ape—*Sivapithecus*. This find greatly changed the debate by showing that *Ramapithecus* and *Sivapithecus* were in fact very similar, representing two species from the same genus, and, more dramatically, that both were more closely related to the orang-utan than to humans (Fig. 1.3b). This meant that *Ramapithecus* could be shifted to the orang-utan side of the evolutionary tree, clearing the way for the more recent divergence between apes and humans that was suggested by the molecular evidence. Molecules had proven their worth in an evolutionary arena. More recent molecular studies have gone even further by showing that human and chimpanzee are more closely related to each other than either is to the gorilla.

The pace at which the science of molecular evolution has developed has greatly accelerated since this time. Nucleotide (DNA) sequences have now replaced proteins as the main source of data, particularly since the invention of the polymerase chain reaction (PCR) in the mid-1980s which allowed segments of DNA to be gathered quickly from very small amounts of starting material. Such is the sensitivity of PCR that it has even been used to obtain DNA from single human hairs and from tissue samples hundreds, or even thousands, of years old. And it is not just at the technical level that gene sequences are transforming biological science—there have also been major advances in theory, with a greater understanding of why genes evolve as they do, and the development of more sophisticated methods for recovering evolutionary information from sequence data. For instance, it is now apparent that DNA sequences not only contain a record of their phylogenetic relationships and times of divergence, but also the signatures of what evolutionary processes have shaped their history and even the size of past populations.

Finally, the success of molecular evolution is reflected by the dramatic rise in the number of scientific papers that include the evolutionary analysis of gene sequences. In the same way, molecular phylogenies are now appearing in journals covering a wide range of biological disciplines which use gene sequences as a primary data source. Molecular evolution has come of age.

1.2 What this book will cover

Modern molecular evolution is a diverse subject, influencing areas as different as ecology and organismal development. We hope to cover as many of these as possible in this book. Despite this diversity, a common theme runs through molecular evolution: that reconstructing the phylogenetic relationships between gene sequences is a crucial first step towards understanding their evolution. The phylogenetic tree can therefore be thought of as the central metaphor of evolution, providing a natural and meaningful way to order data, and with an enormous amount of evolutionary information contained within its branches. In this book we hope to show that molecular phylogenies can be thought of as a sort of navigational aid, guiding us through the sea of gene sequences. As a consequence, we will give most attention to those areas of molecular evolution where phylogenies have proven themselves most successful, although it is important to remember that the power of phylogenies is not solely reserved for molecular data: trees are becoming indispensable analytical tools in many other areas of evolutionary biology. The language of trees is taught in Chapter 2.

Although the first part of this chapter illustrated how molecular phylogenies have transformed systematics, this is by no means the only component of molecular evolution. Perhaps an even more fundamental question is *why* genomes look as they do? There are a multitude of smaller puzzles contained within this. One which has attracted much attention is why so much DNA, in eukaryotes at least, appears to be molecular 'junk', repeated many times. It is difficult to comprehend why evolution, a generator of such wonderful morphological adaptations, would be so wasteful at the molecular level. Related questions concern other aspects of genome organisation. Why are so many genes organised into families? How does recombination influence genome structure? Where do introns come from? In Chapter 3 we describe how genomes are organised and discuss the various evolutionary processes which have given genomes their distinctive architecture.

While the answers to some of these big questions may be lacking, there has been considerable progress in assembling the nuts and bolts of molecular evolution—analysing *how* genes evolve. Such has been the success of this research that it is now possible to describe, in quite detailed mathematical terms, exactly how one DNA sequence evolves into another, and the rates at which these changes take place. These advances have been at two separate but clearly connected levels: what shapes genetic diversity within a single population and how this is translated into evolution over longer timescales. The first level has traditionally been the realm of population genetics, the science which tells us how evolution changes the frequencies of genes in populations. Like many other areas of biological science, population genetics has been transformed by the availability of gene sequence data. In particular, there is a new found role for molecular phylogenies. The basics of population genetics, and the internal revolution initiated by the arrival of gene sequence data, is described in Chapter 4.

Because it requires some more specialist analytical tools, the molecular evolution of genes over longer timescales is suspended until Chapter 7. This area has been dominated by a debate between those who believe that most DNA changes are the outcome of evolution by natural selection—that is, they spread because of their benefit to the organism which houses them, and an opposing view, in which the majority of mutations have no functional relevance, do not affect the evolutionary well-being of the organism in question, and spread simply by good luck. This latter world view, referred to as the **neutral theory of molecular evolution**, was championed by the great Japanese geneticist Motoo Kimura. Although this neutralist–selectionist debate is waning in importance today, it has played a central role historically, initiating new ways to analyse sequence data and focusing our attention on the very essence of the evolutionary process. We therefore discuss many aspects of molecular evolution in the light of this axial dispute over evolutionary mechanism.

Our increased understanding of how genes and genomes evolve has itself led to the development of even better techniques to recover evolutionary information from gene sequences. Anyone sequencing DNA today has at their side a potentially huge array of tools of genetic archaeology, with the phylogenetic tree a multifunctional device. In Chapters 5 and 6 we discuss the pragmatics of how evolutionary information is recovered from gene sequence data. This has a number of elements, starting with sequence alignment (one of the trickiest tasks in sequence analysis), to accurately estimating the evolutionary distance between gene sequences (the true number of substitutions which have accumulated between them) to the multitude of methods for reconstructing phylogenetic trees. Given the key role played by trees in our book, the details of how these trees should be made (Chapter 6) is presented in great detail.

Finally, in Chapter 8, we apply some of the skills learned earlier. In particular, we show how molecular phylogenies may provide answers to a wide range of evolutionary questions. Greatest attention is given to detecting, within trees, the signatures of different evolutionary processes, from speciation to mass extinctions. The uses of phylogenetic trees outlined in this chapter in many ways reflect our own research interests. We are certain that readers will have their own, equally important uses.

The age of complete genomes promises much to many areas in biology. Gene sequence data have already given evolutionary biology a new momentum, providing a fresh perspective to deep-seated problems and opening up new avenues to explore. The uses to which gene sequences can be put will continue to develop in the future. We aim to show you the evolutionary relevance of molecular data, and the relevance evolutionary theory has for the study of genomes.

1.3 Further reading

The genome of *Haemophilus influenzae* is presented by Fleischmann *et al.* (1995), although see the issues of *Science* magazine from 25 October 1996 (Volume 274, pp. 465–688) and 24 October 1997 (Volume 278, pp. 541–768) for discussions of the progress and implications of the genome sequencing projects, and that of 23 August 1996 (Volume 273, pp. 1017–1140) for an introduction into the strange world of *Methanococcus jannaschii*.

A discussion of the 'tree of life' as depicted by rRNA is provided by Olsen and Woese (1993) and Olsen *et al.* (1994), although see Rivera and Lake (1992) for an alternative view. At the other extreme, the story of HIV transmission in the Florida dental cohort is told in detail by Ou *et al.* (1992), with an update by Hillis *et al.* (1994).

The debate surrounding human origins has been told in many places although the paper by Sarich and Wilson (1967) is of immense historical importance, and that of Andrews and Cronin (1982) shows the impact of *Ramapithecus* and *Sivapithecus*. A recent review of the molecular evidence surrounding the phylogeny of hominoid primates is provided by Ruvolo (1997).

Finally, for demonstrations of the multitude of uses to which molecular phylogenies can be put, see the volumes of Avise (1994), Harvey and Pagel (1991) and Harvey *et al.* (1996).

Chapter 2
Trees

2.1 Introduction to trees

All of life is related by common ancestry. Recovering this pattern, the 'Tree of Life', is one of the prime goals of evolutionary biology. This chapter introduces the fundamentals of trees. You may find it useful to read the chapter through once, then refer back to it as you read the rest of the book. All the concepts introduced in this chapter will be discussed in more detail in subsequent chapters; our goal here is to give you some familiarity with trees so that interpreting them eventually becomes second nature.

2.1.1 Tree terminology

Figure 2.1 illustrates the terminology used in this book to describe trees. Unfortunately tree terminology varies greatly among authors, and among different disciplines, such as mathematics and biology. Where possible we will list the commonly used synonyms that you may encounter in the literature.

A **tree** is a mathematical structure which is used to model the actual evolutionary history of a group of sequences or organisms. This actual pattern of historical relationships is the **phylogeny** or **evolutionary tree** which we try and estimate. A tree consists of **nodes** connected by **branches** (also called **edges**). **Terminal nodes** (also called **leaves**, **OTUs [Operational Taxonomic Units]**, or **terminal taxa**) represent sequences or organisms for which we have data; they may be either extant or extinct. **Internal nodes** represent hypothetical ancestors; the ancestor of all the sequences that comprise the tree is the **root** of the tree (see below).

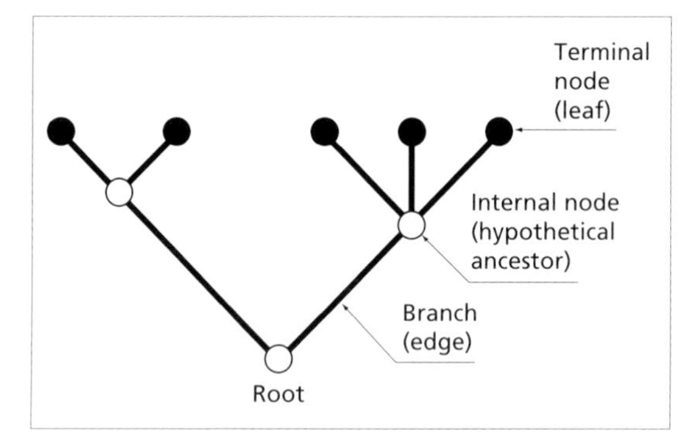

Fig. 2.1 A simple tree and associated terms.

The nodes and branches of a tree may have various kinds of information associated with them. For example some methods of phylogeny reconstruction (e.g. parsimony) endeavour to reconstruct the characters of each hypothetical ancestor; most methods also estimate the amount of evolution that takes place between each node on the tree, which can be represented as **branch lengths** (or **edge lengths**). Trees with branch lengths are sometimes called **weighted trees**.

Box 2.1 Trees are like mobiles

There are many different ways of drawing trees, so it is important to know whether these different ways actually reflect differences in the kind of tree, or whether they are simply stylistic conventions. For instance, the order in which the labels on a tree are drawn on a piece of paper (or computer screen) can differ without changing the meaning of the tree. This is because the edges of a tree can be freely rotated without changing the relationships among the terminal nodes. The diagram below shows the same tree drawn three different ways:

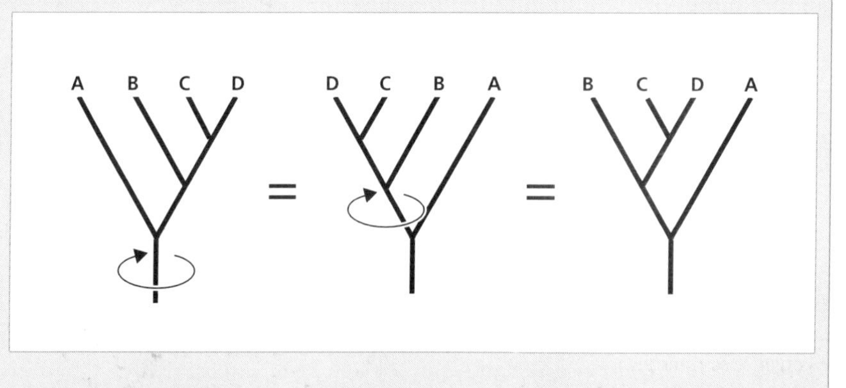

In this sense a tree is just like a mobile; no matter how many times you rotate the 'hanging' objects you do not change how they are connected to one another.

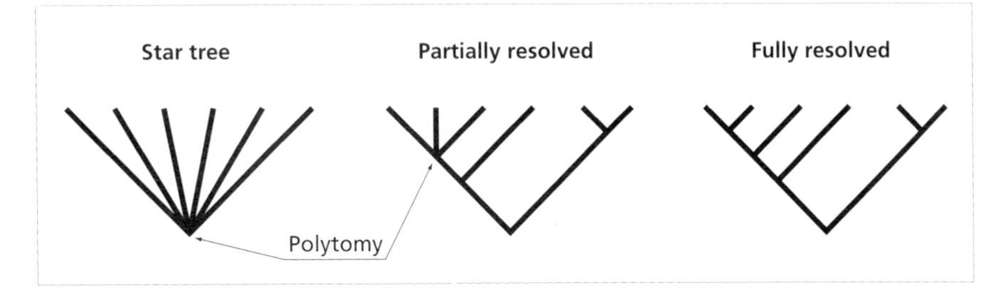

Fig. 2.2 Three trees showing various degrees of resolution, ranging from a complete lack of resolution (star tree) to a fully resolved tree. Any internal node with more than two immediate descendants is a polytomy.

The number of adjacent branches possessed by an internal node is that node's **degree**. If a node has a degree greater than three (i.e. it has one ancestor and more than two immediate descendants) then that node is a **polytomy**. A tree that has no polytomies is fully resolved (Fig. 2.2).

Polytomies can represent two rather different situations (Fig. 2.3); firstly they may represent simultaneous divergence—all the descendants evolved at the same time (a 'hard' polytomy); alternatively, polytomies may indicate uncertainty about phylogenetic relationships—the lineages did not necessarily all diverge at once, but we are unsure as to the actual order of divergence (a 'soft' polytomy). These two interpretations—simultaneous divergence or uncertainty—are obviously quite different. Typically polytomies are treated as 'soft'. It may be thought unlikely that multiple lineages would diverge at exactly the same time; however, if lineages diverge rapidly in time relative to the rate of character evolution then there may be insufficient evidence available to us to ever be able to reconstruct the exact order of splitting, in which case the polytomy is effectively 'hard'.

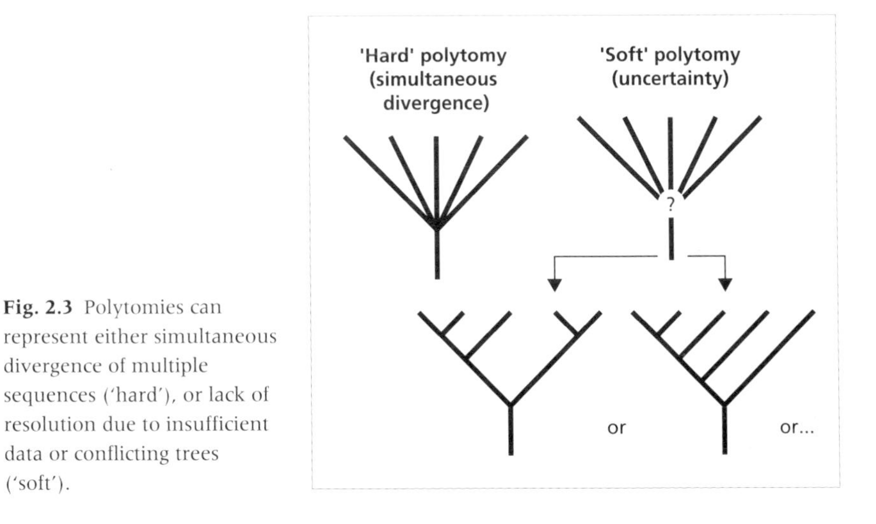

Fig. 2.3 Polytomies can represent either simultaneous divergence of multiple sequences ('hard'), or lack of resolution due to insufficient data or conflicting trees ('soft').

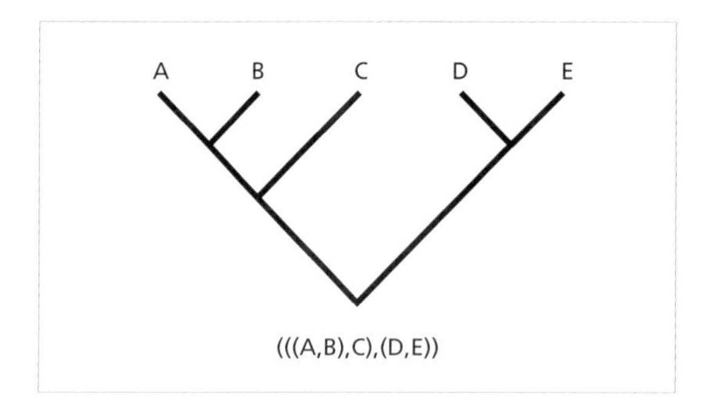

Fig. 2.4 A tree and its shorthand representation using nested parentheses.

2.1.2 A shorthand for trees

Trees can be represented by a shorthand notation that uses nested parentheses. Each internal node is represented by a pair of parentheses that enclose all descendants of that node. This format makes it easy to describe a tree in the body of some text without having to draw it. The format is also used by many computer programs to store representations of trees in data files. Figure 2.4 gives an example of this shorthand.

2.1.3 Cladograms, additive trees and ultrametric trees

Different kinds of tree can be used to depict different aspects of evolutionary

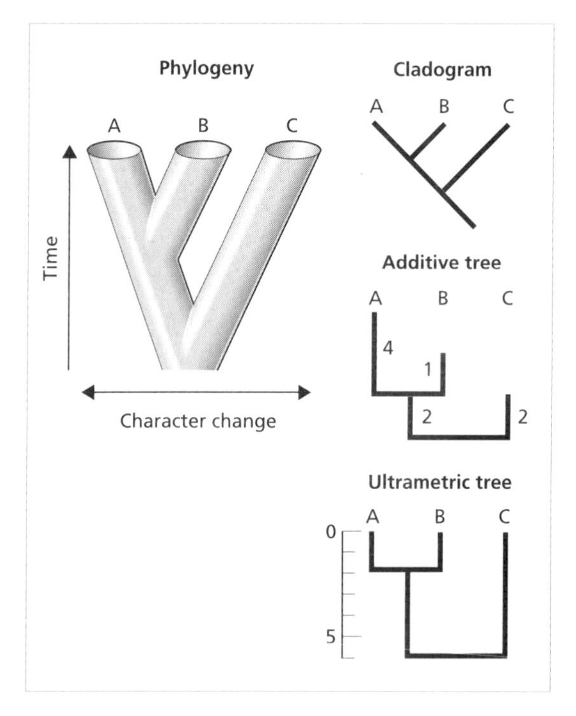

Fig. 2.5 A phylogeny and the three basic kinds of tree used to depict that phylogeny. The cladogram represents relative recency of common ancestry; the additive tree depicts the amount of evolutionary change that has occurred along the different branches, and the ultrametric tree depicts times of divergence.

history. The most basic tree is the **cladogram** which simply shows relative recency of common ancestry, that is, given the three sequences, A, B and C, the cladogram in Fig. 2.5 tells us that sequences A and B share a common ancestor more recently than either does with C. In the biomathematical literature cladograms are often called '*n*-trees'.

Additive trees contain additional information, namely branch lengths. These are numbers associated with each branch that correspond to some attribute of the sequences, such as amount of evolutionary change. In the example shown in Fig. 2.5, sequence A has acquired four substitutions since it shared a common ancestor with sequence B. Other commonly used terms for additive trees include 'metric trees' and 'phylograms'.

Ultrametric trees (sometimes also called 'dendrograms') are a special kind of additive tree in which the tips of the trees are all equidistant from the root of the tree. This kind of tree can be used to depict evolutionary time, expressed either directly as years or indirectly as amount of sequence divergence using a molecular clock.

Additive and ultrametric trees both contain all the information found in a cladogram—the cladogram is the simplest statement about evolutionary relationships that we can make. For some questions knowledge of relative recency of common ancestry is sufficient. However, there are other evolutionary questions (such as determining relative rates of evolution) which require the additional information contained in additive and ultrametric trees.

Box 2.2 What do the horizontal and vertical axes of a tree represent?

It is tempting to think of a tree as being a graphical plot like a scatter plot, in which case the question arises 'what do the horizontal and vertical axes represent?' For cladograms, which have no branch length information, neither axis has any special meaning; you can squash the tree flatter, or stretch it out without changing the relationships among the terminal nodes. Hence, the two cladograms shown below are the same.

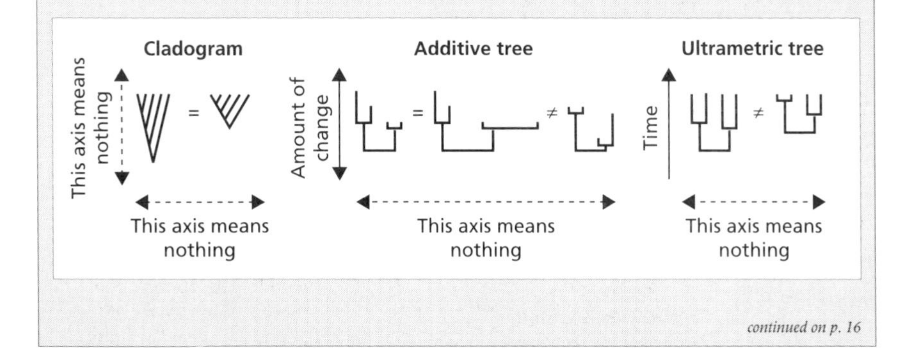

continued on p. 16

Box 2.2 *continued*

For additive trees one of the axes does have meaning; it represents the amount of evolutionary change. In the diagram above if we stretch the tree along the horizontal axis (i.e. left to right) we do not change interpretation of the tree; however, changes in the vertical axis (up and down) change the amount of evolutionary change along the branches, hence the trees are not the same *additive* trees. Similarly, for an ultrametric tree one axis typically represents time whereas the other has no meaning. The two ultrametric trees shown above are different because the two trees specify different divergence times.

A last consideration is that trees can be drawn in a number of orientations, such as 'planted' with the root at the bottom as in the diagram above, 'left-to-right' as in Fig. 2.17, or even 'top-down' with the root at the top. The choice among these representations is entirely arbitrary; in some circumstances it may be more convenient to draw the tree one way rather than another. Just remember that if the tree diagram is rotated then the x- and y-axes in the above diagram need to be rotated as well. Hence, if an additive tree is drawn left to right then the horizontal axis represents evolutionary change and the vertical axis has no meaning.

2.1.4 Rooted and unrooted trees

Cladograms and additive trees can either be rooted or unrooted. A **rooted** tree has a node identified as the root from which ultimately all other nodes descend, hence a rooted tree has direction. This direction corresponds to evolutionary time; the closer a node is to the root of the tree, the older it is in time. Rooted trees allow us to define ancestor–descendant relationships between nodes: given a pair of nodes connected by a branch, the node closest to the root is the ancestor of the node further away from the root (the descendant). **Unrooted** trees lack a root, and hence do not specify evolutionary relationships in quite the same way, and they do not allow us to talk of ancestors and descendants. Furthermore, sequences that may be adjacent on an unrooted tree need not be evolutionarily closely related. For example, given the unrooted tree in Fig. 2.6, the gibbon (B) and orang-utan (O) sequences are neighbours on the tree, yet the orang-utan is more closely related to the other apes (including humans). This is because the root of the tree lies on the branch leading to the gibbon. Had we placed the root elsewhere, say on the branch leading to the gorilla (G), then the gibbon and orang-utan sequences would indeed be closely related.

In the unrooted tree for the apes shown in Fig. 2.6, we could have placed the root on any of the seven branches of the tree. Hence, this unrooted tree corresponds to a set of seven rooted trees (Fig. 2.7).

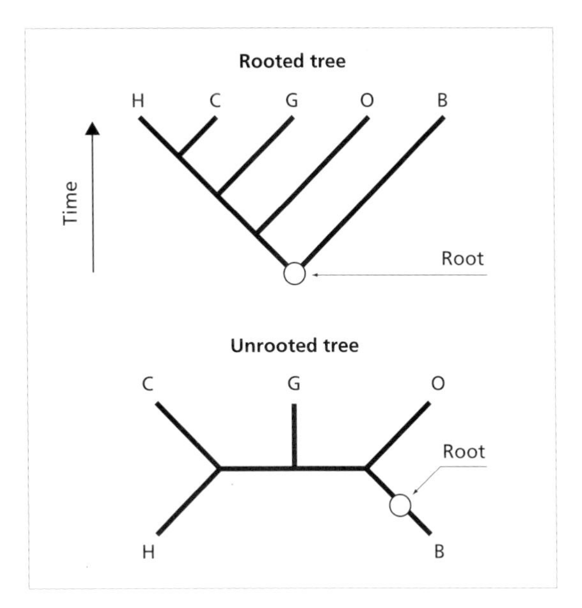

Fig. 2.6 Rooted and unrooted trees for human (H), chimp (C), gorilla (G), orang-utan (O), and gibbon (B). The rooted tree (top) corresponds to the unrooted tree below.

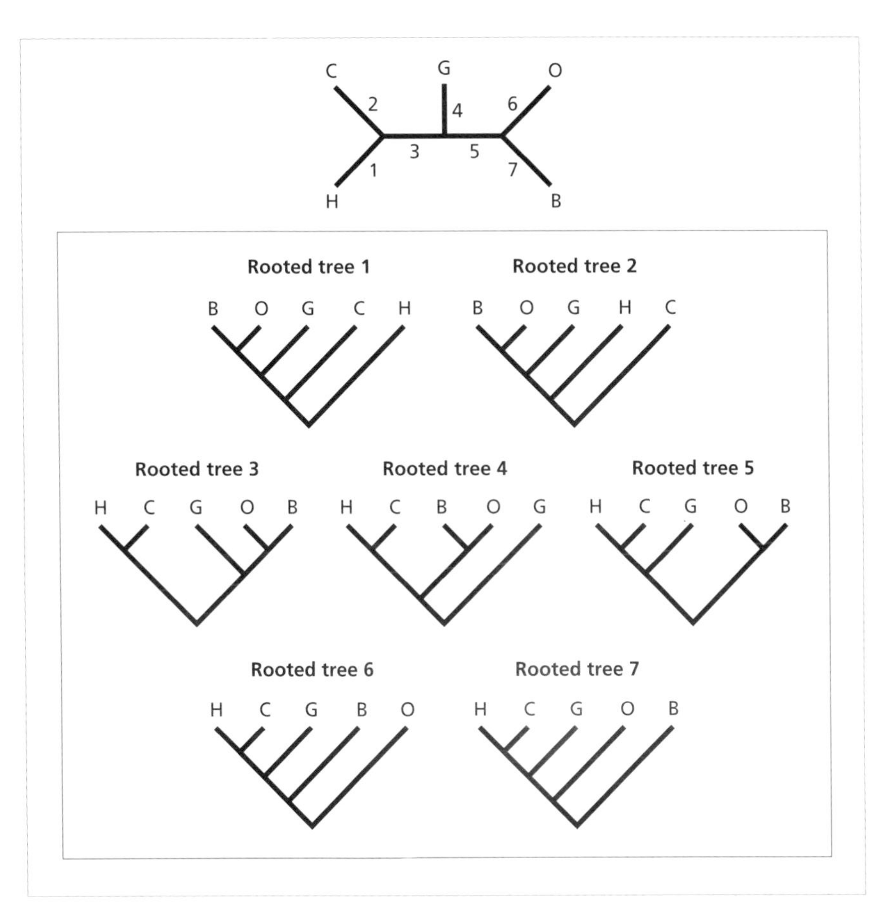

Fig. 2.7 The seven rooted trees that can be derived from an unrooted tree for five sequences. Each rooted tree 1–7 corresponds to placing the root on the corresponding numbered branch of the unrooted tree. (Sequence labels as for Fig. 2.6.)

The distinction between rooted and unrooted trees is important because many methods for reconstructing phylogenies generate unrooted trees, and hence cannot distinguish among the seven trees shown in Fig. 2.7 on the basis of the data alone. In order to root an unrooted tree (i.e. decide which of the seven trees is the actual evolutionary tree) we need some other source of information. Methods of rooting trees are discussed in Chapter 6 (note that this does not apply to ultrametric trees which are rooted by definition).

The number of possible unrooted trees U_n for n sequences is given by

$$U_n = (2n-5)(2n-7)\ \ldots\ (3)(1) \tag{2.1}$$

for $n \geq 2$. The number of rooted trees R_n for $n \geq 3$ is given by

$$\begin{aligned} R_n &= (2n-3)(2n-5)\ \ldots\ (3)(1) \\ &= (2n-3)U_n \end{aligned} \tag{2.2}$$

Table 2.1 lists the numbers of rooted and unrooted fully resolved trees for 2–10 sequences. Note that the number of unrooted trees for n sequences is equal to the number of rooted trees for $(n-1)$ sequences. Note also that the number of trees rapidly reaches very large numbers: for 10 sequences there are over 34 million possible rooted trees. For a relatively modest 20 sequences there are 8 200 794 532 637 891 559 000 possible trees, whereas the number of different trees for 135 human mitochondrial DNA sequences used in the study of the evolution of modern humans (see Chapter 4), 2.113×10^{267}, exceeds the number of particles in the known universe! This explosion in number of trees is a fundamental problem for phylogeny reconstruction, where the goal is to identify which tree of all the possible trees is the best estimate of the actual phylogeny.

2.1.5 Tree shape

Typically, the information in a tree in which we are most interested is the relationship among the sequences, and perhaps the lengths of the branches.

Number of sequences	Number of unrooted trees	Number of rooted trees
2	1	1
3	1	3
4	3	15
5	15	105
6	105	945
7	945	10 395
8	10 395	135 135
9	135 135	2 027 025
10	2 027 025	34 459 425

Table 2.1 Numbers of unrooted and rooted trees for 2–10 sequences.

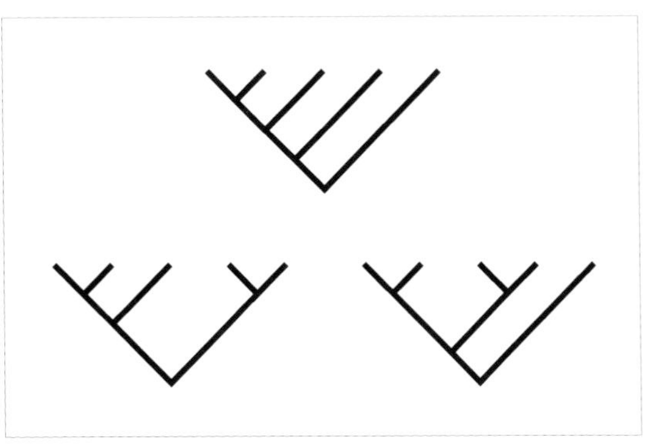

Fig. 2.8 The three possible shapes for a rooted tree for five sequences.

However, other aspects of the tree may also reflect evolutionary phenomena and hence be of interest. Figure 2.8 shows the three possible **shapes** (or **topologies**) for a rooted tree for five sequences. All 105 possible trees (Table 2.1) for five sequences will have one of these three shapes.

2.1.6 Splits

Trees can be represented in a variety of ways other than as graphs. One useful representation is as sets of sets, called **splits** or **partitions**. Each split takes the set of sequences (e.g. {H, C, G, O, B}) and partitions them into two mutually exclusive sets: you can think of a split as the two sets of sequences obtained by chopping ('splitting') the tree at a given branch. For example, the tree shown in Fig. 2.9 has seven branches and hence seven splits. However, all splits comprising a single terminal node on one hand and the rest of the tree on the other are not 'phylogenetically informative' in the sense that all possible trees will contain those splits. Hence, the only informative splits are those

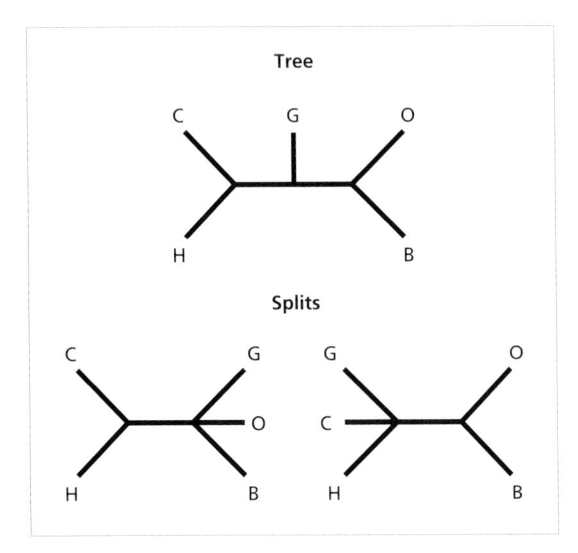

Fig. 2.9 An unrooted tree and its two splits.

resulting from chopping internal branches. The tree shown in Fig. 2.9 has two informative splits: {{C, H}, {G, O, B}} and {{G, C, H}, {O, B}}.

Given these two splits we can combine them to reconstruct the original tree. Notice that there are other possible partitions of the set {H, C, G, O, B}, such as {{H, G}, {C, O, B}}. This split groups humans and gorillas together to the exclusion of the other apes, which is **incompatible** with the split {{C, H}, {G, O, B}}, which groups humans and chimps. Incompatible splits cannot be combined to form a tree.

Another way of representing the splits in Fig. 2.9 is to assign arbitrary letters to each half of a split, such as the letter 'A' to each sequence on the left and the letter 'T' to each sequence on the right. This gives the following table:

Sequence	**Split 1**	**Split 2**
H	A	A
C	A	A
G	T	A
O	T	T
B	T	T

Each split now resembles a single nucleotide site with only the bases A and T. In Chapter 6 you will encounter some methods for reconstructing phylogenies that make use of this relationship between nucleotide sites and splits.

2.2 Reconstructing the history of character change

The tree relating a set of sequences tells us only part of what we want to know. The tree alone does not tell us when a particular evolutionary change, such as a nucleotide substitution, took place, or whether the occurrence of the same amino acid in two sequences is the result of inheritance from a common ancestor or independent evolution. To address these questions we need to be able to reconstruct the history of character change. This problem is addressed

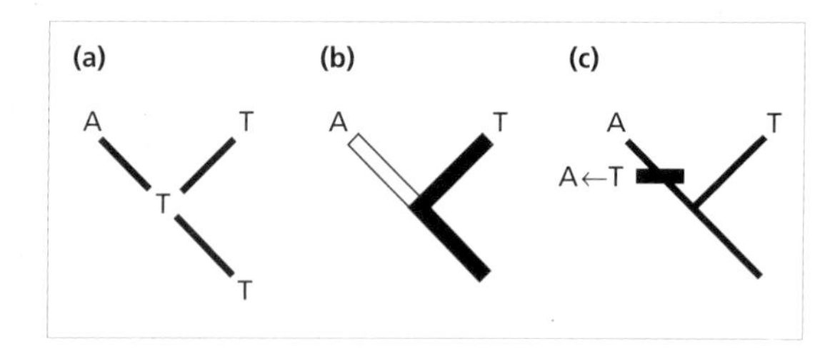

Fig. 2.10 Three equivalent ways of representing the same evolutionary change on the same tree. (a) Each node is labelled by the corresponding nucleotide; (b) each branch is coloured corresponding to the nucleotide at the end of each branch; and (c) indicating on which branch the change took place.

in more detail in Chapter 5, so here we will merely introduce some of the different ways of representing evolutionary change on a tree (Fig. 2.10) and describe some basic terminology.

Given a tree, we can distinguish between ancestral ('primitive') and derived character states. If a sequence has the same base as the common ancestor of all the sequences being studied then it is the primitive or **plesiomorphic** state; otherwise it is a derived or **apomorphic** state. Unique derived character states are **autapomorphies** (*aut* = alone), shared derived states are **synapomorphies** (*syn* = shared) (Fig. 2.11). Given any two character states that are identical (e.g. the same nucleotide base) the similarity between them may be because they have both inherited it directly from their ancestor which also had that state. This is an instance of **homology**. Alternatively, the similarity may have occurred independently in which case it is **homoplasy**. Only homologous similarity directly reflects common ancestry. Homoplasy is

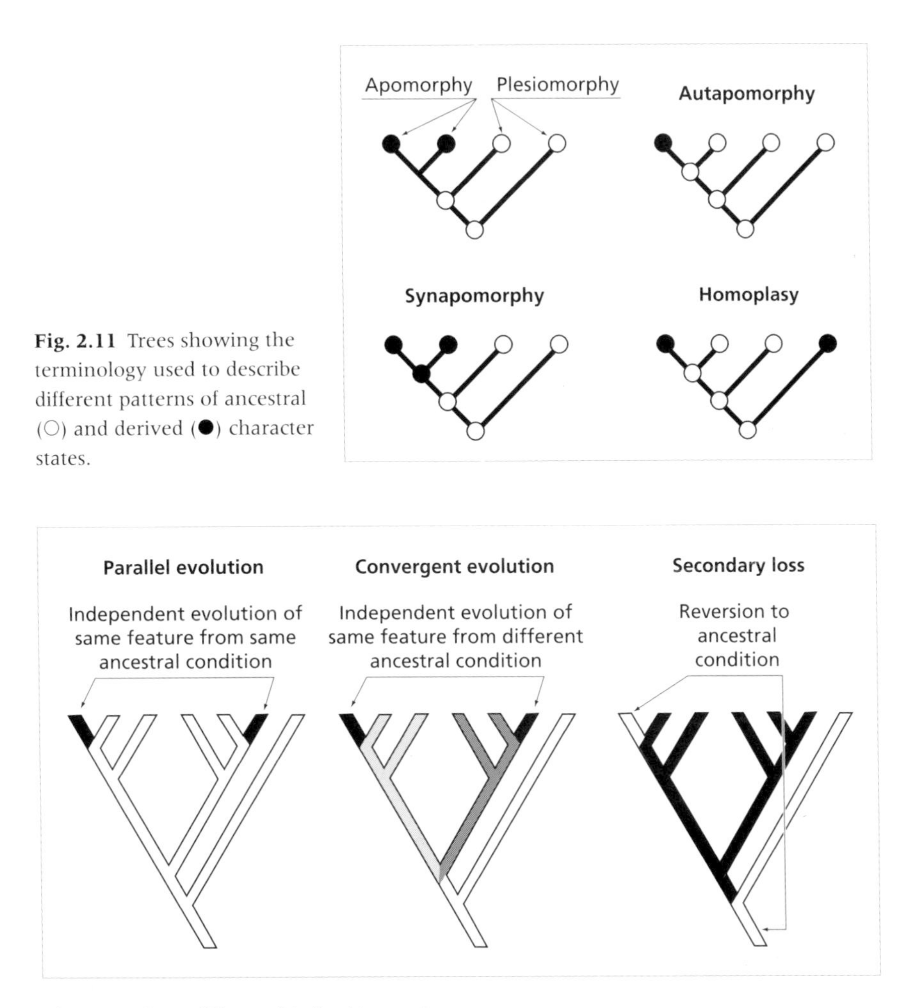

Fig. 2.11 Trees showing the terminology used to describe different patterns of ancestral (○) and derived (●) character states.

Fig. 2.12 Three different kinds of homoplasy.

a poor indicator of evolutionary relationships because the similarity does not reflect shared ancestry. It is sometimes useful to distinguish between different kinds of homoplasy (Fig. 2.12). **Convergence** and **parallel evolution** both result in the independent evolution of the same feature in two unrelated sequences; the difference between the two lies in whether the similarity was acquired from the same (parallelism) or a different (convergence) ancestral condition. Homoplasy may also be due to the **secondary loss** of a derived feature, which results in the apparent reversion to the ancestral condition, such as the loss of legs in snakes and some amphibia.

2.2.1 Ancestors

Phylogenies presuppose ancestors—previously living organisms that are now extinct but which left descendants which comprise modern species. These ancestors (or their sequences) are represented by the internal nodes of a tree. These ancestors are hypothetical, but some methods of phylogenetic reconstruction allow us to infer what they (or their sequences) may have looked like.

All molecular phylogenies include ancestors, but for the most part these remain hypothetical entities represented by the internal nodes of the tree, and inferred solely on the basis of sequences from extant organisms. It used to be thought that the possibility that a sequence being studied was actually an ancestor could be safely ignored, hence all sequences were placed at the tips of evolutionary trees. However, two recent developments have meant that molecular biologists must deal with a problem previously restricted to palaeontology—namely the recognition of ancestors. The first of these developments is the recovery of DNA from extinct taxa; the second is the increasing number of sequences being obtained from viruses such as human immunodeficiency virus (HIV) which evolve sufficiently fast to be tracked in 'real time'.

If all sequences are from extant organisms, then they can be placed at the tips of the tree. However, if some of the sequences are extinct it is possible, if unlikely, that they may have been ancestral to one or more of the extant sequences: is a sequence extracted from an extinct taxon an ancestor to modern taxa or is it on an evolutionary side branch? Cladists have adopted the convention that extinct taxa that lack autapomorphies are candidates for being ancestral, as it is equally parsimonious to treat them as **sister taxa** (i.e. each other's closest relative) or as ancestors (Fig. 2.13). Treating a taxon with autapomorphies as an ancestor would require us to postulate additional evolutionary changes (this invokes the parsimony principle discussed in Chapter 6). Note that under this rule a taxon with no autapomorphies need not be an ancestor, rather there is nothing to refute that possibility.

We can apply the cladistic convention to viral sequences where the virus is evolving sufficiently rapidly for successive samples to show evolutionary change. For example, Fig. 2.14 shows a cladogram for eight HIV sequences

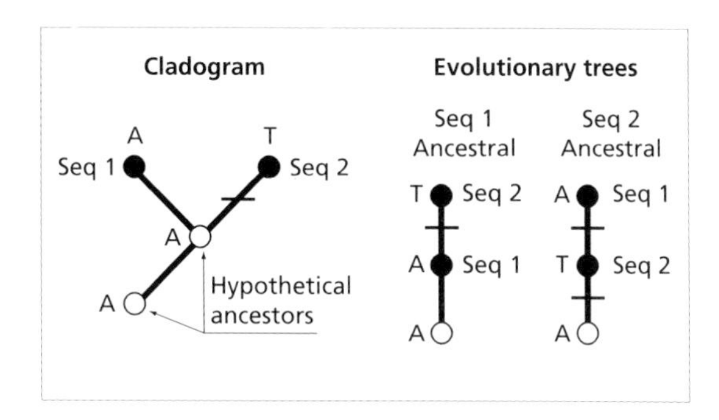

Fig. 2.13 A cladogram for two sequences (Seq 1 and Seq 2) showing the nucleotide at a single site, and two of several possible evolutionary trees derived from that cladogram. We could postulate that either sequence is ancestral to the other. However, postulating Seq 2 to be an ancestor of Seq 1 requires the gain and subsequent loss of T, whereas if Seq 1 is an ancestor no additional substitutions need be postulated. Note that a third phylogeny would be identical to the cladogram (see Box 2.3).

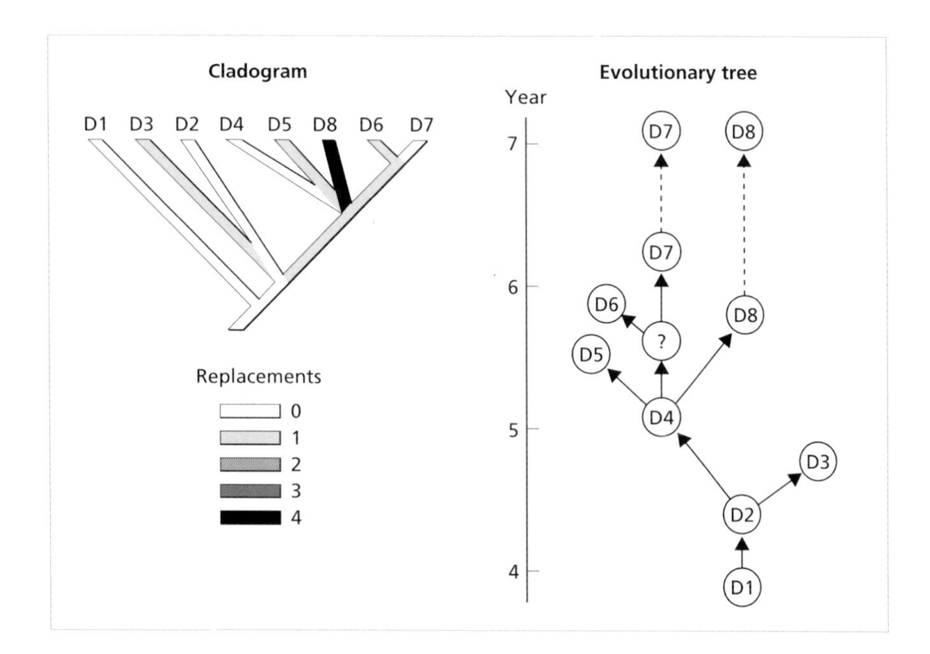

Fig. 2.14 Cladogram and corresponding evolutionary tree for eight V3 loop amino acid sequences for HIV samples taken from a single patient over 3 years. In the cladogram on the left all eight sequences are depicted as terminal nodes; however, four sequences (D1, D2, D4 and D7) have no autapomorphies (i.e. there are no replacements along the branch leading to each sequence) and hence are possible ancestors. The evolutionary tree on the right depicts the same relationships as the cladogram, but the sequences lacking autapomorphies (except D7) are treated as ancestors which is consistent with the order of appearance of the sequences. Modified from Holmes *et al.* (1992).

obtained from a single patient over 3 years. Because the samples were obtained over a period of time it is possible that some of the sequences sampled earlier in time gave rise to later sequences. Indeed, some sequences lack autapomorphies and hence by the cladistic criterion are potential ancestors, a conclusion which is supported by the order of the sequences in time.

Box 2.3 Cladograms and evolutionary trees

In this book we use the term 'cladogram' to refer to an evolutionary tree that has no information on branch lengths (e.g. Fig. 2.5). Within cladistics a distinction is made between a cladogram and an evolutionary tree. In a cladogram the terminal taxa are always at the tips of the tree, no matter if the taxa are extant or extinct, or whether one or more of the taxa are ancestral to any of the others. However, in an evolutionary tree some of the taxa may be ancestral to the others. Given the cladogram ((A, B), C) shown below, there are six different evolutionary trees that are consistent with the cladogram. One of these trees is the cladogram itself; the other five trees have one or more of the taxa A, B and C being ancestral to the others. Note that in all six trees A and B are more closely related to each other than to C.

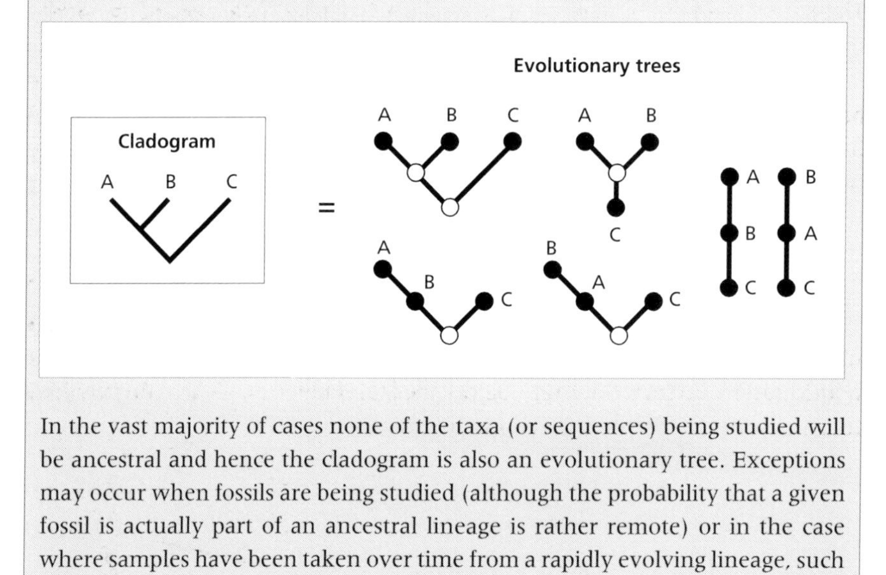

In the vast majority of cases none of the taxa (or sequences) being studied will be ancestral and hence the cladogram is also an evolutionary tree. Exceptions may occur when fossils are being studied (although the probability that a given fossil is actually part of an ancestral lineage is rather remote) or in the case where samples have been taken over time from a rapidly evolving lineage, such as a virus (Fig. 2.14).

2.3 Trees and distances

Measures of sequence dissimilarity may be used to estimate the number of evolutionary changes that occurred in two sequences since they last shared a common ancestor (see Chapter 5). These measures quantify the evolutionary distance between the two sequences. Trees themselves can also be represented by distances, and this link has motivated a range of tree-building methods that seek to convert pairwise distances between sequences into evolutionary trees. We shall describe these measures in Chapter 5. However, in order for a distance measure to be used to build phylogenies it must satisfy some basic requirements: it must be a metric, and it must be additive.

2.3.1 Metric distances

Let d (a, b) be the distance between two sequences, a and b. A distance d is a **metric** if it satisfies these properties:

1	$d(\mathrm{a}, \mathrm{b}) \geq 0$	(non-negativity)
2	$d(\mathrm{a}, \mathrm{b}) = d(\mathrm{b}, \mathrm{a})$	(symmetry)
3	$d(\mathrm{a}, \mathrm{c}) \leq d(\mathrm{a}, \mathrm{b}) + d(\mathrm{b}, \mathrm{c})$	(triangle inequality)
4	$d(\mathrm{a}, \mathrm{b}) = 0$ if and only if $\mathrm{a} = \mathrm{b}$	(distinctness)

The first property is *non-negativity*; two sequences must have a non-negative distance. The second property is *symmetry*; two sequences have the same dissimilarity regardless of the direction in which the dissimilarity is measured. These two properties may seem trivial, but not all measures of similarity meet these seemingly obvious requirements.

The third property is the *triangle inequality*, which states that the dissimilarity between any two sequences cannot exceed the sum of the dissimilarities between each sequence and a third. This condition is equivalent to ensuring that it is possible to represent the distances between the three sequences as a triangle (Fig. 2.15), hence the name. The last condition (*distinctness*) requires that sequences that are different must have a non-zero dissimilarity.

Of these conditions, 1, 2 and 4 are generally true for all measures of sequences dissimilarity calculated directly from sequences. However, indirect measures of sequence dissimilarity such as those obtained from DNA–DNA hybridisation or from immunological measurements need not always obey these conditions, particularly condition 2.

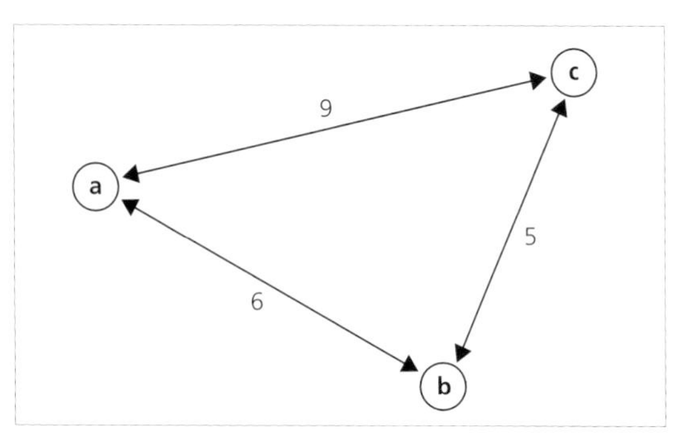

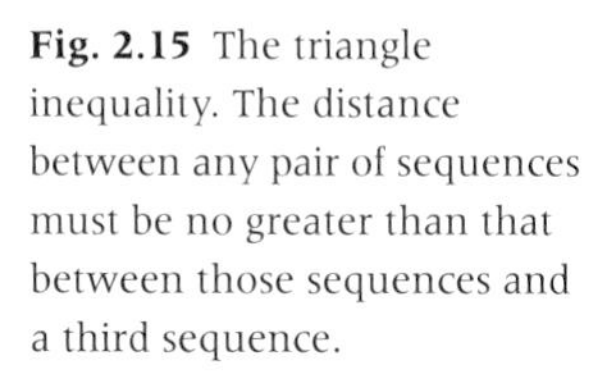

Fig. 2.15 The triangle inequality. The distance between any pair of sequences must be no greater than that between those sequences and a third sequence.

2.3.2 Ultrametric distances

A metric is an ultrametric if it satisfies the additional criterion that:

5 $d(\text{a, b}) \leq \text{maximum}\ [d(\text{a, c}), d(\text{b, c})]$

This criterion implies that the two largest distances are equal, so that they define an isosceles triangle (Fig. 2.16).

Ultrametric distances have the very useful evolutionary property of implying a constant rate of evolution. Indeed the 'relative rate' test for a molecular clock (see Box 7.2, Chapter 7) is a test of how far the pairwise distances between three sequences depart from ultrametricity. Furthermore, if distances between sequences are ultrametric then the most similar sequences are also the most closely related.

2.3.3 Additive distances

Being a metric (or ultrametric) is a necessary, but not sufficient condition for being a valid measure of evolutionary change. A measure must also satisfy the *four-point condition*:

6 $d(\text{a, b}) + d(\text{c, d}) \leq \text{maximum}\ [d(\text{a, c}) + d(\text{b, d}), d(\text{a, d}) + d(\text{b, c})]$

This is equivalent to requiring that of the three sums $d(\text{a, b}) + d(\text{c, d})$, $d(\text{a, c}) + d(\text{b, d})$, and $d(\text{a, d}) + d(\text{b, c})$, the two largest are equal.

2.3.4 Tree distances

An additive distance measure defines a tree. Perhaps the easiest way to see this is to consider the distances shown in Fig. 2.17. Sequence d is equidistant from all other sequences; sequence c is equidistant from a and b. If we take any three sequences the distances between them define an isosceles triangle (the two largest distances are equal), hence the distances shown in Fig. 2.17

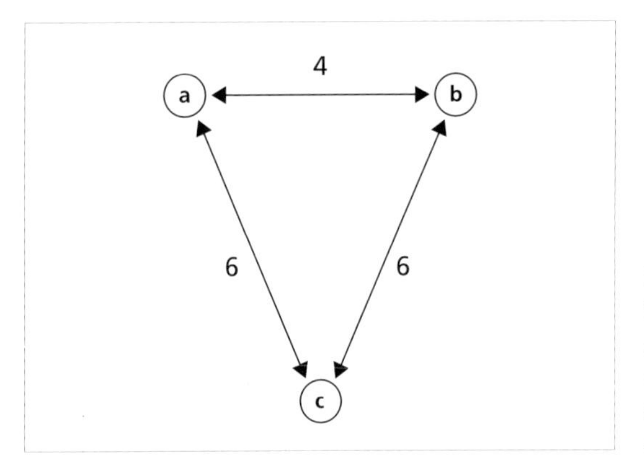

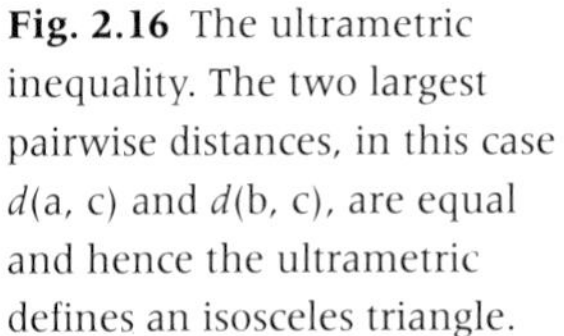

Fig. 2.16 The ultrametric inequality. The two largest pairwise distances, in this case d(a, c) and d(b, c), are equal and hence the ultrametric defines an isosceles triangle.

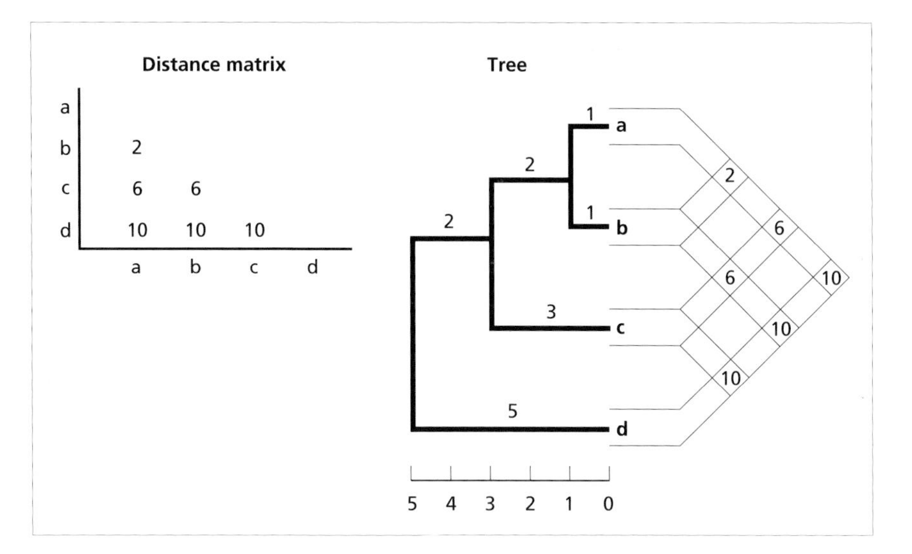

Fig. 2.17 An ultrametric distance matrix between four sequences a–d and the corresponding ultrametric tree. For any two sequences, the value in the distance matrix corresponds to the sum of the branch lengths along the path between the two sequences on the tree.

are ultrametric. These same distances can be represented by the ultrametric tree shown in Fig. 2.17. If we trace the shortest path between any pair of sequences in the tree, and add up the corresponding branch lengths, we obtain the same values as those in the distance matrix. For example, travelling from

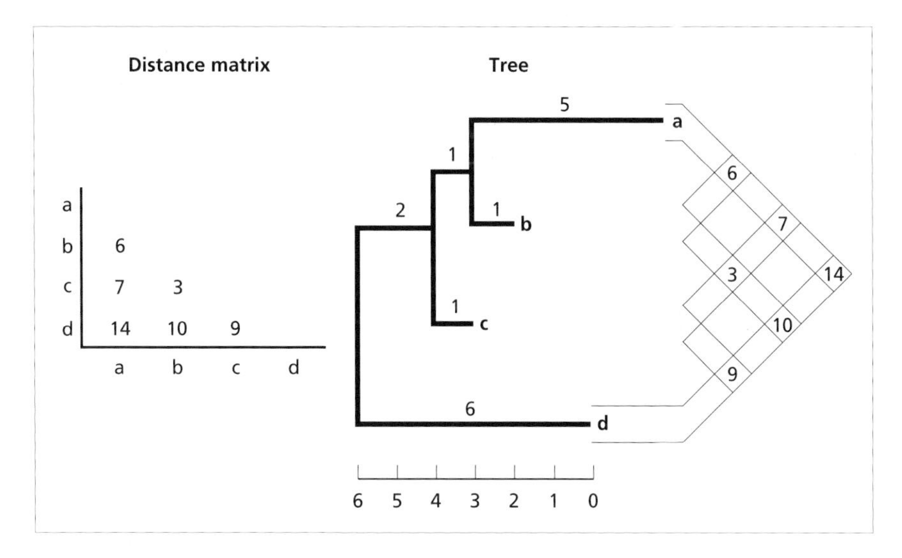

Fig. 2.18 An additive distance matrix between four sequences and the corresponding additive tree. For any two sequences, the value in the distance matrix corresponds to the sum of the branch lengths along the path between the two sequences on the tree.

sequence a to sequence d and adding branch lengths we obtain the value of $1 + 2 + 2 + 5 = 10$, hence $d(a, d) = 10$.

When distances are not ultrametric but only metric they can still be represented by a tree, in this case an additive tree (Fig. 2.18). This additive tree again represents the additive distances exactly. Notice that sequences b and c are the most similar ($d(b, c) = 3$) but are not the most closely related. Similarity and evolutionary relationship will only coincide exactly if the distances are ultrametric. This has important implications for using distances to reconstruct trees (Chapter 6).

The distances obtained from the tree are **tree distances** (also called 'patristic distances'), to distinguish them from **observed distances** which are obtained directly from the sequences themselves. In the examples shown in Fig. 2.17 and Fig. 2.18, the observed and tree distances match exactly. For real data this is rarely the case, indicating that the observed distances cannot be completely accurately represented by a tree. The discrepancy between observed and tree distances can be used to measure how good the fit is between the observed distances and the best tree representation of those distances (see Chapter 6).

2.4 Organismal phylogeny

Although the main subject of this book is molecular evolution, a major use of DNA sequences is the reconstruction of the evolutionary history of the organisms from which those sequences are obtained.

2.4.1 Clades and classification

Phylogenies form the basis of **classification**, which is the formal naming of groups of organisms. Cladistic classifications recognise only **monophyletic** groups or **clades**. A monophyletic group includes all the descendants of an ancestral taxon, whereas a non-monophyletic group omits some of those descendants (Fig. 2.19). A good example of a non-monophyletic group is the 'apes', which is not monophyletic as it excludes humans.

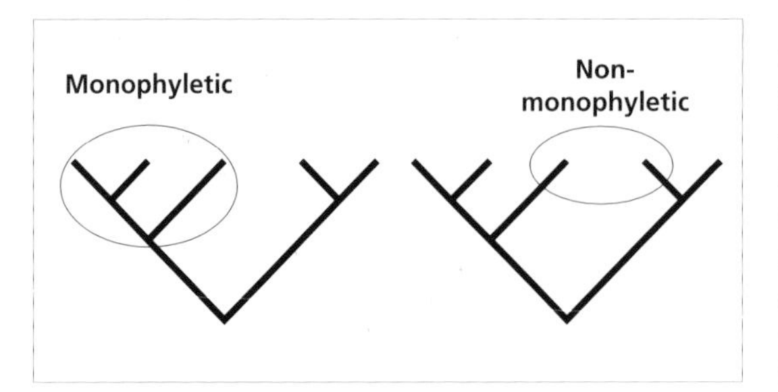

Fig. 2.19 The difference between monophyly and non-monophyly. A monophyletic group includes all descendants of their common ancestor, whereas in a non-monophyletic group one or more descendant is not included.

Many authors distinguish between two kinds of non-monophyly: **paraphyly** and **polyphyly** (Fig. 2.20). Paraphyletic groupings are based on shared primitive characters (plesiomorphies), and hence typically exclude one or more taxa that have autapomorphies. The paradigm example is the 'reptiles' as classically defined, which excludes birds because of their novel anatomy and behaviour, even though crocodiles are more closely related to birds than they are to other reptiles. Polyphyletic groups are typically assemblages of taxa that have been erroneously grouped on the basis of convergent characters, such as 'vultures'. The New and Old World vultures look strikingly similar but have evolved independently from different ancestors (storks and birds of prey, respectively).

Cladistic classifications have often been criticised as being limited in that they tell us little about the organisms themselves beyond who their nearest relatives are. For example, advocates of more traditional approaches to classification like to be able to reflect the evolutionary innovation shown by birds compared to their closest living relatives (the crocodiles) by elevating birds to their own class and consigning their dowdy relatives to the non-descript group the 'reptiles'. However, rigorous and objective alternatives to cladistic classifications have been hard to construct. Cladistic classifications also have the great advantage of being immune to variation in rates of evolution. For

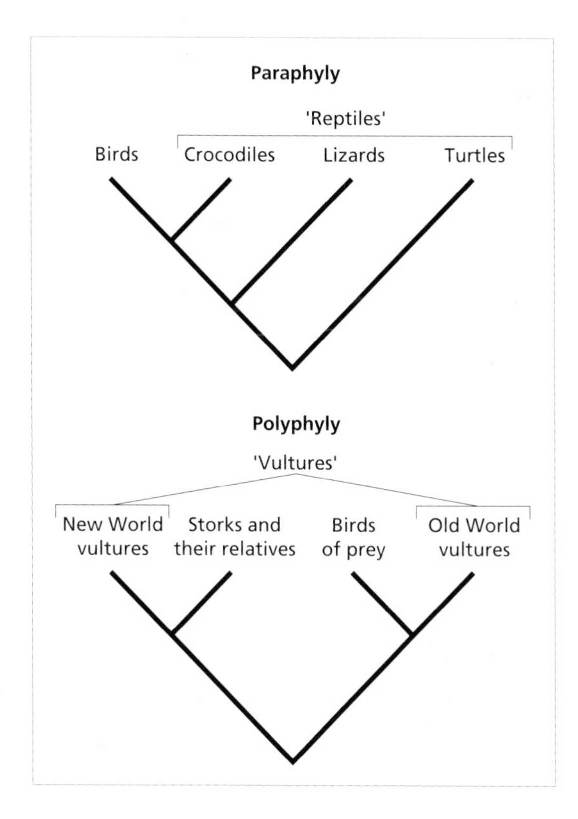

Fig. 2.20 Two kinds of non-monophyletic groups. 'Reptiles' are a paraphyletic grouping that is based on the absence of the apomorphic ('derived') characters possessed by birds. 'Vultures' are a polyphyletic grouping comprising two groups of birds that have independently evolved similar morphology and habits from different ancestors.

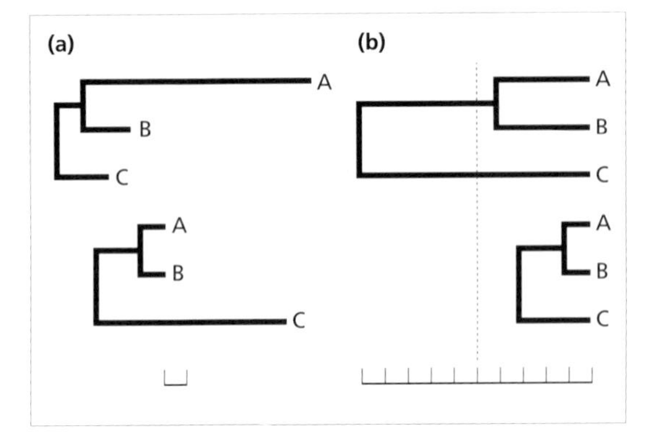

Fig. 2.21 Examples of variation in rate of evolution among genes from the same organisms. For all four trees the cladistic group AB is preserved. Dashed line is an arbitrary threshold for placing species in different higher taxonomic groups.

example, in Fig. 2.21(a) we have additive trees for two different genes from the same organisms A–C. In the first case the gene from species A has evolved much more rapidly than its homologous gene in B and C. Classifying the three species based on similarity would lead us to group B and C together. However, a second gene indicates that A and B are more conservative than C, in which case we might prefer to group A and B to the exclusion of C. If similarity is our criterion for delimiting taxonomic groups then we would have to choose between these two genes, essentially an arbitrary choice. Note however that the cladistic relationship remains the same in both trees. Figure 2.21(b) shows a different case where the rate of evolution for a given gene is constant. This might lead us to base taxonomic groups on amount of genetic divergence. However, using this method another gene evolving at a slower rate might lead to a different classification. Again, the cladistic groupings have not changed.

2.4.2 Gene trees and species trees

The naïve expectation of molecular systematics is that phylogenies for genes match those of the organisms, hence obtaining the first necessarily gives us the second. However, there are a number of reasons why this need not be so. The first is that gene duplications may result in a species containing a number of distinct but related sequences. In the example shown in Fig. 2.22 three species A–C each have two copies of the same gene, α and β. A phylogeny for all six genes allows us to correctly recover the organismal phylogeny ((A, B), C) from either the α or β genes. However, if we were unfortunate enough to sequence only genes 1, 3 and 5, and were unaware that they were part of a larger gene tree, we would infer that the organismal tree was ((A, C), B) because gene 3 from species C is more closely related to gene 1 from species A than is gene 5 from species B, even though species B is actually closer to species A than is species C.

This example illustrates that organismal phylogeny can be correctly inferred only if we have the complete set of genes, or if we restrict ourselves to

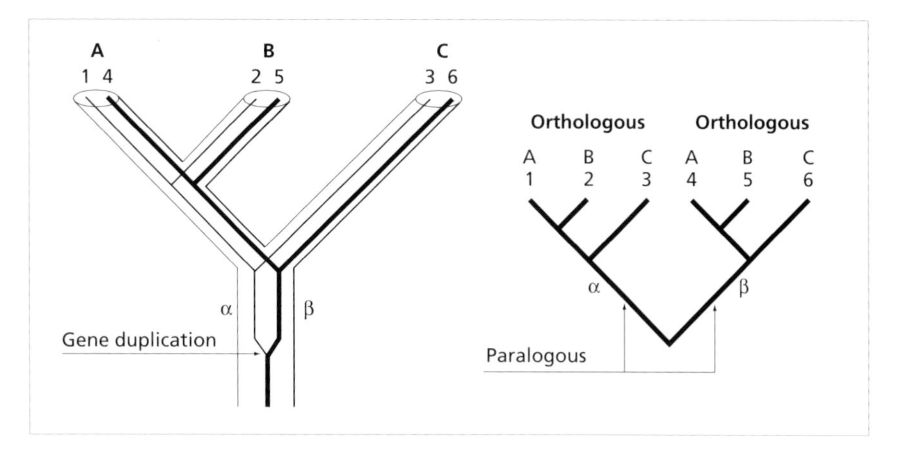

Fig. 2.22 Phylogeny for three species A–C and six genes that stem from a gene duplication resulting in two paralogous clades of genes, α and β. The α genes 1–3 are orthologous with each other, as are the β genes 4–6; however each α gene is paralogous with each β gene as they are separated by a gene duplication event, not a speciation event.

a set of genes that have not themselves undergone a duplication. That is, we require **orthologous** genes. Two homologous genes are orthologous if their most recent common ancestor did not undergo a gene duplication, otherwise they are termed **paralogous**. In Fig. 2.22 genes 1–3 are orthologous, as are genes 4–6, but any pair of α and β genes are paralogous.

2.4.3 Lineage sorting and coalescence

Another process that complicates the relationship between organismal and gene phylogeny is **lineage sorting**. Even if we restrict our attention to orthologous genes for the reason given above, the presence of ancestral polymorphism coupled with the differential survival of those alleles can result in allele phylogeny not matching organismal phylogeny. If we start with a pair of orthologous alleles and travel down the tree (i.e. backwards in time) we will eventually encounter their most recent common ancestor. This is the point at which the two gene lineages **coalesce** (Fig. 2.23) and the time at which this occurs is the **coalescence time**.

In the example shown in Fig. 2.23, alleles 3 and 4 have a recent coalescence point which lies within their organismal lineage B. However, alleles 1 and 2 have a more ancient coalescence time which pre-dates the age of their lineage A, that is, they are older than species A. Furthermore, even though alleles 1 and 2 are both found in the same species, they are not each other's closest relative. The presence of a paraphyletic pair of alleles in lineage A may have consequences later on in evolutionary time. Imagine that shortly after species A and B diverged, and while alleles 1 and 2 were still both extant, species A itself speciated into species A_1 and A_2 (Fig. 2.24). In this case, species A_1 inherited

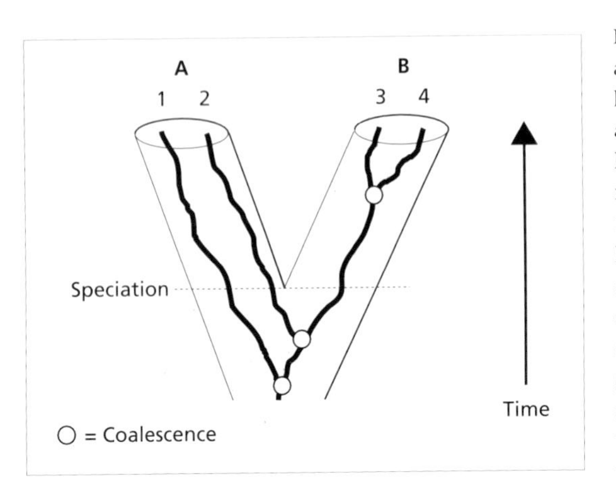

Fig. 2.23 A gene tree for four alleles (1–4) in two organismal lineages, A and B. The points at which pairs of allele lineages join (coalesce) are marked by open circles. Alleles 3 and 4 coalesce within lineage B, but alleles 1 and 2 are older than lineage A. Note especially that alleles 1 and 2 do not form a monophyletic group—2 is more closely related to 3 and 4 than it is to the other allele (1) found in the same species.

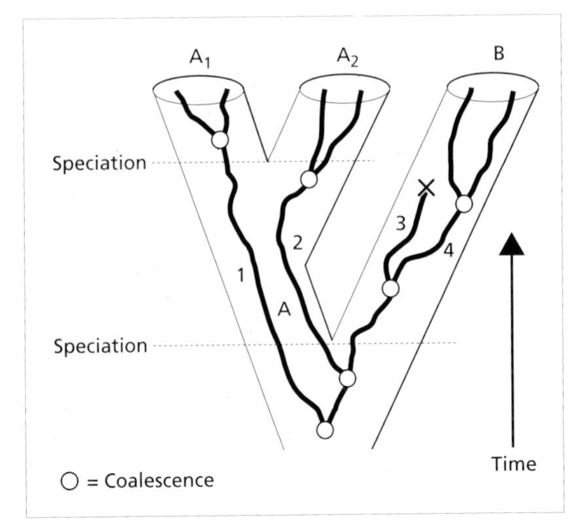

Fig. 2.24 One possible extrapolation into the future of the gene and species trees shown in Fig. 2.23. In this instance species A diverged into species A_1 and A_2. The two alleles (1 and 2) present in A when it speciated were inherited by A_1 and A_2 respectively. Allele 3 has gone extinct.

allele 1 and species A_2 inherited allele 2. Put another way, the two allele lineages 1 and 2 were sorted among the descendants of A. Note that even though all three species have monophyletic suites of alleles, the alleles found in A_2 are actually more closely related to species B than to its sister species A_1. Were we to use the phylogeny of these alleles to infer the phylogeny of the three species A_1, A_2 and B we would incorrectly conclude that the species tree was (A_1, (A_2, B)). This hypothetical example illustrates the problem of lineage sorting. If the alleles present in a lineage prior to that lineage speciating are not monophyletic then the distribution and relationships of these alleles need not accurately reflect the phylogeny of the organisms themselves.

Lineage sorting is likely to be a problem for organismal phylogenetics if the time it takes for alleles within a lineage to coalesce is greater than the interval between successive speciation events.

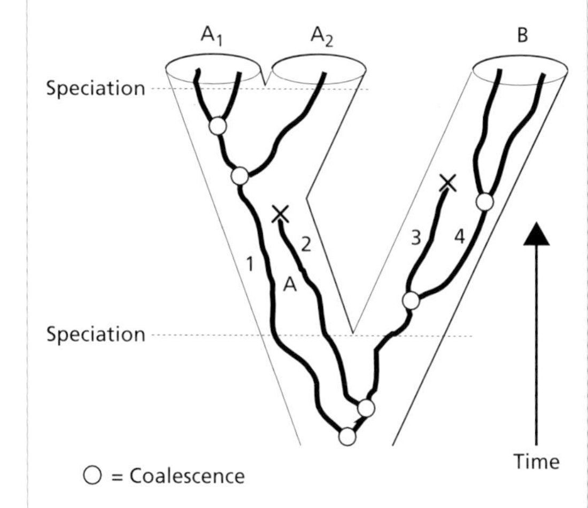

Fig. 2.25 The same situation as in Fig. 2.24 but lineage A speciating later in time, by which time allele 2 has gone extinct. Consequently species A_1 and A_2 inherit a monophyletic set of alleles.

Figure 2.25 shows an alternative extrapolation of Fig. 2.23 in which species A splits into two daughter species later than in Fig. 2.24, after allele 2 has gone extinct. Consequently, when A speciates its descendants receive a monophyletic set of alleles. In this case, allele phylogeny faithfully reflects species phylogeny.

The key difference between Fig. 2.24 and Fig. 2.25 is the length of time between successive speciations of the same lineage. Due to a combination of chance and selection, allele lineages will either persist, radiate or go extinct. The longer the interval between speciation events the greater the chance that these processes will result in lineages with a monophyletic set of alleles. The importance of gene trees and coalescence times for modern population genetics is discussed in more detail in Chapter 4.

2.5 Consensus trees

Often we want to compare trees derived from different sequences, or from the same sequences using different methods. Given two or more different trees we can ask 'what do these trees agree on?' **Consensus trees** are trees that represent the commonality (if any) among a set of trees. For example, consider the two trees for hominoids shown in Fig. 2.26. The two trees are very similar, but tree 1 groups humans and chimps together, whereas tree 2 groups the chimp and gorilla. Both trees agree that humans and African apes are more closely related to each other than each is to the orang-utan, and that the great apes and humans form a clade that excludes the gibbon. We can summarise the agreement between these trees in a consensus tree, which contains a polytomy indicating that there is conflict concerning our relationships with the African apes. This is an example of a soft polytomy (see section 2.1.1); the consensus tree is not indicating that the African apes

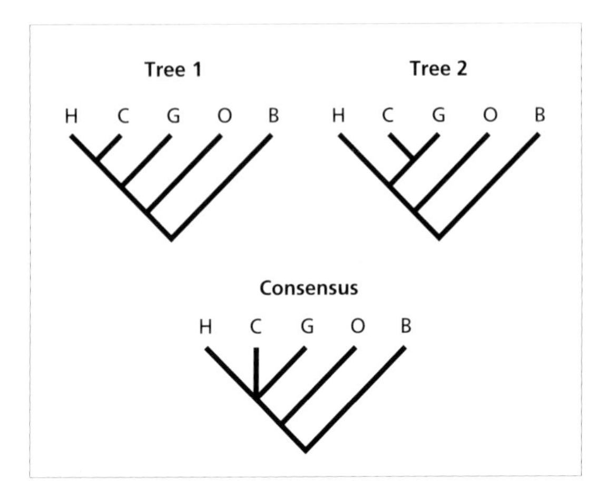

Fig. 2.26 Two different trees for humans (H), chimps (C), gorillas (G), orang-utans (O) and gibbons (B), and their consensus tree.

and humans speciated simultaneously, rather we have insufficient evidence to determine the exact order of speciation. There is a range of different consensus methods, three of which are discussed in Box 2.4.

Box 2.4 Types of consensus tree

A consensus tree summarises information common to two or more trees. There is a range of different methods which differ in what aspect of tree information they use, and how frequently that information must be shared among the trees to be included in the consensus. We discuss three commonly used methods here. The **strict consensus** tree includes only those groups (or splits, see section 2.1.6) that occur in all the trees being considered. Among the three trees below, only the split {{A, B, C}, {D, E}} is common to all three trees, and so the strict consensus of these trees contains just that split.

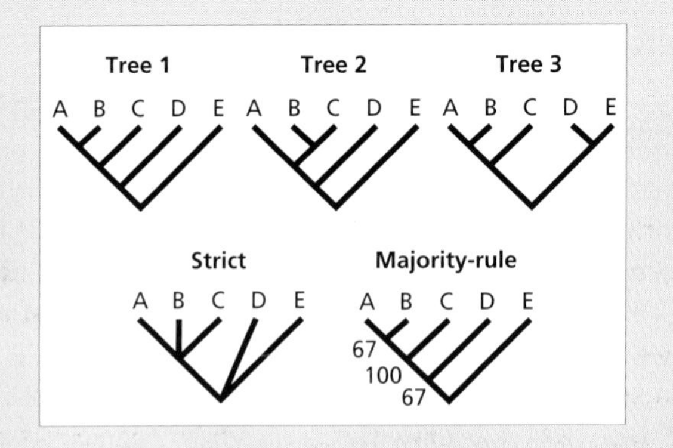

Notice, however, that there are some splits such as {{A, B}, {C, D, E}} that are found in two of the three trees. We can relax the requirement that a split be in

continued

Box 2.4 *continued*

all trees; for example, we could retain those splits found in a majority of trees, and this is exactly what the **majority-rule consensus** does. Any split in more than half the trees is included in the consensus tree, so the two splits shared by two of the three trees are also included. Note that any split in the strict consensus tree will also be in the majority-rule tree. The splits in the majority-rule tree are usually labelled by what percentage of trees that split occurs in.

Strict and majority-rule consensus methods are two examples of methods that use splits. However, trees that have no splits in common (and hence will give completely unresolved strict and majority-rule consensus trees) may still have points of similarity. For instance, the two trees below share no splits, yet both agree that if we consider just sequences B, C and D, B is more closely related to C than either is to D.

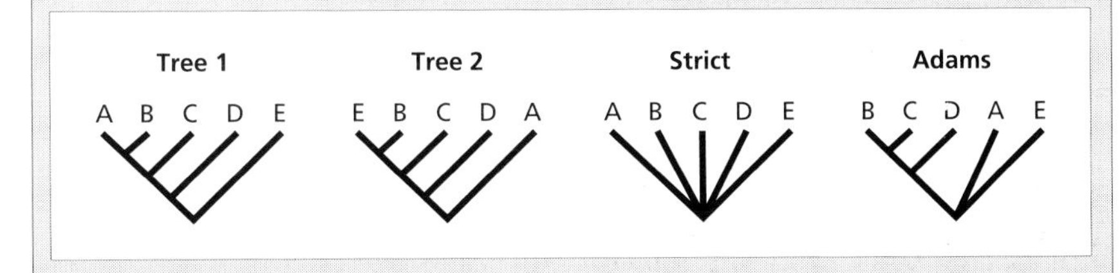

The strict consensus tree for these trees is a star tree, but this somewhat overstates the differences between the two trees. The **Adams consensus** tree captures the information that both tree 1 and tree 2 have the subtree ((B, C), D). Although Adams consensus trees can sometimes be a little difficult to interpret, they are very useful in situations where one or more sequences have very different positions on different trees, but there is a subset of sequences upon whose relationships the different trees agree.

2.6 Networks

So far in this chapter we have assumed that the evolutionary relationships among sequences and organisms are best represented by a tree. In other words, we are using the tree to model reality. However, the actual evolutionary history may not be particularly tree-like, in which case analyses that assume a tree may be seriously misleading.

For example, the metaphor of a 'family tree' is itself rather misleading, as anyone will know who has drawn one. A tree has single root and branches outwards such that the branches never meet, whereas in a family tree or pedigree every time a male and female organism mate their branches fuse. Generally the history of each individual gene can be adequately represented by a tree; however, in cases where a gene has undergone recombination a

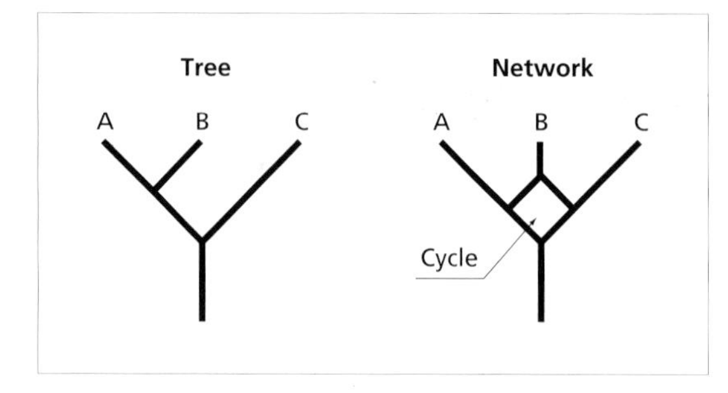

Fig. 2.27 A tree and a network. Networks contain cycles, whereas trees do not.

network may be more appropriate. A network (Fig. 2.27) contains one or more cycles (a set of nodes where it is possible to trace a path starting and ending at the same node without visiting any other node more than once).

2.7 Summary

1 Evolutionary relationships can be represented by a variety of trees. Cladograms depict relative recency of common ancestry, additive trees incorporate branch lengths, ultrametric trees can be used to represent evolutionary time.
2 Trees may be either rooted or unrooted, but only rooted trees have an evolutionary direction.
3 The number of possible trees increases rapidly with increasing number of sequences.
4 Evolutionary trees can depict ancestor–descendant relationships.
5 Distances satisfying the 'four-point condition' define a corresponding tree.
6 Gene trees may differ from species trees.

2.8 Further reading

Maddison and Maddison (1992) give an excellent introduction to trees and phylogenies. Barthélemy and Guénoche (1991) provide a detailed and elegant discussion of the kinds of trees, and the relationship between distances and trees. See Poinar and Poinar (1995) for the recovery of DNA from amber, Austin *et al.* (1997) for a sceptical review of the authenticity of geologically ancient DNA, and Smith (1994) on the problem of ancestors. The HIV example is taken from Holmes *et al.* (1992). For the distinction between hard and soft polytomies, see Maddison (1989). Swofford (1991) provides an excellent review of consensus methods. For a discussion of classification see chapter 14 in Ridley (1996).

Chapter 3
Genes: Organisation, Function and Evolution

3.1 Levels of genetic organisation

Perhaps no science has made such rapid progress in recent years and has more potential to change the way we live as the science of genetics. Determining the laws of heredity was one of the greatest achievements of biological endeavour and today this knowledge is allowing us to make great advances in medicine, through the identification of genes which make us susceptible to illness, in agriculture, with the development of highly productive plants and animals, and in evolution where the genetic similarities between species are being used to reconstruct their evolutionary history.

The first to realise that differences in the physical characteristics of organisms could be caused by underlying differences in 'heritable factors' was an Augustinian monk called Gregor Mendel who published his findings in 1866, just 7 years after the appearance of Darwin's *On The Origin of Species*. These heritable factors which determine much of the physical make up of organisms, including the size, shape and colour of the pea plants Mendel studied, are called **genes**. Genes themselves are composed of a more fundamental molecule called **deoxyribonucleic acid** or **DNA**. DNA has two extremely important properties: first, it contains the instructions for how organisms should be put together in the enormous variety of ways that characterises life on earth, and second it can be copied or **replicated** so that these instructions are passed on to successive generations. As evolutionists we are interested in another property of DNA: that the process of replication is not perfect so that errors, or **mutations**, continually arise which provide the raw material upon which evolution works. Furthermore, because mutations gradually accumulate through time, so that species which have diverged recently usually differ by fewer mutations than those which went their separate ways in the distant past, we can use the sequence of mutations to retrace the pathways of evolutionary change. Genes are therefore valuable archives of evolutionary information.

In this chapter we will outline how genetic information is organised and how this organisation evolved, as well as discussing some of the ways in which genes manage to perform their extraordinarily diverse range of functions.

3.1.1 DNA, proteins and chromosomes

At a biochemical level, DNA is known as a **nucleic acid** composed of a continuous string of four basic molecules called **nucleotides**. Each nucleotide contains a phosphate group, a sugar (deoxyribose) and one of four **bases**—**adenine (A)**, **cytosine (C)**, **guanine (G)** and **thymine (T)**, which are linked together in a long sequence. The four bases can be placed into two groups depending on their chemical structure: adenine and guanine form one group called **purines**, while cytosine and thymine are members of another called **pyrimidines**. There are about 3.3×10^9 bp (base pairs) in the human genome and it is their precise linear order which determines the function of any particular gene. Mutations in genes are simply changes in the running order of bases.

An example of a DNA sequence is shown in Fig. 3.1. The sequence in (a) is part of the *pol* (polymerase) gene of HIV-1 that produces the enzyme reverse transcriptase, which the virus uses to replicate itself. The sequence in (b) is a mutant form, or an **allele**, of this gene where an A base has changed to G base. This mutation is of enormous importance because it enables HIV-1 to develop resistance to the drug AZT (zidovudine) which is often used to treat people with this virus infection.

Genetic information can also be carried by another nucleic acid called **ribonucleic acid (RNA)** which has similar properties to DNA, although ribose replaces deoxyribose as the sugar used, and the base thymine is replaced by another called **uracil (U)**. RNA has a critical role in the processing of genetic

(a)

Position 70

Amino acid	Glu	Asn	Pro	Thr	Lys	Trp	Lys	Lys	Lys
DNA	GAA	AAT	CCA	ACT	AAA	TGG	AAA	AAG	AAA

(b)

DNA	GAA	AAT	CCA	ACT	A**G**A	TGG	AAA	AAG	AAA
Amino acid	Glu	Asn	Pro	Thr	Arg	Trp	Lys	Lys	Lys

Fig. 3.1 A DNA sequence. Part (a) shows the DNA sequence from part of the *pol* gene of HIV-1 and the amino acids this sequence encodes. Part (b) shows an A to G mutation at amino acid 70 (lysine changes to arginine) which confers resistance to the drug AZT.

information (see section 3.2.2) and some organisms, such as many of the viruses which cause serious disease in humans, have genomes composed solely of RNA. Unlike DNA, some RNA molecules (called **ribozymes**) are able to undertake certain biochemical (catalytic) reactions without the aid of proteins. This has led to the idea that the first living systems on earth may have been composed of RNA rather than DNA—an environment called the 'RNA world'.

Returning to DNA, we find that a molecule of this nucleic acid is actually composed of two sequences (or **strands**) of nucleotides which are wound against each other in opposite directions (anti-parallel) so that they resemble two springs coiled against each other: the bases of each strand are located on the inside of the molecule, and connected to each other by hydrogen bonds, while the outside of the molecule is composed of a backbone of sugars and phosphates (Fig. 3.2). This structure is called a **double-helix**. The bases on the two strands are joined following certain rules: the C nucleotides on one

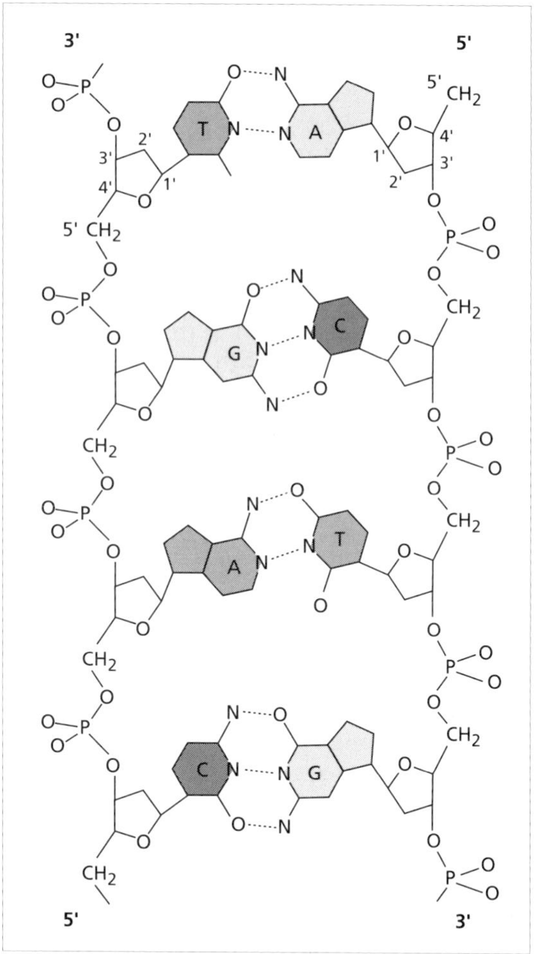

Fig. 3.2 The DNA double-helix. A DNA sequence, which specifies the properties of each gene, is read (transcribed) in a particular direction—referred to as 5′ to 3′—along the molecule. Bases on each strand of the helix are linked by a sugar–phosphate backbone and the bases are joined between strands by hydrogen bonds which bind A to T (or U) bases and C to G bases. From Dickerson (1983) with permission (illustration by Irving Geis).

strand pair with G nucleotides on the other, while the A nucleotides pair with T nucleotides (or U nucleotides in the case of RNA). This is known as Watson–Crick base pairing (although other forms of base pairing exist), because it was James Watson and Francis Crick who in 1953 first proposed that DNA takes on the shape of a double helix, a discovery for which they were later awarded the Nobel prize along with Maurice Wilkins.

As we shall see a little later in this chapter, although genes are composed of DNA, not all DNA forms genes. This is especially true of eukaryotes where most DNA appears to be **non-coding DNA** and may have no function. In contrast, **coding DNA** produces **proteins**—more complex molecules that perform many of the essential tasks of life, **regulatory sequences** that control how proteins are made, or special RNA molecules, the most important of which are the rRNAs and tRNAs which are used in protein synthesis. These will be discussed in more detail in section 3.2.2.

Proteins can be classified according to their function. Some, like actin and collagen, make up the physical structure of cells and tissues. Others, called **enzymes**, catalyse chemical reactions, as in the case of the polymerases which are important in DNA replication (see section 3.2.1). Another set of proteins are used in the transportation and storage of chemicals in the body, such as haemoglobin and myoglobin, which transport and store oxygen, respectively, whilst others, like the hormones, act to regulate biological processes. Finally, some proteins are designed to recognise foreign organisms and operate as part of the immune system, a good example being the immunoglobulins.

Each protein is itself composed of smaller units called **amino acids**, and each amino acid is determined by a specific sequence of DNA (how this is done will be described in a moment). For example, in the case of the *pol* gene of HIV-1, viruses which are susceptible to AZT have a lysine (Lys) amino acid at position 70 in the gene, while resistant strains have an arginine (Arg) (Fig. 3.1). There are 20 naturally occurring amino acids, signified by different letters, which have varying physical and chemical properties (Table 3.1).

Proteins are produced by a sequence of these amino acids linked together by **peptide bonds**, where the α-carboxyl carbon of one amino acid joins with the α-nitrogen of another, to form a molecule called a **polypeptide**. Polypeptide sequences vary greatly in length, from just tens of amino acids (the mature insulin protein in humans has only 51 amino acids) to many hundreds. Although it is possible to think of proteins as comprising a simple linear polypeptide sequence (known as the **primary sequence**), they usually take on far more complex three-dimensional structures—that is, they can be folded in different ways—which play a vital role in determining their function. The smallest of these structures are called **motifs** and represent runs of highly conserved amino acids. One large group of motifs are those specifying C_2H_2 zinc fingers, 21 to 26 amino acids in length, which are important components of DNA-binding proteins. If a motif, or more usually a group of motifs, can be folded independently then it is called a **module** (C_2H_2 zinc fingers also

Table 3.1 The 20 amino acids and their letter codes. Information about polarity taken from Li (1997).

Amino acid	One-letter code	Three-letter code	Polarity
Alanine	A	Ala	Non-polar
Arginine	R	Arg	Positive charge
Asparagine	N	Asn	Polar
Aspartic acid	D	Asp	Negative charge
Cysteine	C	Cys	Polar
Glutamine	Q	Gln	Polar
Glutamic acid	E	Glu	Negative charge
Glycine	G	Gly	Polar?
Histidine	H	His	Positive charge?
Isoleucine	I	Ile	Non-polar
Leucine	L	Leu	Non-polar
Lysine	K	Lys	Positive charge
Methionine	M	Met	Non-polar
Phenylalanine	F	Phe	Non-polar
Proline	P	Pro	Non-polar
Serine	S	Ser	Polar
Threonine	T	Thr	Polar
Tryptophan	W	Trp	Non-polar
Tyrosine	Y	Tyr	Polar
Valine	V	Val	Non-polar

correspond to modules). Modules can be thought of as the structural and functional core of proteins. For example, the proteins encoded by *HOM/Hox* genes, which are important factors in the development of animal body plans (see section 3.3.2), contain a module (comprising a single motif approximately 60 amino acids in length) called the **homeodomain**, which is also found in other genes involved in the control of developmental processes.

Even if amino acid sequences differ they can produce similar folded structures. The most common of these **secondary structures**, caused by regularly occurring patterns of chemical bonding, are the α-helix and β-sheet, which are important in determining the overall structure of most proteins. These small structural regions can be joined to produce a more complex folded architecture, the **tertiary structure**, in which amino acids far apart in the primary sequence can be placed close together. Related proteins tend to have similar tertiary structures. Finally, two or more tertiary structures can come together to produce a **quaternary structure**, composed of several different polypeptide chains, each with a slightly different function. A good example of such a complex multicomponent protein is haemoglobin. A molecule of human haemoglobin has four haem groups which bind oxygen, each associated with a particular polypeptide chain. In normal human adult haemoglobin two of the four polypeptide chains, labelled α, have 141 amino acid residues, while

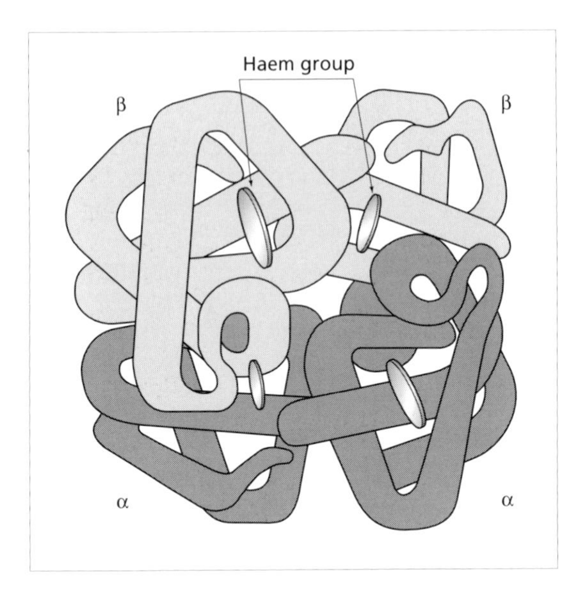

Fig. 3.3 Schematic view of the structure of human haemoglobin. The protein is made up of two α chains (141 amino acids) and two β chains (146 amino acids) which are shaded differently. Each of the four chains contains a haem group, shown by a disc, which binds oxygen. From Griffiths *et al.* (1993), with permission from WH Freeman.

the other two, labelled β, have 146 residues (Fig. 3.3). The related protein myoglobin, which stores oxygen in the muscles, has a single polypeptide chain of 153 amino acid residues.

So far we have discovered that genes are composed of DNA and usually produce proteins. There is, however, one more level of genetic organisation which needs to be considered—how DNA actually resides in the cell. This involves much larger genetic structures called **chromosomes**. The complete set of chromosomes, and hence all of an organism's genetic material, is also referred to as its **genome** and the position a gene occupies on the chromosome is its **locus**. A major part of modern genetic analysis involves finding the positions of gene loci on chromosomes. Most eukaryotes have more than two chromosomes, and sometimes several hundred. For example, the 12 068 kb genome of the yeast *Saccharomyces cerevisiae* is organised into 16 chromosomes. Bacteria and archaea usually have small genomes, with genes generally located on a single circular chromosome. Furthermore, in eukaryotes chromosomes are located within a membrane-bound nucleus whereas bacterial cells are simpler, and because there is no nucleus, the chromosome-bound DNA lies within the main body of the cell.

Chromosomes consist of DNA and proteins that bind tightly to this DNA. This complex of DNA and protein is called **chromatin**, and in eukaryotes it takes on two different forms: **euchromatin**, the most common configuration and where most genes are found, and **heterochromatin**, which is permanently in a condensed and inactive state in specific chromosomal locations, such as near the centromeres and telomeres, and which contains much fewer numbers of genes. Packaging is actually so tight that some 1–2 m of DNA can fit into a cell nucleus of only 1 μm in diameter!

In sexually reproducing organisms chromosomes can be subdivided into **autosomes**, which carry the genes used by both sexes, and a smaller set of chromosomes that are specific to each sex—the sex chromosomes. In humans, for example (Fig. 3.4), there are 22 autosomes and two sex chromosomes, designated X and Y (in birds these are called the W and Z chromosomes).

Chromosomes also take on distinctive shapes, typically with a long (q) and a short (p) arm separated by a restriction called a **centromere** (the region which enables chromosomes to segregate in mitosis and meiosis—see section 3.2.1). The ends of the chromosome are called the **telomeres**. Although they can be easily seen with microscopes, chromosomes are often treated with special fluorescent chemicals which show up bands on their surface. The most common of these **chromosome banding** techniques is Giemsa, which produces the so-called G bands, and this is often used in chromosome classification (sometimes even to reconstruct phylogenetic trees) and gene mapping. Other banding techniques make use of quinacrine hydrochloride (Q bands) and reversed Giemsa (R bands). As we shall discover in Chapter 7, G and R bands also correlate with major elements of genome organisation.

3.1.2 The genetic code

At this stage we can see that genetic information is organised at a number of levels: from the nucleic acids DNA and RNA, to the more complex proteins, and finally to the large cellular structures that are the chromosomes. The next question we need to address is how information flows from one level to another, and particularly how the base sequence of DNA can produce such a wide variety of proteins? The crucial concept here, and one of the most important in molecular biology, is that there is a **genetic code** which specifies how a combination of any of the four bases produces each of the 20 amino acids. The critical part of the code, which was deciphered in the 1960s, is that each amino acid is specified by a consecutive and non-overlapping sequence of three of the four bases. These triplets of bases are called **codons** and, with four bases, there are 64 possible codons. The genetic code is shown in Table 3.2.

Because there are only 20 amino acids, but 64 possible codons, the same amino acid is often encoded by a number of different codons, which usually differ in the third base of the triplet. Because of this repetition the genetic code is said to be **degenerate** and codons which produce the same amino acid are called **synonymous codons**. Three amino acids—arginine, leucine and serine—are encoded by six different codons, while another five can be produced by four codons which differ only in the third position ('**fourfold degenerate**' sites). A further nine amino acids are specified by a pair of codons which differ by a transition substitution at the third position ('**twofold degenerate**' sites), while isoleucine is produced by three different codons and methionine and tryptophan by only a single codon. Three of the 64 codons do not specify amino acids at all. These are the **stop codons** and they determine

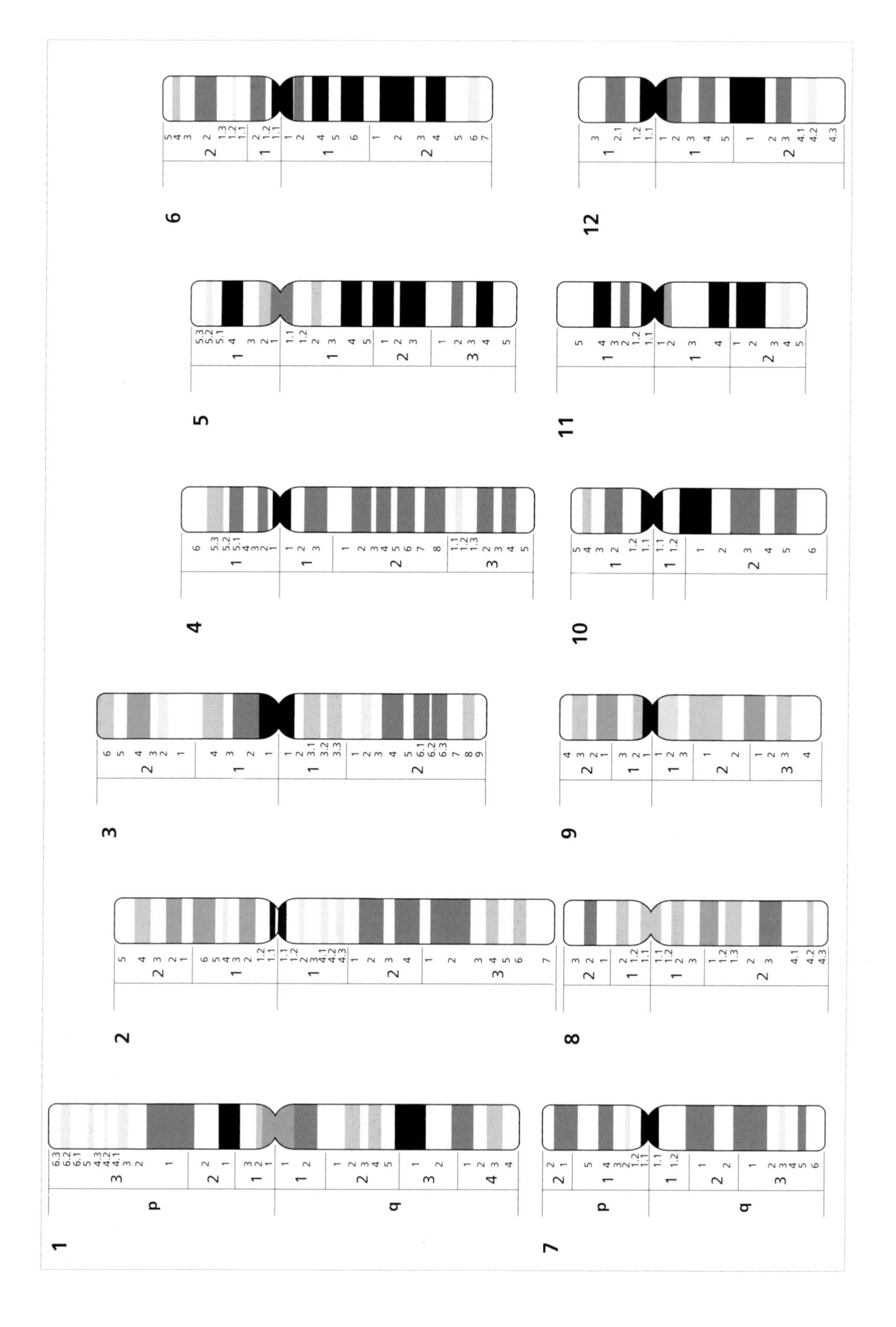

1
p
q
2
3
4
5
6
7
p
q
8
9
10
11
12

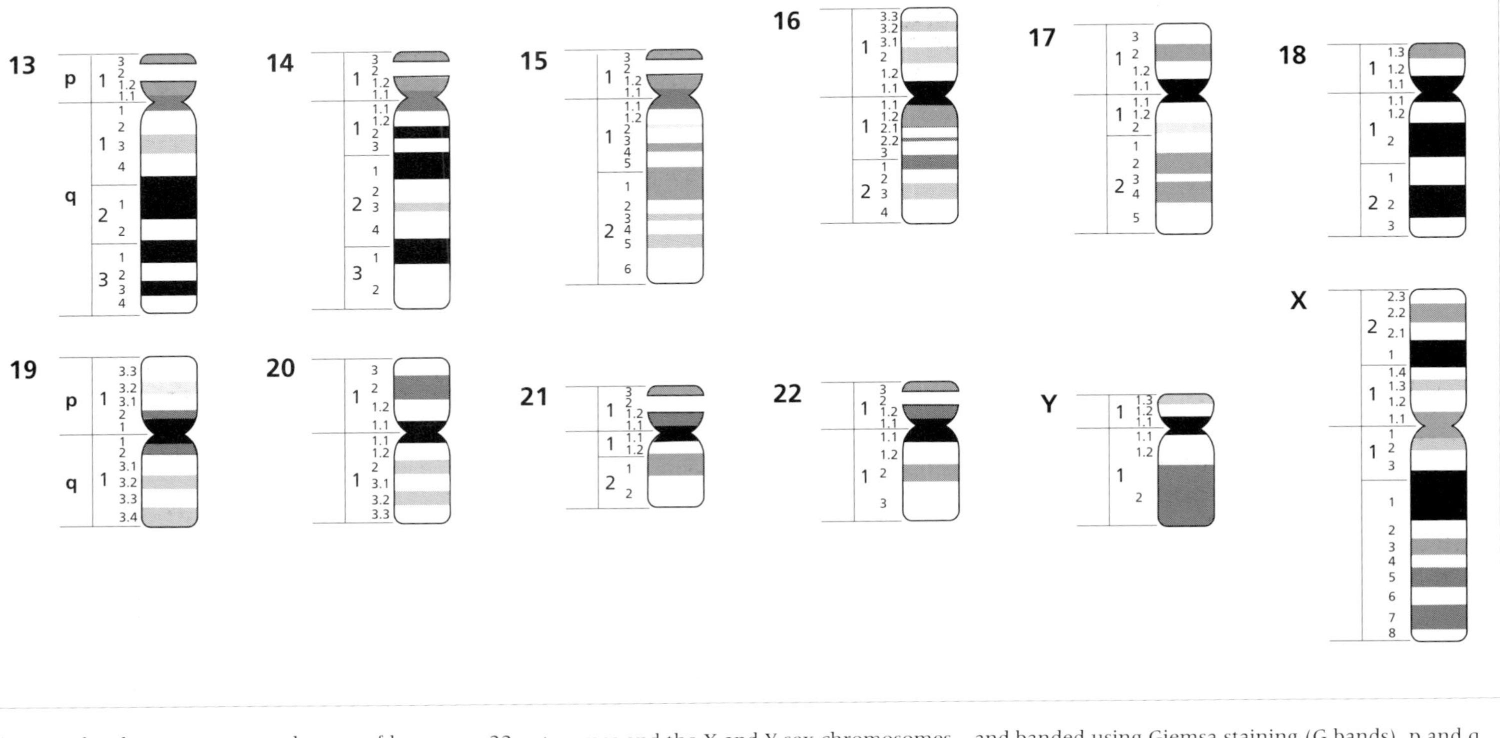

Fig. 3.4 The chromosome complement of humans—22 autosomes and the X and Y sex chromosomes—and banded using Giemsa staining (G bands). p and q refer to the short and long arms of the chromosome, respectively. These arms can then be divided into regions, marked by horizontal lines and assigned numbers, and then to individual bands, which are also numbered. For example, band 8p22 refers to band 2, region 2 on the p (short) arm of chromosome 8. Adapted from Watson *et al.* (1987).

Table 3.2 The genetic code.

Codon	Amino acid	Codon	Amino acid	Codon	Amino acid	Codon	Amino acid
UUU	Phe	UCU	Ser	UAU	Tyr	UGU	Cys
UUC	Phe	UCC	Ser	UAC	Tyr	UGC	Cys
UUA	Leu	UCA	Ser	UAA	Stop	UGA	Stop
UUG	Leu	UCG	Ser	UAG	Stop	UGG	Trp
CUU	Leu	CCU	Pro	CAU	His	CGU	Arg
CUC	Leu	CCC	Pro	CAC	His	CGC	Arg
CUA	Leu	CCA	Pro	CAA	Gln	CGA	Arg
CUG	Leu	CCG	Pro	CAG	Gln	CGG	Arg
AUU	Ile	ACU	Thr	AAU	Asn	AGU	Ser
AUC	Ile	ACC	Thr	AAC	Asn	AGC	Ser
AUA	Ile	ACA	Thr	AAA	Lys	AGA	Arg
AUG	Met	ACG	Thr	AAG	Lys	AGG	Arg
GUU	Val	GCU	Ala	GAU	Asp	GGU	Gly
GUC	Val	GCC	Ala	GAC	Asp	GGC	Gly
GUA	Val	GCA	Ala	GAA	Glu	GGA	Gly
GUG	Val	GCG	Ala	GAG	Glu	GGG	Gly

where the polypeptide sequence ends. UAG, also known as the amber codon, was the first stop codon to be discovered. The other stop codons are UGA (the opal codon) and UAA (the ochre codon). The beginning of a polypeptide sequence is also usually marked by a specific codon, although there is no codon which only has this function. In most genes, this **initiation codon** is AUG (methionine) although GUG and UUG are sometimes used in bacteria.

For a long time it was believed that the genetic code was universal and there has been much speculation about how this code evolved. However, it is now clear that some genetic systems use slightly different codes, most notably the mitochondrial genomes of various species (mitochondrial DNA is discussed in more detail in the next section). Perhaps the most common deviation from the universal code, which is found in the nuclear genome of mycoplasma and all mitochondrial genomes except those of plants, is that the stop codon UGA instead encodes the amino acid tryptophan (Table 3.3).

3.1.3 Mitochondria and chloroplasts

Not all DNA in eukaryotes is stored within the cell nucleus. In the cytoplasm that surrounds the nucleus there are structures, called **organelles**, which contain their own DNA and which are present in about 10^2 to 10^4 copies per cell. One type of organelle, found in most eukaryotes, are the **mitochondria**

Table 3.3 Changes from the 'universal' genetic code. Adapted from Jukes and Osawa (1991).

Universal code	Change	Genome: Nuclear	Genome: Mitochondrial
UGA (Stop)	Trp	*Mycoplasma*	All except plants
AUA (Ile)	Met		Yeast
			Metazoans (except echinoderms)
	Met to Ile		Echinoderms, platyhelminths
AGR (Arg)	Ser		Metazoans (except vertebrates)
	Stop		Vertebrates
AAA (Lys)	Asn		Flatworms, echinoderms
UAR (Stop)	Gln	*Acetabularia* Ciliated protozoans (except *Euplotes*)	
CUN (Leu)	Thr		Yeasts
CUG (Leu)	Ser	*Candida cylindracea*	
UAA (Stop) UAG (Stop)	Glu	Diplomonads (except *Giardia*)	
UGA (Stop)	SeCys*	Vertebrates, bacteria (in special enzymes)	

N = any nucleotide, R = any purine. *Selenocysteine.

which help degrade fats and sugars, including those important in respiration. The other type of organelle, found in plants and algae, are the **chloroplasts**. These contain chlorophyll and are where photosynthesis takes place.

Organelle genomes usually consist of a single DNA molecule and each gene is normally present only once. Unlike nuclear genes, both mitochondria and chloroplasts only encode proteins which are required for their essential functions—respiration and photosynthesis, respectively—and for the machinery needed to express these genes. Non-coding DNA is rare in mitochondria and chloroplasts.

Mitochondria originated more than a billion years ago when a free-living bacterium, the closest living relatives of which are the α-proteobacteria, entered a eukaryotic cell. This **endosymbiosis** event was beneficial to both organisms and mitochondria are now an essential feature of many eukaryotic cells. Bacterial endosymbiosis is also thought to be the origin of chloroplasts, with the cyanobacteria (blue–green algae) as the most likely ancestors and it may even be that eukaryotes themselves are the result of a fusion between a bacterium and an archaeon.

Gene content, genome size and genome organisation can vary widely among mitochondria, with only four genes common to all those studied so far. Mitochondrial genome (mtDNA) sizes range from only 6 kb (kilobases) up to more than 2000 kb, with the human version being some 16 kb in length (Fig.

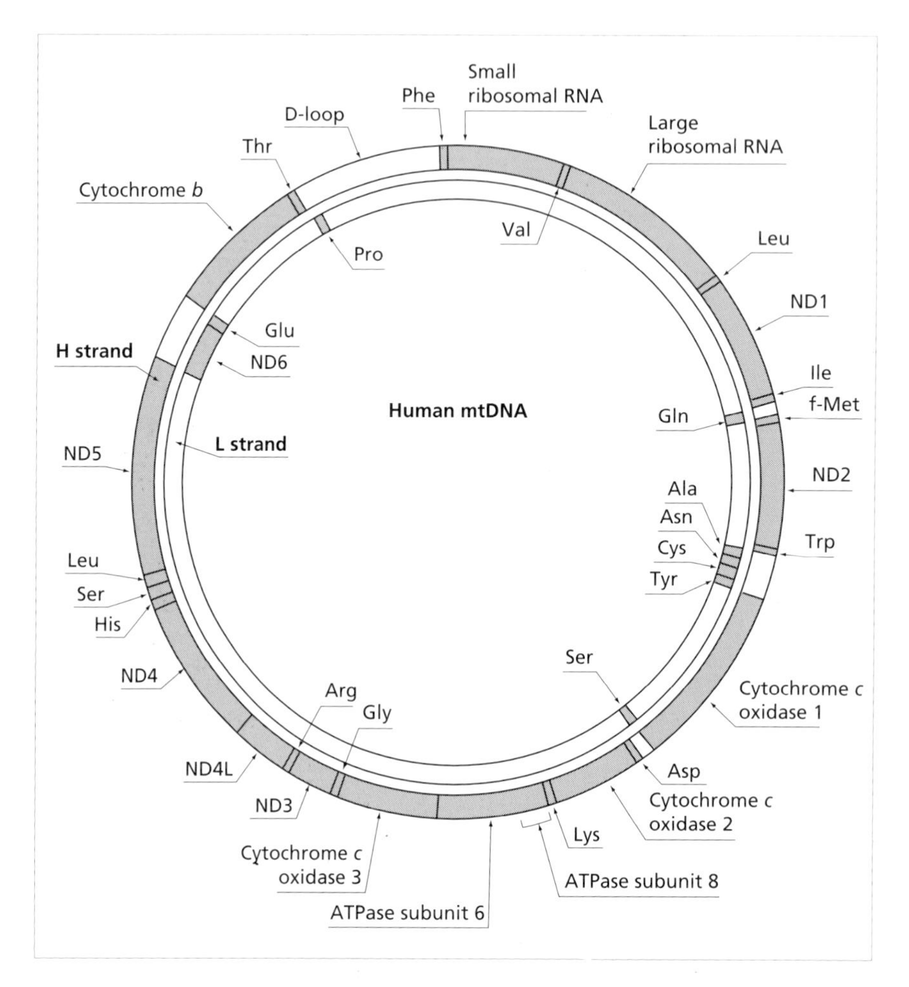

Fig. 3.5 The human mitochondrial genome (mtDNA). Different genes (proteins or tRNAs) are located on the two strands of DNA—the outer H (heavy) strand and the inner L (light) strand. The H strand is transcribed in a clockwise direction, the L strand anticlockwise. The three-letter amino acid codes specify each tRNA type. Adapted from Darnell *et al.* (1990).

3.5). In animals and plants, mitochondria are maternally inherited through the egg cytoplasm, the sperm cell usually having insufficient room for organelles. This maternal form of inheritance, along with the fact that mtDNA does not appear to undergo recombination (at least in higher animals) and in mammals evolves about tenfold faster than nuclear DNA, make it an extremely important study tool in molecular population genetics and systematics. However, these are not the only features which make mtDNA genomes different from their nuclear counterparts. Other deviations include a slightly different genetic code, modified tRNA and rRNA genes and, in many plant and fungal mtDNAs, self-splicing **introns**—sequences which are removed from the gene before the final protein product is synthesised (see sections 3.1.4 and 3.2.2).

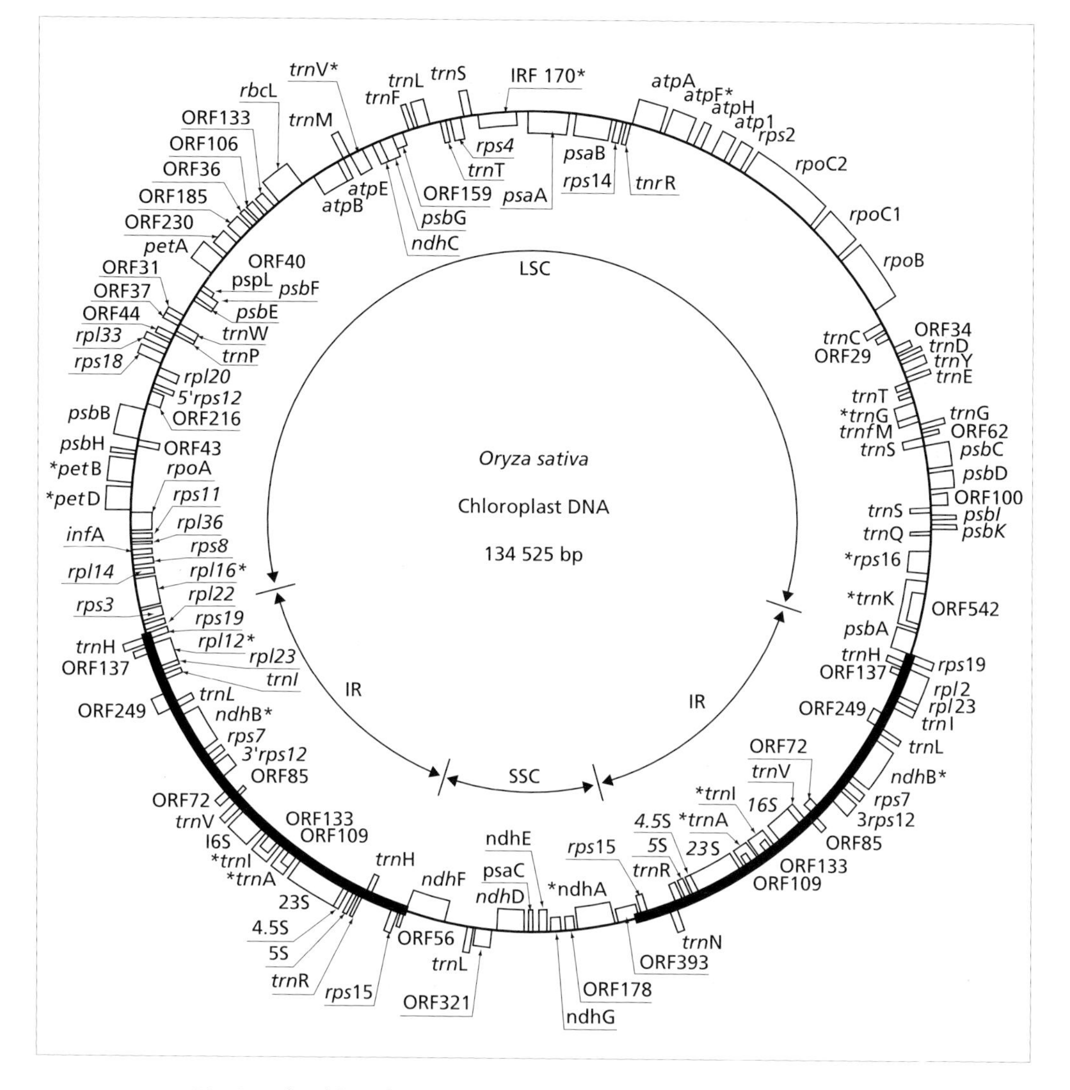

Fig. 3.6 The chloroplast genome (cpDNA) of rice (*Oryza sativa*). Like mtDNA, genes are located on two strands, with those on the outside (A) strand transcribed anticlockwise and those on the inside (B) strand transcribed in the clockwise direction. Split genes are denoted by asterisks and tRNAs by 'trn'. From Kendrew (1994) and originally Hiratsuka *et al.* (1989).

The chloroplast genome (cpDNA) is usually much larger than that of mitochondria, ranging in size from 120 to 220 kb. Much of the size variation is due to two inverted repeat (IR) regions which separate a large single copy (LSC) region and a small single copy (SSC) region. Although most chloroplasts carry the same genes there are some species-specific differences, sometimes because genes have migrated to the nuclear genome and been lost from the chloroplast genome. Figure 3.6 depicts the chloroplast genome of rice (*Oryza sativa*). This, like other chloroplast DNAs, consists of about 100 protein-coding genes, most of which are involved in protein synthesis or photosynthesis, about 30 tRNA genes and four rRNA genes. As with mitochondrial genomes, introns have also been found in some chloroplast genes.

Plant chloroplast genes evolve some four- to fivefold more slowly than those in the nucleus (which in turn evolve about the same rate as mammalian nuclear genes—see Chapter 4), but about threefold faster than plant mtDNA genes. They have also been frequently used in molecular systematic studies, particularly sequences from the large subunit of ribulose-1,5-biphosphate carboxylase—*rbcL* (see Chapter 7). However, the phylogenetic analysis of cpDNA is made more complicated by intragenome recombination, strong codon bias (particularly for A- and T-ending codons) and, in non-coding regions, patterns of nucleotide substitution that are strongly influenced by the base composition at adjacent sites.

3.1.4 The structure of genes

Now that we have seen how genetic information is organised at the DNA, protein and chromosome levels, and how these levels interact with each other, it is necessary to examine the structure of genes in a little more detail. Because producing a fully functional organism is a complex task, genes come in many forms. Those which produce a protein, a tRNA, or an rRNA are often referred to as **structural genes**, while those which control when and how genes are expressed are often referred to as **regulatory genes** (or 'regulatory sequences', although some of these may also encode proteins). Some, **housekeeping genes**, need to be expressed in all cell types, such as those involved in protein synthesis (see section 3.2.2). Other, **tissue-specific genes** are only expressed in a particular cell or tissue type. Insulin, for example, which performs a number of functions including regulating the uptake and metabolism of glucose, is only expressed in pancreatic β-cells.

Whatever their function, all genes comprise a coding region which specifies a polypeptide chain or an RNA molecule (Fig. 3.7) and which can vary dramatically in length. In the case of protein-coding genes, a single coding region usually produces a single protein. But this is not always the case. In some circumstances, particularly in bacteria and viruses, a coding sequence can be translated starting from the first, second or third bases in the codon (that is, using different **reading frames**)—and so produce multiple proteins. An extreme example of this is hepatitis B virus (HBV), one of the commonest and most serious viral infections of humans, and which has 63% of its 3.2 kb genome translated in more than one reading frame.

The coding regions of genes are normally flanked by regulatory sequences which control when the gene is turned on and off and how much of the gene product is made. Such control of **gene expression** governs two processes, **transcription** and **translation**. These processes are key aspects of protein

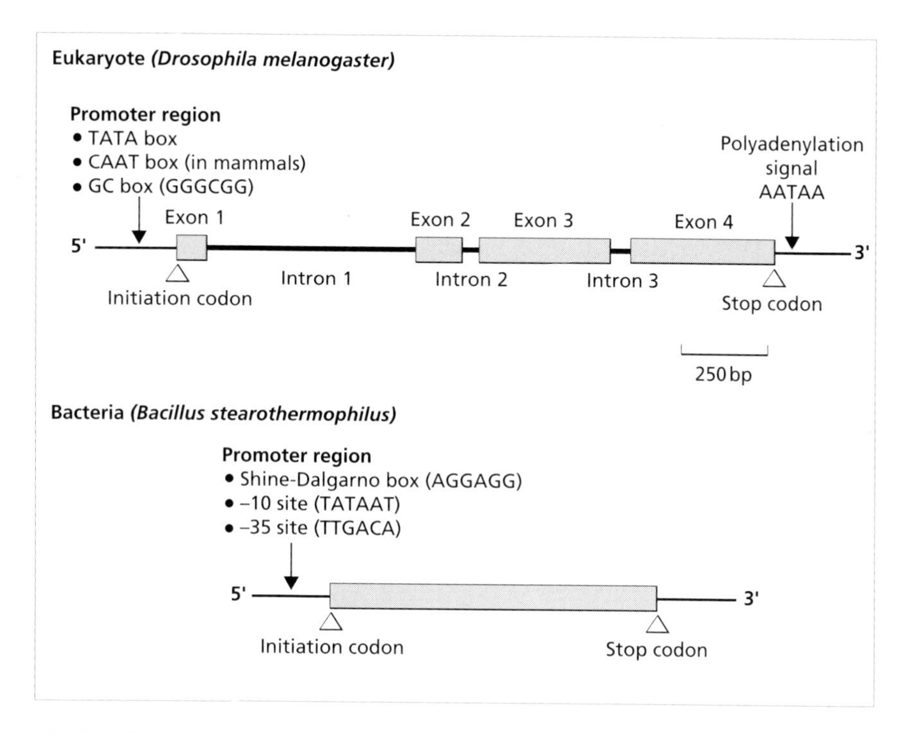

Fig. 3.7 The structure of a typical protein coding gene—alcohol dehydrogenase (*Adh*)—from a eukaryote (*Drosophila melanogaster*) and a bacterium (*Bacillus stearothermophilus*). Although both genes contain a coding region (shaded grey), the *Drosophila Adh* gene also contains spliceosomal introns (thick lines) which separate those parts of the gene which encode amino acids—exons. Spliceosomal introns are not present in Bacteria or Archaea. Some of the conserved motifs typically found in the promoter regions of eukaryote and bacterial genes are also shown, as are the positions of the initiation and stop codons and the polyadenylation signal. The coding regions (and introns) have been drawn to scale, while the 5′ and 3′ flanking regions have not for ease of presentation.

synthesis and are therefore described in more detail in section 3.2.2 (although you may want to glance at that section now to make the most of the next two paragraphs). One of the most important groups of these regulatory sequences are the **promoters** which rest on the 5′ side ('upstream') of coding regions—often called the **promoter region** (Fig. 3.7). These contain sites where RNA polymerases, enzymes crucial to the transcription process, can bind. Another important set of regulatory sequences found in eukaryotes are **enhancers** which increase the rate of transcription. These are normally located further upstream than promoters.

Because of their functional importance, promoters have sequences that are often highly conserved between species. For example, most promoters in eukaryote and bacterial genes contain the sequence motif TATA—the **TATA box**. In bacteria this sequence is located about 10 bp before the site where transcription starts (15–30 bp in the case of eukaryotes) and forms part of a slightly larger motif with the consensus sequence TATAAT. This is known as the '**–10 site**' (or the 'Pribnow box') and allows the RNA polymerase to be correctly positioned for transcription. A second conserved motif in bacteria, with the consensus sequence TTGACA, is located about 35 bp before the transcription start site (the '–35 site') and also functions in the control of transcription. Finally, the **Shine–Dalgarno box**, a ribosome binding site with the consensus sequence AGGAGG, is found about 7 bp before the initiation codon (i.e. the ATG) in bacteria. In some bacterial genes, a set of structural genes with similar functions can be under the control of a single promoter. This is called an **operon** and means that the expression of all the genes can be coordinated to exactly when they are required. Operons are also found in the Archaea, the genes of which generally resemble those of bacteria in structure.

As well as containing the TATA box, the promoter regions of some eukaryotic genes also contain a 'CAAT box' located ~40 bp before the transcription initiation site and a 'GC box', which has the consensus motif GGGCGG, located ~110 bp before the transcription start site. Both these motifs are binding sites for transcription factors (proteins which regulate transcription). Conserved motifs are also found at the 3′ ends of genes, where they signify the end of transcription and increase the stability of the RNA transcripts produced. For example, the **polyadenylation signal** is marked by a conserved sequence motif, usually AATAA.

As well as their promoters, another, and highly important, difference between the genes of eukaryotes and those of Bacteria and Archaea is that the former often contain **introns**—sequences which are discarded ('spliced') from the coding regions of genes during protein synthesis (again see section 3.2.2). The remaining pieces of DNA, which together will encode the finished protein, are called **exons** (Fig. 3.7).

Introns occur frequently within eukaryotic genes and make up most of the length of very large genes. The number, size and organisation of introns also varies greatly from gene to gene: histones (important DNA-binding proteins) have no introns while the chicken pro-α_2-collagen gene has over fifty. The virus SV40 contains an intron of only 31 bp, whilst one of more than 210 000 bp is found in the human dystrophin gene (mutations in which cause Duchenne muscular dystrophy—a genetic disease characterised by the enlargement of the muscles), which is one reason why this gene has a total length of more than two million bases! Some introns are so large that they even contain genes within them, an example being the alcohol dehydrogenase (*Adh*) gene of *Drosophila* which is located within an intron of the much larger *outspread* gene, and introns have been found within other introns—'twintrons'! Although introns are removed from genes, and so are usually considered to be non-functional, there is a strong conservation of sequence around the intron–exon boundaries in eukaryotes: introns nearly always begin and end with the nucleotides GT and AG, respectively, probably because this allows them to be accurately spliced. Intron splicing is itself an extremely important process because a single gene can make different proteins if it is spliced at different places.

Introns can be placed into different classes depending on how they are spliced and their structure. Most introns in eukaryotes are **spliceosomal introns** (or 'nuclear introns') because they are spliced by a **spliceosome** consisting of proteins and RNA. Spliceosomal introns are found in animals, plants, fungi and the protist eukaryotes (that is, those that are not plants, animals or fungi) most closely related to these 'higher eukaryotes'. Their absence from Bacteria, Archaea, the more divergent protists, as well as mitochondria and chloroplasts, suggests that they have only recently entered eukaryotes (the evolution of spliceosomal introns is discussed more fully in Box 3.1). Other, **protein-spliced introns**, are found in some tRNA and rRNA genes.

Box 3.1 The evolution of introns

The discovery that eukaryotic genes are not always intact continuous sequences, but are usually split into segments by introns which do not code for proteins, was one of the biggest surprises of modern genetics. Not surprisingly, there has also been great speculation about how introns evolved.

There are two competing hypotheses for the origin of spliceosomal introns (Fig. B3.1). The first, proposed by Walter Gilbert, is that introns mark the boundaries between ancient genes, which encoded distinct proteins, that now exist only as exons. Throughout evolution these once independent proteins have gradually been put together in different combinations to produce the more complex proteins we see today by **exon shuffling**, a process which introns may have aided by acting as hotspots for recombination. Because of the antiquity of proteins, introns are also thought to be an ancient invention, arising before the divergence of Archaea, Bacteria and eukaryotes. This is the **introns-early** theory (or the 'exon theory of genes'). An alternative hypothesis proposes that introns invaded eukaryotic genes only recently, hence their absence from the genes of Archaea, Bacteria and ancient eukaryotes, and have been increasing in number since this time. This is the **introns-late** theory.

A crucial prediction of the introns-early hypothesis is that spliceosomal introns mark out structural or functional units within proteins, because these correspond to the ancient independent proteins. For example, introns occur at the same places in all known globin genes, including myoglobin and the plant leghaemoglobins (even though the latter have an extra intron), and their points of insertion have been claimed to mark out natural structural domains of this protein (although this is debated). However, although introns are found to lie between intact codons (rather than splitting single codons in two) more frequently than if they inserted randomly, in most cases they do not separate functionally distinct parts of proteins.

continued on p. 54

Box 3.1 *continued*

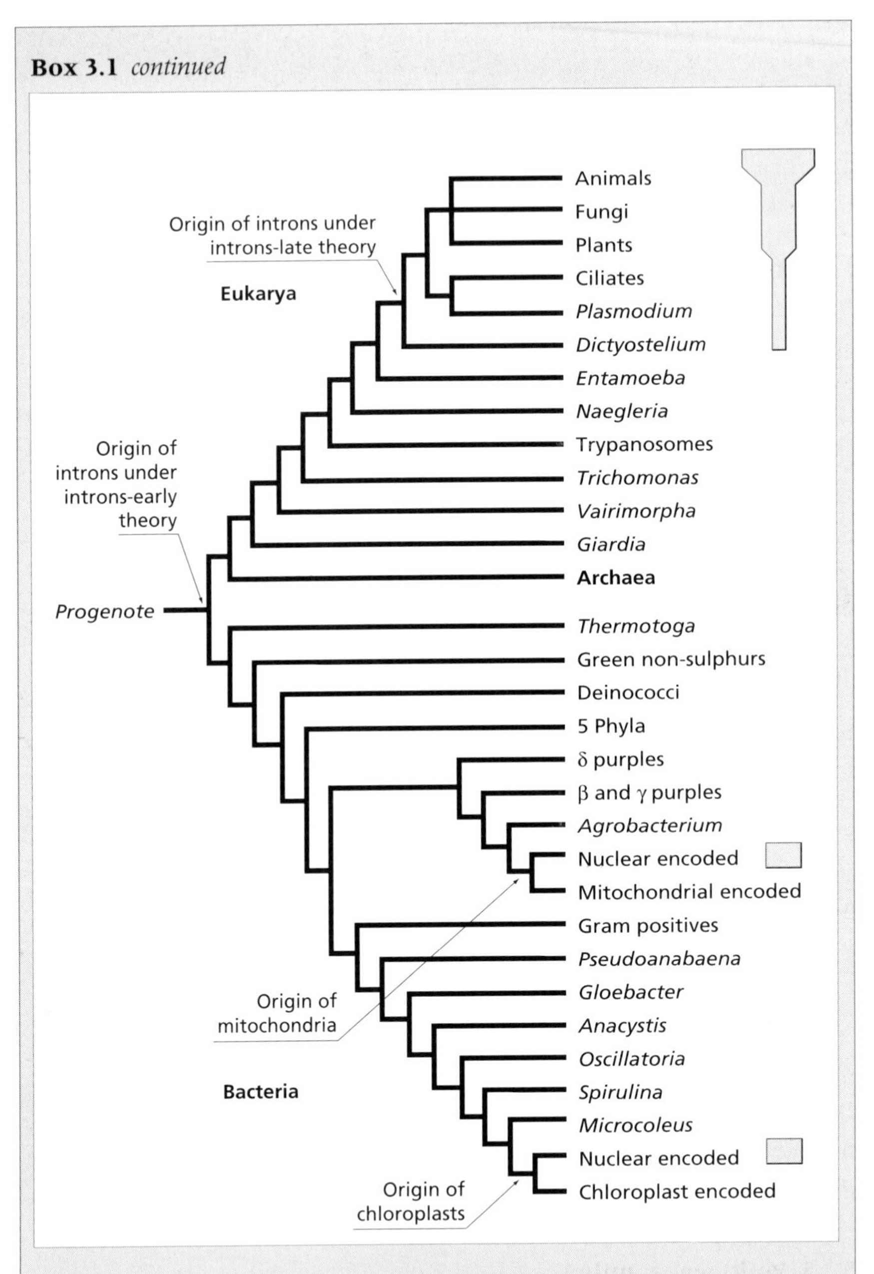

Fig. B3.1 Competing hypotheses for the origin of introns. The shaded bars on the right of the tree indicate the presence of spliceosomal introns, and the width of the bar their frequency. The progenote is usually taken to refer to the ancestor of all living cells. Adapted from Palmer and Logsdon Jr. (1991), with permission.

A far more serious problem for the introns-early hypothesis is that spliceosomal introns are *absent* from Archaea and Bacteria. Supporters of this

continued

Box 3.1 *continued*

theory attempt to explain this by massive intron loss, perhaps because there was selection to produce a streamlined genome with minimum replication time. However, intron loss cannot explain why spliceosomal introns *are* found in genes which have migrated from mitochondria and chloroplasts to the nuclear genome, but *not* in the mitochondria and chloroplasts themselves, nor in the bacterial groups thought to have given rise to them: the proteobacteria (α-purple bacteria), thought to be the ancestors of mitochondria, and the cyanobacteria (blue–green algae) which gave rise to the chloroplasts. This suggests that spliceosomal introns first appeared in eukaryotic genes, including those derived from organelle genomes, *after* they acquired mitochondria and chloroplasts. It is even possible that spliceosomal introns are descendants of group II self-splicing introns which may have invaded eukaryotes along with the bacteria that eventually became mitochondria and chloroplasts.

Whilst exon shuffling probably has led to the formation of new proteins, this seems to have only happened during the evolution of the later eukaryotes.

A variety of introns are able to splice without the aid of proteins, instead relying on the autocatalytic properties of RNA. One such class of self-splicing introns are the **group I introns** many of which are also mobile because they encode proteins, such as DNA endonucleases, that allow them to move ('transpose') around the genome. Group I introns are found in the genomes of some mitochondria and chloroplasts, the rRNA genes of some eukaryotes, such as in the protist *Tetrahymena thermophila*, and in T4 bacteriophage. **Group II introns**, which can also self-splice although by different mechanisms, are less common than the group I introns, being only found in scattered mitochondria and chloroplasts as well as in the bacterial ancestors of these organelles (α-proteobacteria and cyanobacteria, respectively) and contain sequences similar to those of reverse transcriptase. Finally, and least well studied, are the **group III introns**. These are found in a small number of protist eukaryotes, most notably *Euglena gracilis*, and appear to be group II introns with the central portion removed.

3.1.5 Multigene families

If we look at the distribution of genes within genomes we find that many genes, rather than existing as individual copies producing a single protein, are in fact part of a larger family of related genes. Some of these **multigene families** are located in a specific region of a single chromosome and can be repeated many times, whilst others are dispersed throughout the genome. Arranging genes into families was an important evolutionary innovation because it often allows proteins with similar functions to be encoded by clustered genes that

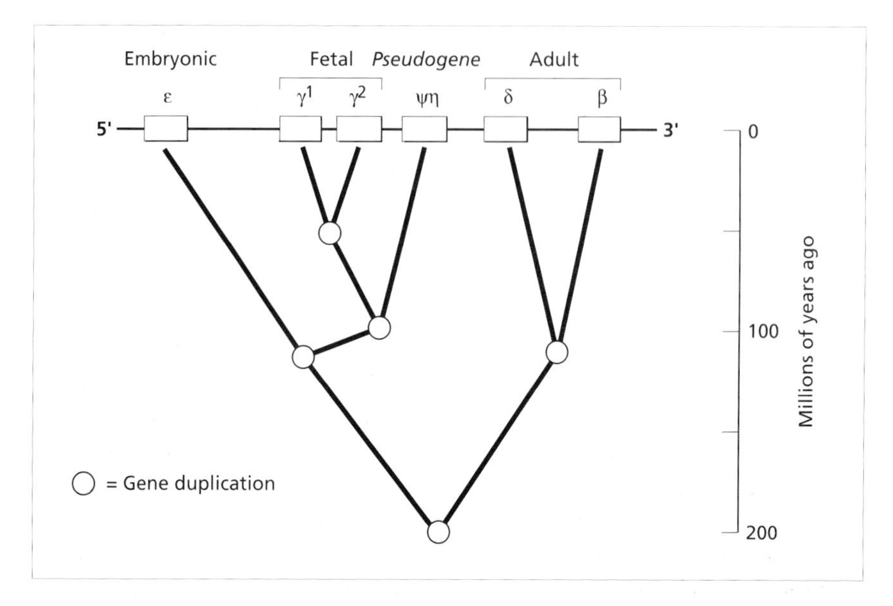

Fig. 3.8 Evolution of the human β-globin gene family. The family has six genes and is located on chromosome 11. Starting with an ancestral gene some 200 million years ago, the family has grown by a series of gene duplications. The newly formed genes then evolve new functions so that different amounts of oxygen can be carried to the embryo, the fetus and the adult. A single pseudogene, ψη, has also been produced.

can be regulated in an efficient manner. For example, invertebrate haemoglobins usually consist of a single globin gene which produces a single polypeptide chain, whereas vertebrates have a variety of multipolypeptide globin genes, produced by a process of **gene duplication** (in which a direct copy of a gene is made through various mechanisms), which are adapted to the varying oxygen requirements of different development stages (Fig. 3.8). More information about the structure and evolution of multigene families, especially by gene duplication, is provided in section 3.3.2.

Although the importance of genes is that they encode proteins or RNAs, not all genes are functional: some have acquired mutations, such as those that block the initiation of transcription, prevent correct RNA splicing or introduce premature stop codons, which inactivate them. These dead genes are called **pseudogenes**. Most pseudogenes also arise through gene duplication but as only one of these copies is usually needed, the other may acquire inactivating mutations and gradually diverge from its functional ancestor. Pseudogenes of this type are found in many multigene families. For example, the human α-globin gene family contains three pseudogenes; ψα1 and ψα2, which are derived from the adult α-globin gene, and ψζ which arose from the embryonic ζ gene (the Greek character ψ is used to signify a pseudogene). As we can see from Fig. 3.8, a single pseudogene, ψη, is present in the human β-globin gene family.

Other pseudogenes, like the $\psi\alpha3$ pseudogene from the mouse α-globin cluster, look less like their functional precursors because they lack introns and the sequences involved in gene regulation, such as promoters. These **processed pseudogenes** have probably been produced by the reverse transcription (from RNA to DNA) of the mature mRNA transcript of a gene (which will itself lack introns and promoter sequences—see section 3.2.2). Furthermore, because processed pseudogenes are sometimes found on a different chromosome from their functional ancestor, it is clear that they can be transmitted throughout the genome.

Finally, it is worth noting that genes in eukaryotes are often located in regions of the genome which have high concentrations of pairs (dinucleotides) of G and C bases. These are called **CpG islands** (where the p denotes the phosphate which links the two bases). In most cases, the CpG dinucleotide is highly susceptible to mutation through **methylation**, where C mutates to T, so that CpG pairs are usually rare. However CpG islands, which account for about 2% of the mammalian genome, have approximately tenfold more CpG than the rest of the genome because they are protected from methylation by a special DNA repair system. Most importantly, CpG islands are found at the 5′ ends of most mammalian genes so that they are useful in mapping the position of gene loci.

3.2 How genes function

3.2.1 DNA replication

In order for the genetic information stored in DNA to be inherited between generations it must first be replicated, so that new copies of the two strands are made. Accurate replication is therefore the critical step in the reproduction of living organisms. In eukaryotes, DNA replication is a complex process because it must be coordinated across different chromosomes. To achieve this, replication starts at a number of different places, which are recognised by sequence-specific binding proteins. Starting replication from a number of locations simultaneously also makes the process a lot quicker, and it only takes a few hours in humans. In contrast, the smaller genomes of bacteria are replicated from a single point of origin.

The first step in DNA replication is the unwinding of the double helix. This is undertaken by proteins called helicases. After the two DNA strands have been separated, new copies of each are made by proteins called **DNA polymerases**, creating a **replication fork** (Fig. 3.9). Four types of DNA polymerase are used to replicate nuclear DNA in eukaryotes—α, β, δ and ε—whereas three types of DNA polymerases are used in bacteria—I, II and III. Mitochondrial DNA is replicated by a single γ polymerase. The DNA polymerases synthesise the new DNA strands from pools of deoxyribonucleoside triphosphates (dNTPs), which contain the bases to be inserted into the growing

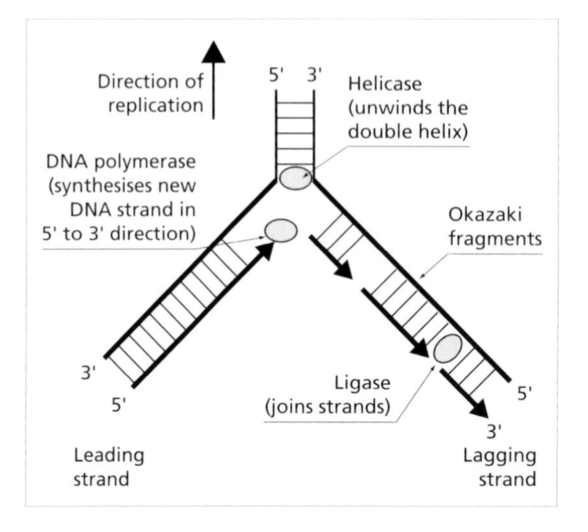

Fig. 3.9 Schematic representation of DNA replication. A replication fork is shown, which diverges from the origin of replication. One of the new strands—the 'leading strand'—is synthesised continuously, whereas the other, 'lagging strand', is synthesised in smaller sections called Okazaki fragments. Three of the most important proteins in DNA replication—DNA polymerase, helicase and ligase—are shown as shaded circles although in reality a variety of other proteins are utilised, including multiple polymerases. See Kendrew (1994) for more details.

molecule. Each original strand acts as a **template** for the production of a new strand so that replicated DNA molecules will consist of one parental strand, from the original DNA molecule, and one newly synthesised strand. Finally, the template and the newly copied strand are joined together, by enzymes called ligases, to produce two copies of the mature double-stranded DNA molecule.

DNA replication is closely tied to processes at the chromosomal and cellular levels (Fig. 3.10). In most eukaryotes, chromosomes reside in cells either as single sets (N; **haploid**) or pairs ($2N$; **diploid**) (chromosomes that form a pair are **homologous**, while those from different pairs are **non-homologous**). The cells with a haploid number of chromosomes are the sex cells, otherwise known as the **gametes**, which in animals make up the eggs and sperm. These fuse during sexual reproduction to form a diploid **zygote**. To produce a new organism the zygote then goes through a number of cellular divisions to produce both the diploid body cells (the **soma**) in a process known as **mitosis**, or more haploid gamete cells through **meiosis**. The key difference between mitosis and meiosis is therefore that the latter requires two cellular divisions, as opposed to the single division in mitosis, so that the gametes only have a haploid number of chromosomes. A second important difference between meiosis and mitosis is that homologous chromosomes pair up during meiosis thereby allowing recombination to take place. The details of this extremely

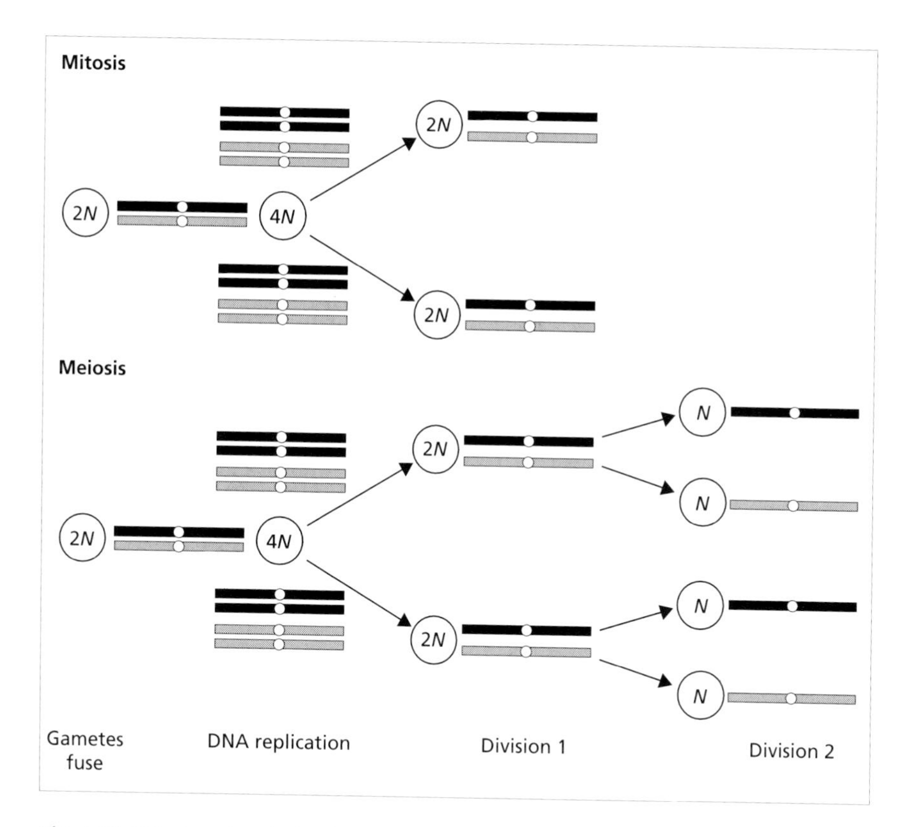

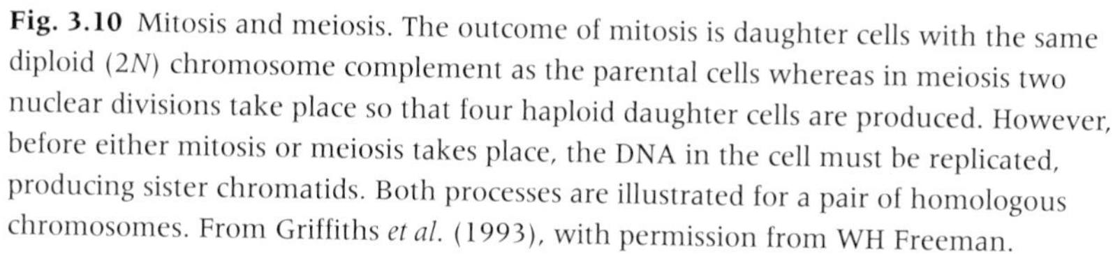
Fig. 3.10 Mitosis and meiosis. The outcome of mitosis is daughter cells with the same diploid ($2N$) chromosome complement as the parental cells whereas in meiosis two nuclear divisions take place so that four haploid daughter cells are produced. However, before either mitosis or meiosis takes place, the DNA in the cell must be replicated, producing sister chromatids. Both processes are illustrated for a pair of homologous chromosomes. From Griffiths *et al.* (1993), with permission from WH Freeman.

important evolutionary process are described in section 3.2.4. Despite these differences, mitosis and meiosis are similar in that before either occurs the DNA in every chromosome must be replicated. The outcome of replication is two copies of each chromosome, and thus of each double helix, known as **sister chromatids**. This double chromosome structure then divides into two daughter chromosomes, with each going to a different daughter cell, although a second division, resulting in four haploid daughter cells, takes place in meiosis.

3.2.2 Protein synthesis

Not only must DNA be able to copy itself, but it must also possess an efficient system for the production of proteins. Controlling which genes are expressed to produce proteins, and when, is one of the most complex of all molecular processes. The essence of this problem is that, with only a few exceptions, each cell of the body has a highly specialised role but an identical set of genes.

How does the body know which gene to turn on or off in each cell? This problem has been solved in many different ways. For example, the signal to activate a gene, so that its protein product can be produced, may come from a hormone lying outside the cell, or from special regulatory genes. Changes in when and how genes are expressed is an important aspect of evolutionary innovation.

Protein synthesis involves two major steps; the **transcription** of the DNA from the gene required into RNA and the **translation** of this RNA into protein (Fig. 3.11). In most eukaryote genes, transcription is undertaken by an enzyme called **RNA polymerase II**, although two other types of RNA polymerases are sometimes used: RNA polymerase I synthesises rRNA, while RNA polymerase III synthesises tRNAs as well as the small nuclear and cellular RNA molecules. Bacteria, on the other hand, use a single RNA polymerase for transcription. Transcription results in the production of a special RNA, called **messenger RNA (mRNA)**, which contains the protein-coding sequence of

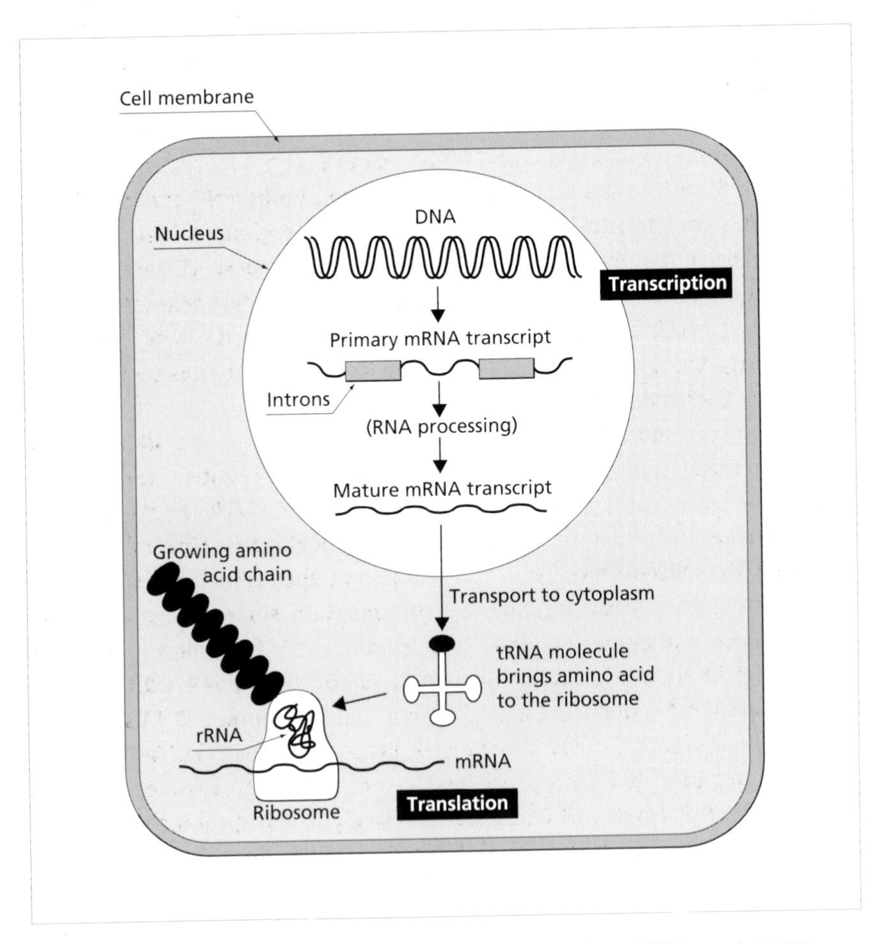

Fig. 3.11 A cartoon of protein synthesis in eukaryotes. Based on Griffiths *et al.* (1993).

the gene required along with 5′ and 3′ flanking sequences which control the process (see below). The terms 5′ and 3′ refer to the direction transcription takes—5′ to 3′—and DNA sequences are usually shown in this orientation.

Transcription is initiated by special proteins, known as **transcription factors**, which bind to particular sites in the gene to start the process. There are many different types of these binding sites, an important example that we have already encountered being promoters. Sometimes a transcription factor is itself only produced in a single cell type. For example, the immunoglobulins contain a binding site for a transcription factor which is only produced in B cells, so that immunoglobulins are only synthesised in B cells.

Because DNA is a double-stranded molecule, only one of the two strands acts as the template for transcription. This is the **sense** (or coding) strand, although which strand takes on this role differs among genes. The other or **complementary DNA** strand is then referred to as the **antisense** (or the nonsense) strand. Although transcription uses the sense strand as a template it actually produces an mRNA molecule which has the same nucleotide sequence as the complementary strand so that A, C, G and T bases in the sense strand become U (T), G, C and A, respectively, in the mRNA molecule.

Although the mRNA molecule contains the complete sequence of the protein to be produced, not all the mRNA is used to make the protein: two flanking regions, one of less than 100 bp (in eukaryotes) before the initiation codon–the 5′ untranslated region (5′-UTR), and another of between 50 and 200 bp after the stop codon—the 3′ untranslated region (3′-UTR), are not translated because they contain control signals and do not specify amino acids. Furthermore, a number of modifications are made to the ('primary') mRNA molecule before it is ready to produce proteins. This is known as **mRNA processing** (Fig. 3.12), and takes a slightly form in eukaryotes and bacteria. We will deal with the former.

First, the 5′ end of the molecule is modified through the addition of a special chemical 'cap' (composed of 7-methylguanosine residues) and a stretch of A bases, known as a **poly(A) tail**, is added to the 3′ UTR (a process called 'polyadenylation'). Next, the introns are cut ('spliced') away from the mRNA molecule by spliceosomes, and what is left of the mRNA molecule—the exons—are pasted back together. Furthermore, in some genes, and most notably those of angiosperm (plant) mitochondria and chloroplasts, individual nucleotides in the mRNA molecule can be changed by **RNA editing**. This means that mRNA sequences can be different from the genomic DNA sequences which produced them, so that care must be taken if mRNA and DNA sequences are combined in a phylogenetic analysis. Once all these processing steps have finished, the ('mature') mRNA molecule is ready to start producing proteins.

Once synthesised and processed, the mRNA molecules pass through pores in the cell nucleus into the cytoplasm where they are translated into protein. The synthesis of proteins takes place at special cellular organelles called **ribosomes** which are located on the surface of the endoplasmic reticulum.

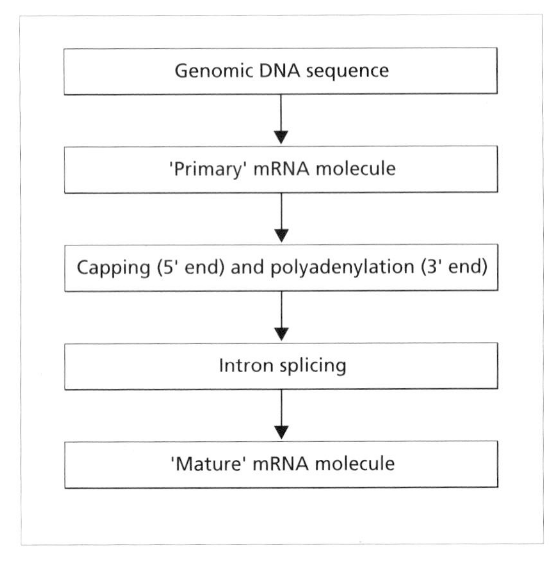

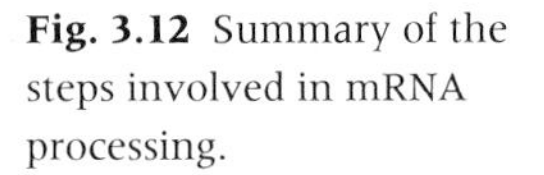
Fig. 3.12 Summary of the steps involved in mRNA processing.

The ribosome is composed of a set of proteins and **ribosomal RNAs (rRNA)**. Each amino acid to be inserted into the new protein is brought to the ribosome by a **transfer RNA (tRNA)** molecule that links up with a specific codon in the mRNA molecule, so that a polypeptide is gradually constructed. The part of the tRNA molecule that recognises an mRNA codon is called the **anticodon**, although some tRNAs can recognise more than one codon. tRNAs, like rRNAs, are never translated into proteins themselves and so are synthesised from special sets of genes (we will come across these genes again later).

Once the process of translation is finished, the completed polypeptide chains are passed into the lumen of the endoplasmic reticulum where they fold into their characteristic shapes. Many proteins undergo further post-translational modification before they become fully functional, such as **glycosylation** in which carbohydrates are attached to their surface. Finally, the finished proteins can be released to other parts of the body if they are required outside of their cell of manufacture.

Not all genes are synthesised in the way described above. For example, mitochondrial and chloroplast genes are synthesised on their own internal ribosomes and protein synthesis is simpler in bacteria where there is no nucleus or other membrane-bound structures within their cells.

Finally, the process of protein synthesis shows how genetic information flows from DNA to RNA to protein. This was originally thought to be such an important discovery that it was called the **central dogma of molecular biology**. However, in the 1970s the enzyme **reverse transcriptase** was discovered which changes the relationship by synthesising DNA from an RNA template. We will hear more about reverse transcriptase on pp. 80–82.

3.2.3 Mutation

Although DNA replication is a very efficient and accurate system, it does not function correctly on every occasion. Sometimes errors, or **mutations**, can creep into the process. While many of the mutations we see in DNA sequences arise through this faulty replication, others occur spontaneously, and environmental factors such as ultraviolet and chemical radiation can also damage DNA. For example, higher mutation rates than normal were found in families living near the Chernobyl nuclear reactor in Belarus which suffered a serious accident in 1986 and contaminated a large area with caesium-137. If mutations are inherited between generations they will provide the raw material of evolutionary change. Only mutations which occur in the sex cells (otherwise known as the **germ-line**) have the possibility of being inherited. Errors which take place in other (somatic) cells of the body are not passed on.

Copies of the same gene sequence which differ by mutation are called **alleles**. If these alleles also differ in amino acid sequence they may generate slightly different forms of protein. Understanding the forces which generate and maintain allelic variation in populations is the realm of population genetics and is dealt with in Chapter 4.

There are many different types of mutation (Fig. 3.13). Most involve the replacement, or **substitution**, of one base for another in the DNA sequence. These **point mutations** can be placed into two categories. **Transitions** occur when a purine nucleotide (A and G) is substituted for another purine, or a pyrimidine (C and T) is replaced by another pyrimidine. **Transversions** occur when a pyrimidine is substituted for a purine, or vice versa. If point mutations do not change the overlying amino acid, as is often the case if they occur in the third position of codons, they are referred to as **synonymous** (or **silent**) mutations, while those that change the amino acid are called **non-synonymous** (or **replacement**) mutations (see Chapter 5). If a second mutation occurs in a gene that restores the original phenotype (although not necessarily the original DNA sequence) this is referred to as a **back mutation**.

Another class of mutations are those that lead to **insertions** and **deletions** of nucleotides (collectively called **indels**). Although indels are most likely to occur in non-coding DNA, they sometimes occur in coding regions, often with serious deleterious effects. For example, the most common mutation responsible for the genetic disease cystic fibrosis (which affects between 1 in 2000–4000 Caucasians) is a three basepair deletion which removes the amino acid phenylalanine at position 508 (ΔF-508) in the cystic fibrosis transmembrane conductance regulator (*CFTR*). Indels are particularly likely to be deleterious if they change the nucleotide sequence so that it no longer fits the triplet codon structure, because bases have been inserted or deleted which are not a multiple of three. Such **frameshifts** will often cause major changes in amino acid sequence and are the cause of a number of genetic diseases, including

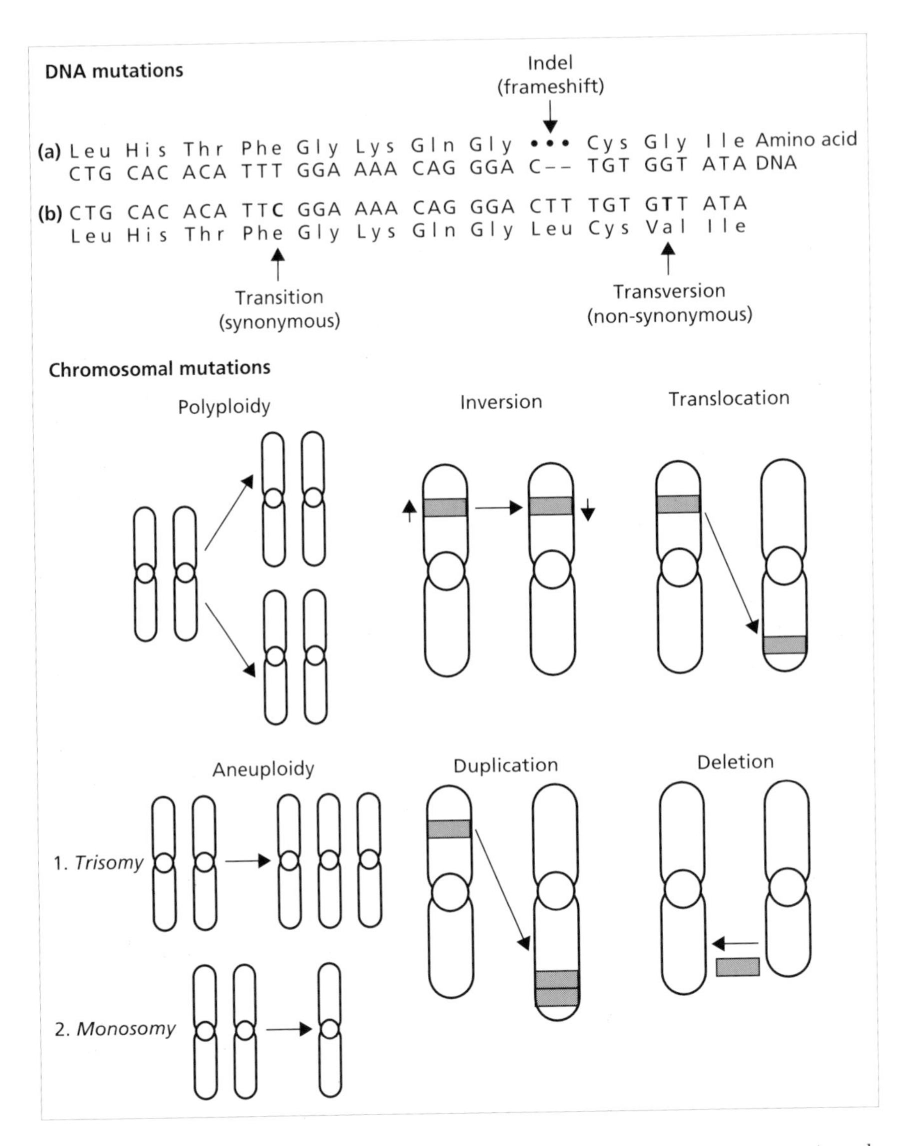

Fig. 3.13 Different types of DNA and chromosome mutations. Two DNA sequences (a and b) are shown with point mutations and insertions and deletions (indels) between them. At the chromosomal level, mutations such as polyploidy and aneuploidy alter chromosome number whilst inversions, translocations, duplications and deletions change chromosome structure.

β-thalassaemia (anaemia caused by the unbalanced synthesis of globin genes), haemophilia and Duchenne muscular dystrophy.

Mutations can also occur at the chromosomal level, which are often visible as disease syndromes. The largest group of chromosomal mutations are those where the number of chromosomes in the cell is altered, frequently because of errors during meiosis. The most common examples are **polyploidy**, where the cell nucleus contains multiple sets of chromosomes because the first meiotic division has not occurred, or cases where the chromosome number changes by smaller steps—**aneuploidy**. Specific cases of aneuploidy are **trisomy**, where there is one more chromosome than the diploid number (and the cause of Down's syndrome), or **monosomy** where there is one fewer chromosome.

Other mutations affect chromosome structure rather than number. Some involve a break in the chromosome, which may occur spontaneously (for unknown reasons), or be induced by mutagens. Some chromosomal regions appear to be more susceptible to breakage than others. An example is the 'fragile sites', one of which leads to an important genetic disease called **fragile X syndrome** (see p. 80). In most cases the broken ends of chromosomes rejoin and the break is successfully repaired, although genetic material can be deleted in the process (giving rise to a chromosomal 'deficiency') or perhaps lead to more complex structural rearrangements. For example, if the broken part flips end-for-end before rejoining the rest of the chromosome in the reverse direction, the result is a chromosome **inversion**. Inversions may be harmless if no genetic material is lost and no important genes disrupted at the breakpoints of the inversion. They also appear to have been relatively frequent events: an analysis of primate chromosomes, for instance, has shown that two inversions occurred on the branch leading to humans after its divergence from the great apes. On other occasions part of the broken chromosome may join another chromosome—a process known as **translocation**. Finally, if breaks occur in the chromosome during replication then it is possible that there will be a **duplication** of part of the chromosome.

Because mutations can have such disastrous affects on phenotype, it is not surprising that systems have evolved to repair damage to DNA. **DNA repair** can take place in a variety of ways. Occasionally, errors can be repaired directly as they occur, through the action of specialised enzymes. However, such **direct repair** is relatively rare. A more common mechanism is **excision repair** where the damaged DNA sequence, which may be single bases or many nucleotides, is excised (or cut) from the molecule during replication. The missing bases are then resynthesised by a DNA polymerase which uses the undamaged complementary strand as a template. Another important mechanism of repair that operates during DNA replication is **proofreading**, where mistakes are recognised by DNA polymerases and removed by proteins called exonucleases. Some repair pathways are able to recognise errors after DNA replication has been completed. One such mechanism is **mismatch repair** which detects mismatches—cases where A does not bind with T and C does not bind with G—that have arisen during DNA replication. In the bacterium *Escherichia coli* (the O157:H7 strain of which causes about 20 000 cases of food poisoning in the USA each year) the mismatch repair system reduces the mutation rate from ~1 bp per 10^8 bp replicated to ~1 bp per 10^{10} bp. Finally, **homologous recombination** (see below) can replace damaged DNA with the undamaged sequence present on the homologous chromosome.

3.2.4 Recombination

Although mutations arise on individual chromosomes, they can move between homologous pairs, and even to other chromosomes, by a process of **recombination**. This is one of the most important aspects of sexual reproduction because, by shuffling mutations, it means that progeny resemble neither of their parents.

Perhaps the most important type of recombination in eukaryotes takes place during meiosis. In meiosis, unlike mitosis, newly replicated (duplicated) homologous chromosomes form pairs. As they lie side-by-side they form a complex consisting of four chromatids. At this four-strand stage the chromatids can break and genetic material is exchanged between non-sister pairs (i.e. chromatids from the other chromosome), before reuniting again. These **crossing-over** events result in the formation of **chiasmata**, in which the chromatids form a cross-like structure (Fig. 3.14). Each chromosome pair generally has one or more chiasmata and these are important because they appear to allow chromosomes to segregrate properly to the daughter cells. The further apart two genes are on the chromosome, the greater the chance of crossing-over between them. Because homologous chromosomes do not pair during mitosis, crossing-over is rare at this stage, so that the daughter cells produced by mitosis are genetically identical to those of their parents.

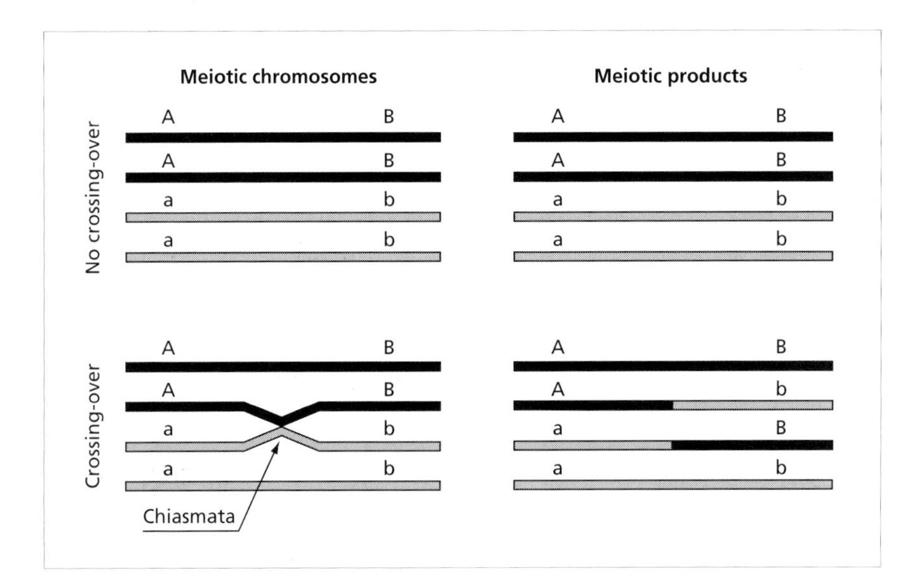

Fig. 3.14 Crossing-over between non-sister chromatids on duplicated homologous chromosomes (one shaded black, the other grey) during meiosis. Two loci (A and B) with two alleles at each are shown. In the top example, no crossing-over takes place but in the bottom one crossing-over leads to a new combination of alleles in the meiotic products. Adapted from Griffiths *et al.* (1993).

Most recombination in eukaryotes occurs between genomic regions which have a great deal of sequence similarity—**homologous recombination**. However, this does not always have to mean the same gene, as recombination between similar, but non-homologous, genes can also take place. One example of such a process is **unequal crossing-over**, which is one of the most important mechanisms determining the structure and evolution of multigene families and a way in which gene duplication can take place. Unequal crossing-over occurs when there is a misalignment between genes during meiosis so that recombination produces gametes with different copy numbers of genes. Misalignment is particularly likely in multigene families because each member usually has a high degree of sequence similarity, so that matching truly homologous genes can be difficult. For example, the Lepore mutation of β-haemoglobin is caused by an unequal crossing-over, and hence a fusion, between the first 50–80 amino acids of the δ-globin gene and the last 60–90 residues from the adjacent β-globin gene, so that the full δ and β genes are absent from descendant gametes (Fig. 3.15). Because unequal crossing-over leads to a variation in copy number between gametes, an 'anti-Lepore' mutant is also produced which carries a mirror image of the Lepore variant along with the δ and β genes.

Unequal crossing-over is also responsible for red–green colour blindness in humans. In this case, individuals with normal colour vision have one red-light and three green-light receptor long-wavelength opsin genes (which are found on the X chromosome). However, those with red–green colour blindness have a fused red and green-light gene, the outcome of an ancient unequal crossing-over event.

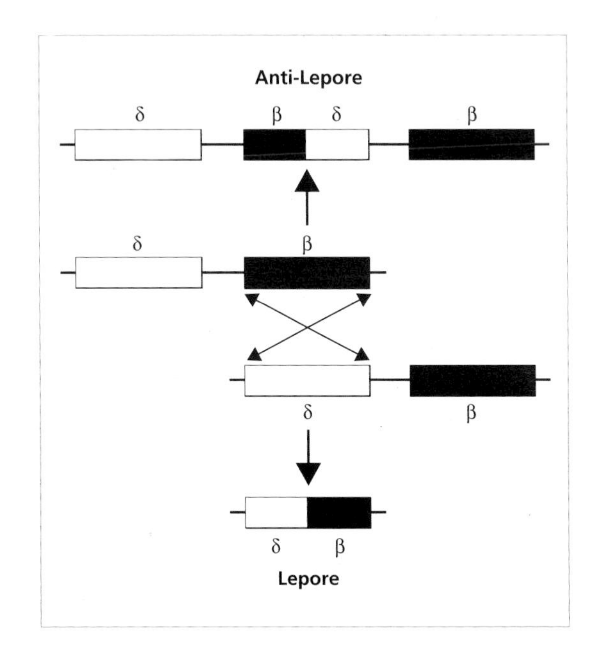

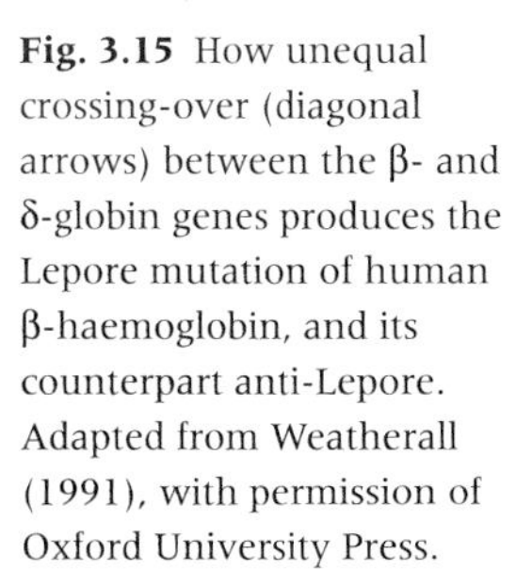
Fig. 3.15 How unequal crossing-over (diagonal arrows) between the β- and δ-globin genes produces the Lepore mutation of human β-haemoglobin, and its counterpart anti-Lepore. Adapted from Weatherall (1991), with permission of Oxford University Press.

Another special form of recombination which takes place in multigene families and contributes greatly to their evolution is **gene conversion**. This occurs when the DNA sequence of one gene is replaced (or 'converted') by the DNA sequence from another. This can occur over DNA sequences stretching from a few basepairs to many kilobases, and between different genes located on the same or different chromosomes (or between alleles of the same gene). Although gene conversion does not alter copy number, it can be biased, so that certain genes or alleles are far more likely to act as donor sequences (the origin of the converting sequences) than others.

Although the precise mechanisms of gene conversion are uncertain, it is clearly dependent on the extent of similarity among sequences: the more similar the sequences, the greater the chance of conversion between them. For example, gene conversion is more frequent in coding regions compared with non-coding regions because sequence similarity is greater in the former, and newly duplicated genes, which will be very similar in sequence, are more likely to experience gene conversion than those that diverged longer ago. This also means that anything which interrupts sequence similarity, like the insertion of a transposable element (see pp. 80–85), may prevent gene conversion from taking place.

A good example of gene conversion is the fetally expressed γ-globin genes of primates (γ^1 and γ^2—part of the β-globin gene family—see Fig. 3.8), which arose by gene duplication within the last 50 million years. Since this time more than 20 conversion events of varying lengths have been documented to have occurred between the two genes, with γ^1 the more frequent donor sequence. These conversion events often begin or end in a stretch of repeated TG sequence located in the second intron: this may act as a 'hot-spot' for gene conversion because the DNA here forms a structure that increases the rate of recombination. For example, in some human γ^2 globins, sequences on the 5′ side of the TG repeat are more similar to those in the adjacent (paralogous) γ^1 gene than they are to the orthologous γ^2 globins of other species, strongly suggesting that the γ^1 gene has converted this part of the γ^2 gene. The expected pattern of sequence relationships is restored on the 3′ side of the TG sequence, with more similarity among orthologous γ^2 genes than to the γ^1 gene (Fig. 3.16).

Although it is usually thought of as a way of homogenising sequences, gene conversion can sometimes enhance genetic diversity. The most spectacular example of this involves the immunoglobulin (Ig) genes of the chicken where gene conversion from a large pool of 80–100 pseudogenes to a single functional gene generates the vast array of different peptides needed by the immune system.

In bacteria recombination can also occur between genes from different species. This **interspecific recombination** (or **horizontal gene transfer**), which appears to have been a frequent occurrence in many bacterial species, can have a number of important consequences. For example, strains of the

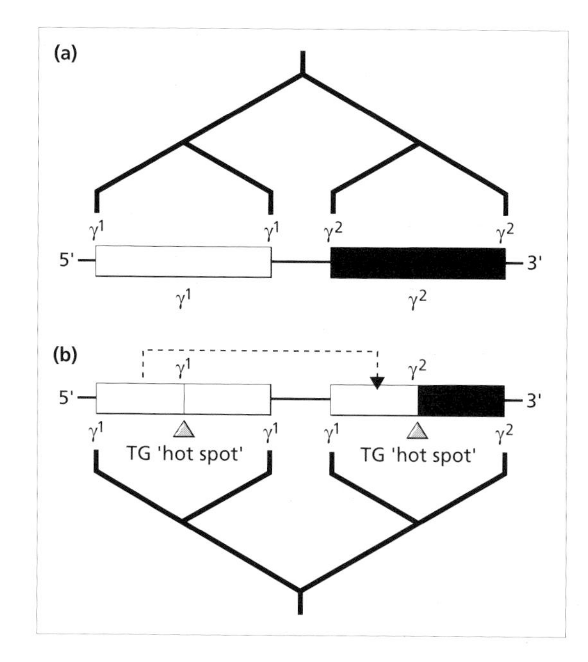

Fig. 3.16 Gene conversion in the duplicated γ-globin genes of primates. Part (a) depicts the expected relationships between the duplicated γ^1 and γ^2 genes from different species: each gene evolves independently so that γ^1 sequences are more closely related to other γ^1 sequences than they are to γ^2 sequences. However, in part (b) a gene conversion event occurs such that the 5′ part of γ^1 is superimposed on the 5′ part of γ^2. This means that the 5′ region of γ^2 is more closely related to γ^1 than it is to the γ^2 sequence found in other species. The boundary for this conversion event is marked by the repeated TG 'hotspot' sequence.

bacterium *Neisseria meningitidis* (a major cause of meningitis and septicaemia) that are resistant to penicillin acquired this property by the horizontal transfer of parts of the *penA* gene (which encodes a penicillin-binding protein) from two other species, *N. flavescens* and *N. cinerea*, which have more natural resistance. The large-scale exchange of genes between bacterial species also means that the evolutionary relationships between them cannot always be represented as simple bifurcating trees, and are better described by an interconnecting network (see Chapter 2).

Overall it is clear that recombination events are frequent in many organisms—with, for example, perhaps one crossing-over for every 5 kb in some yeast chromosomes—and therefore a major source of genetic diversity. Furthermore, some DNA sequences appear to have inherently higher rates of recombination than others. One example of such a hot spot is the ***Chi* site** which is found every 5–10 kb in the genome of *E. coli*. However, rates of recombination are much lower in other genomic regions, and especially in areas of heterochromatin such as at the centromeres, the telomeres or on the small fourth chromosome of *Drosophila* (see Chapter 7), and some viruses with single-stranded RNA genomes do not recombine at all.

Finally, despite its fundamental importance, the molecular mechanisms of recombination are still only partially understood, although in bacteria such as *E. coli*, it is known that certain proteins, specifically *RecA*, are involved in recombination and that this process is associated with DNA repair.

3.3 Genome organisation and evolution

One of the most startling revelations of the DNA sequence revolution is the extraordinary variety of ways in which genomes are organised. For example, the vast majority of DNA in the genomes of Bacteria and Archaea produces proteins (88% in the case of *E. coli*), yet 97% of the vertebrate genome is composed of non-coding DNA and may therefore have no function (although this is hotly debated). Furthermore, most of the eukaryote genome is made up of DNA sequences that are repeated very many times and, as we saw above, many genes are arranged into multigene families. A roadmap of eukaryotic genome organisation is shown in Fig. 3.17. In this section we will describe this organisation in more detail, especially that of the repetitive part, and discuss some of the evolutionary mechanisms that have given rise to it. The evolutionary processes responsible for another important aspect of genome organisation—differences in base composition—are discussed in Chapter 7.

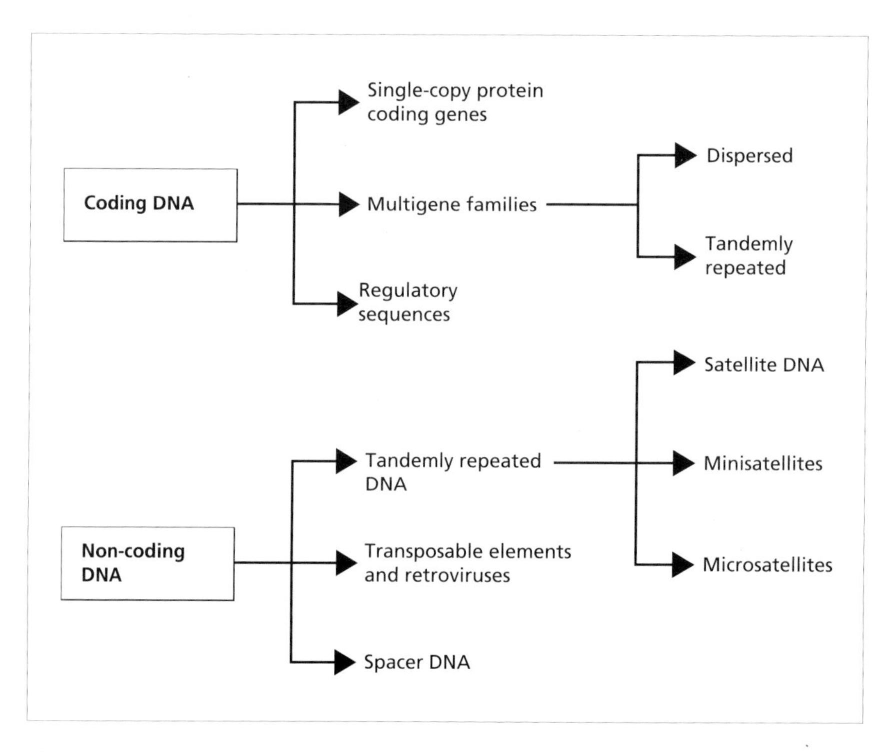

Fig. 3.17 A roadmap of the eukaryote genome. Non-coding regions within genes, such as introns, are not considered separately here.

3.3.1 Species differ in genome size and gene number

One of the first indications that genomes are highly flexible entities was the finding that their sizes can vary greatly between species. This can be quantified as the amount of DNA per haploid genome, called the **C-value** (Fig. 3.18). Variation here can be dramatic: for instance, there is approximately 200 times more DNA in the protist *Amoeba dubia* (670 000 000 kb) than in humans (3 300 000 kb). C-values also vary within taxonomic groups: while there is only a twofold variation in genome size among mammalian species (the largest known genome is found in the aardvark, the smallest in a Muntjak deer), there is a tenfold variation within anuran amphibians (frogs and toads), at least a hundredfold variation between insects, and an enormous 350-fold variation among bony fish. The puzzle for evolutionary biologists is that the

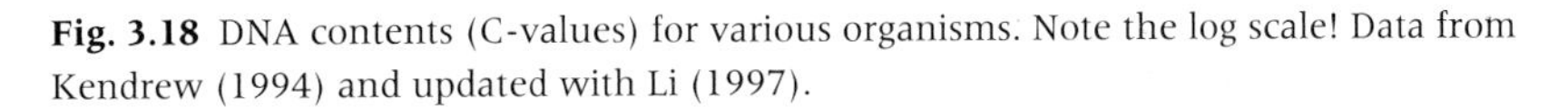

Fig. 3.18 DNA contents (C-values) for various organisms. Note the log scale! Data from Kendrew (1994) and updated with Li (1997).

amount of DNA is not associated with genetic or morphological complexity—some 'simple' organisms, such as amoebae, can have enormous genomes. And why are genomes so large if most DNA is redundant? This has become known as the **C-value paradox**. As we shall see in the next few sections, most of this variation in genome size is due to the presence of non-coding repetitive DNA.

The number of genes found in different species also varies considerably (Table 3.4). Although eukaryotes usually have many more genes than Bacteria and Archaea, and vertebrates have more genes than invertebrates, even the simplest unicellular bacteria require thousands of different genes, while complex multicellular organisms may require as many as 100 000. Furthermore, these differences in gene number cannot explain the huge variation in C-values.

Table 3.4 Estimates of the number of protein coding genes in various free-living organisms. Modified from Bird (1995), with permission.

Species	Gene number
Archaea	
*Methanococcus jannaschii**	1738
*Archaeoglobus fulgidus**	2436
Bacteria	
*Mycoplasma genitalium**	470
*Helicobacter pylori**	1590
*Haemophilus influenzae**	1743
*Bascillus subtilis**	4100
*Escherichia coli**	4288
Eukarya	
Oxytricha similis (ciliated protozoan)	12 000
*Saccharomyces cerevisiae** (yeast)	5885
Dictyostelium discoidium (slime mould)	12 500
Drosophila melanogaster (fruit fly)	12 000–16 000
Caenorhabditis elegans (nematode)	17 800
Fugu rubripes (fish)	50 000–100 000
Mus musculus (mouse)	80 000
Homo sapiens (human)	60 000–80 000

* Complete genome sequences available.

3.3.2 The evolution of multigene families

The most obvious way in which gene number can change between species is through **gene duplication**, where direct copies of genes (or parts of genes) are made. As we have already seen, this is an important process in the evolution of multigene families because it is the first step in creating genes with new functions, and in doing so increases the family copy number (and genome size). Because of its fundamental importance as a source of genetic diversity, we will discuss it in some detail.

Gene duplication can occur in a number of different ways. One mechanism, which we have encountered already, is unequal crossing-over. This is particularly important in increasing the copy number of multigene families. Gene duplication can also occur if complete genomes are duplicated, as appears to have been the case in *S. cerevisiae*. One way this can take place is through **polyploidy** (see section 3.2.3). Although polyploidy is much more common in plants than in animals (approximately 50% of angiosperms are polyploid), because plants can also propagate vegetatively without the need to make gametes, it has occurred fairly regularly in amphibians like the toad *Xenopus laevis*, where there are four copies of each chromosome, so that the species is **tetraploid**. Because four is an even number it is possible for tetraploids to have normal meiosis. Other species in the genus *Xenopus* have chromosome numbers varying from 20 to 108, suggesting that polyploidy has been relatively frequent in this group. Polyploidy may also be important in speciation. An example is the salmonid family of fishes which have twice as much DNA as related species and which are thought to have originated through polyploidy.

Another way in which gene duplication can occur is through **transposition**. This takes place when a DNA sequence at one position in the genome is copied and inserted into a new position elsewhere, sometimes on a different chromosome. Transposition can occur by a number of different mechanisms and these will be outlined in section 3.3.3 when we meet the main beneficiaries—the **transposable elements**, which are a major component of eukaryotic genomes.

Once a gene has been created by duplication, a number of things could happen to it. It could be that only a single copy of the gene is required so that the new copy is automatically redundant. In this case the newly duplicated gene is free of selective constraints and so will accumulate mutations, eventually becoming a pseudogene, or even being deleted from the genome. Another possibility is that natural selection favours a gene with a slightly altered function which could be fulfilled by the new copy, as we have already seen in the case of the haemoglobin genes and their different oxygen carrying capacities.

Although duplication is the most important way in which genes with new functions can arise, in a few rare cases genes have acquired new functions without duplication. In other words, different proteins can be produced by the *same* amino acid sequence. Perhaps the best examples of this are the crystallins, proteins that play a structural role in the eye lenses of animals, and which in some cases can also operate as enzymes: for example, ε-crystallin, found in the eye lenses of some birds and crocodiles, is also the enzyme lactate dehydrogenase (LDH). It seems that the protein's original function was as an enzyme but it was then recruited into a new structural role through changes in gene expression.

The importance of gene duplication in the evolution of multigene families can be illustrated by the family of genes which control the development of body plans in animals. The most important genes of this type in vertebrates are those in the *Hox* family. For invertebrates, like *Drosophila*, the homologous set of genes are called the homeotic gene complex (*HOM*). Like many other developmental genes in eukaryotes, both *Hox* and *HOM* contain a highly conserved protein module known as a homeodomain (or a homeobox if we refer to its underlying DNA sequence) (see section 3.1.1).

Mutations in the *HOM/Hox* genes can drastically affect the organisation of body parts, sometimes making them grow in the wrong places: for example, the *antennapedia* mutant in *Drosophila* causes leg-like structures to grow in place of antennae. Placing these genes in a multigene family helps control their expression: in some of the *HOM* clusters, genes at the 3′ end control development of the anterior body parts of the embryo, while those positioned at the 5′ end control the posterior sections. The *HOM/Hox* genes probably arose early in the evolution of metazoans and were perhaps one of the most important innovations in the development of multicellular organisms. Indeed, there is also a remarkable conservation of other important developmental genes in vertebrates and invertebrates. For example, mutations in the *eyeless* gene of *Drosophila* and the homologous *Pax6* gene of humans both affect the pattern of eye development.

Although the *HOM/Hox* genes are related, they differ in how they are organised. In all vertebrates examined so far there are multiple clusters of *Hox* genes. In the mouse, a well-studied example, each of its four clusters (denoted *Hoxa* to *Hoxd*) is located on a different chromosome and extends for over 100 kb. In contrast, there are two clusters of *HOM* genes in *D. melanogaster*, Bithorax and Antennapedia, which are found on the same chromosome. One of the most interesting of all *Hox* gene clusters is found in amphioxus—the common name of a number of marine animals belonging to the genus *Branchiostoma*. The importance of amphioxus is that it is the closest relative of the vertebrates—'the most vertebrate-like invertebrate'—and so represents a key organism in trying to understand the morphological innovations that have accompanied the evolution of the vertebrates (which possess such novel characteristics as cranial ganglia and teeth). Work by Peter Holland and colleagues has shown that amphioxus has a single cluster of at least 10 *Hox* genes (spanning 270 kb), each of which is homologous to a different *Hox* gene in vertebrates, so that the origin of the vertebrates coincided with a series of gene duplications (Fig. 3.19).

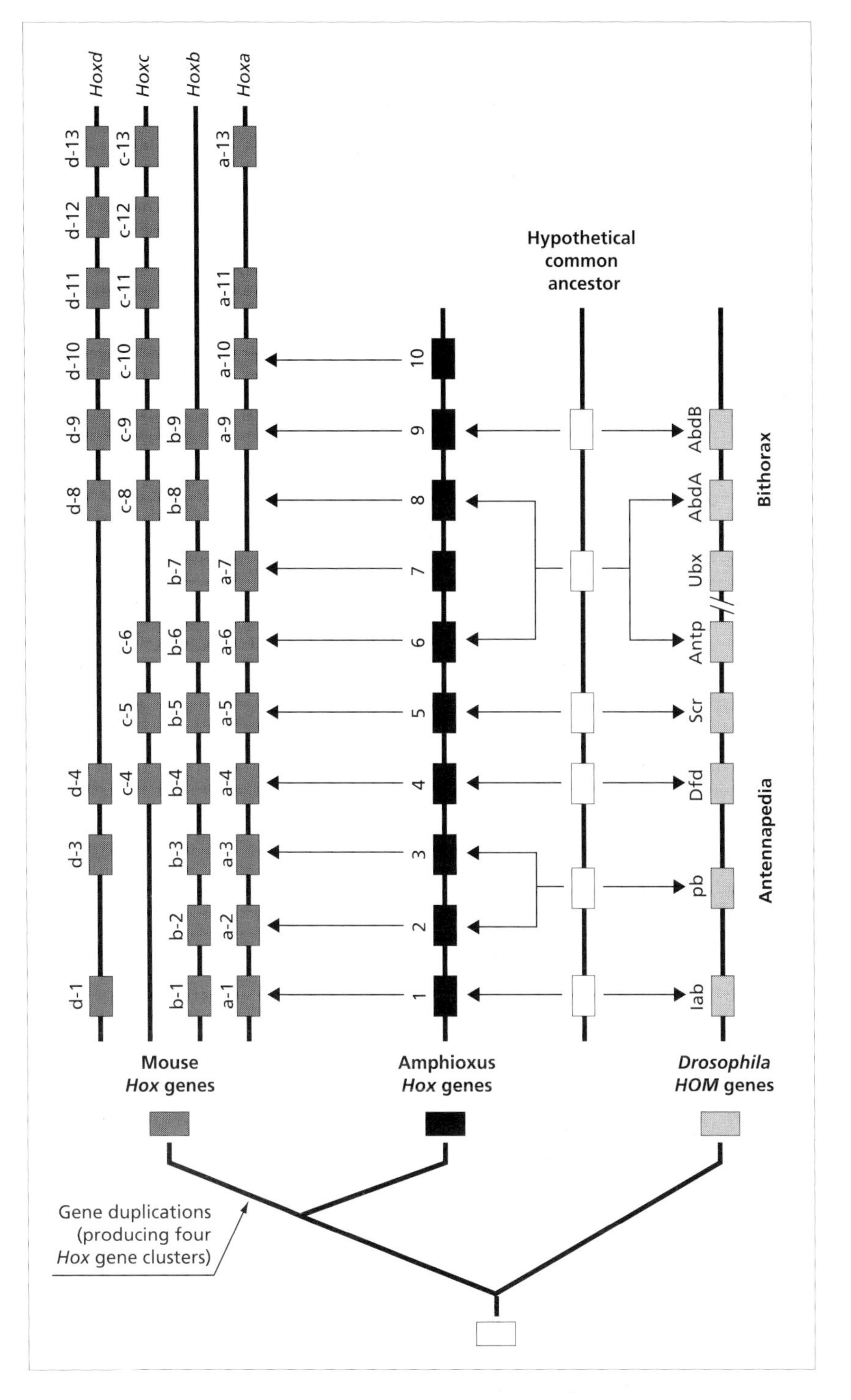

Fig. 3.19 Evolution of *Hox* and *HOM* gene clusters by gene duplication. The hypothetical *HOM/gene* cluster in the ancestor of chordates and arthropods (white boxes) expands to produce the eight *Drosophila HOM* genes (separated into the Bithorax and Antennapedia complexes as shown by the / / symbol), and the *Hox* gene cluster of amphioxus. Further gene duplications in higher vertebrates then produce four *Hox* gene clusters, illustrated here by the mouse. Adapted from Garcia-Fernàndez and Holland (1994), with permission.

It may be that other important evolutionary transitions were also accompanied by large and perhaps rapid increases in gene number brought about by extensive gene duplications.

The *HOM/Hox* genes, along with the globins we encountered previously, represent one type of multigene family—in which there is only a single copy, or just a few copies, of the family in the genome and where each member gene usually has a separate function. In the case of the vertebrate *Hox* genes, the family members are also **dispersed** over a number of chromosomes. Other multigene families however are repeated side by side many times, so that they contain multiple copies of genes with the same function. These are known as **tandem arrays**. Tandem arrays probably evolved because the host cell required large amounts of the protein they produce. A good example is the **rDNA array** which codes for ribosomal RNA (rRNA), part of the ribosome. Because of the crucial role played by ribosomes in protein synthesis, rRNAs are found in every species and are often present in huge quantities in each cell (and in mitochondria and chloroplasts), making up the majority of cellular RNA.

The rDNA arrays found in the nuclear genomes of eukaryotes produce three types of rRNA: 18S, also known as the small subunit ($\approx$1800 bp), 28S, the large subunit ($>$4000 bp), and 5.8S rRNA ($\approx$160 bp) (where S stands for Svedberg unit, a measure of sedimentation rate) (Fig. 3.20). In bacteria, the equivalent rRNA types are 16S, 23S and 5S, although numerous variations are found. As well as these coding sequences, the rDNA array also contains a number of 'spacer' sequences which contain the signals needed to process the rRNA transcript: an external transcribed spacer (ETS) and two internal transcribed spacers (ITS-1 and ITS-2) (other **spacer DNA** sequences which separate genes lack any kind of function—see Fig. 3.17). A group of genes and spacer sequences together makes up an rRNA transcription unit (because a single RNA transcript is produced) and these units are tandemly repeated in many species, with each rDNA array separated by a non-transcribed spacer (NTS). The size of the rDNA array varies greatly between taxa, from a single copy in the protist *Tetrahymena* to 19 300 copies in the lizard *Amphiuma means*.

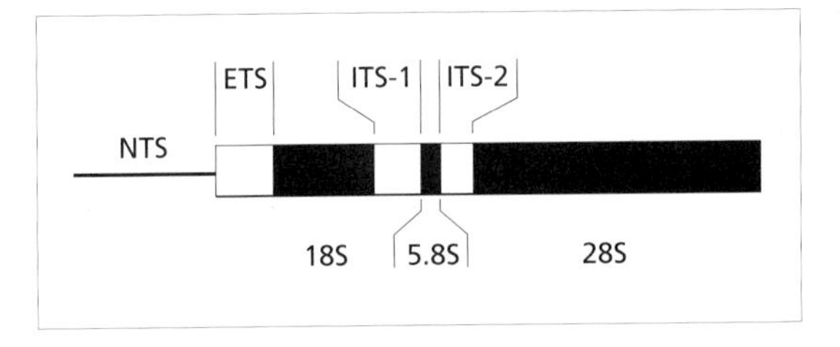

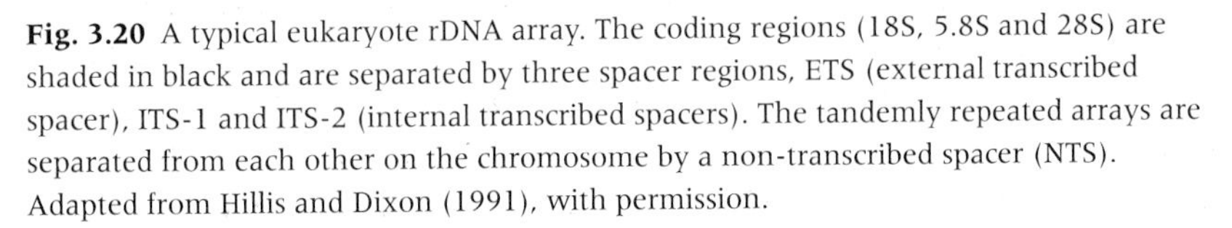

Fig. 3.20 A typical eukaryote rDNA array. The coding regions (18S, 5.8S and 28S) are shaded in black and are separated by three spacer regions, ETS (external transcribed spacer), ITS-1 and ITS-2 (internal transcribed spacers). The tandemly repeated arrays are separated from each other on the chromosome by a non-transcribed spacer (NTS). Adapted from Hillis and Dixon (1991), with permission.

About 200 tandemly repeated copies are found on the X and Y chromosomes of *D. melanogaster*, while in humans there are approximately 300 copies on five chromosomes.

rDNA sequences have been used frequently in molecular systematics because they include both highly conserved (18S) and highly variable sequences (NTS), and so can reconstruct the phylogenetic relationships between both very distant and very closely related species. Most striking of all is that because homologous rDNA genes are found in Archaea, Bacteria and eukaryotes they can be used to construct a tree of all cellular life (see Chapter 1).

Despite their popularity as phylogenetic tools, rDNA arrays do not always evolve in a simple manner. In particular, when rDNA sequences are compared within and between species an interesting pattern emerges: there is often a high degree of sequence similarity within species yet a great deal of divergence between them. This is unusual because, as we will see in Chapter 7, a low level of within species variation is normally associated with a low level of between species divergence. This unusual evolutionary pattern arises because there are genetic mechanisms, particularly those of unequal crossing-over and gene conversion, which transfer DNA sequences between genes so that they evolve together. This process is known as **concerted evolution** and is one of the most important acting on multigene families because it means that mutations can spread to all members, even if they reside on different chromosomes. While concerted evolution allows genes to evolve together, it also greatly complicates the phylogenetic analysis of multigene families because it becomes difficult to discern which genes are really homologous, so that orthologous and paralogous genes can be mixed (see Chapter 2). Furthermore, the point of sequence divergence no longer corresponds to the time of the last gene duplication or speciation event, but rather to the time of the last gene conversion or unequal crossing-over.

In sum, multigene families are a simple and highly effective way of creating genetic and functional flexibility. However, the complex processes which shape their evolution also mean that they are often composed of 'mosaics' of sequences, each with a different phylogenetic history, rather than strictly homologous genes gradually diverging through time.

3.3.3 Non-coding repetitive DNA sequences

Not all the repetitive DNA sequences in the genomes of eukaryotes produce proteins or RNAs. Other types of repetitive DNA, usually present in much higher copy numbers, do not encode products used by the cell (Table 3.5). However, the fact that these sequences are repeated many times and are non-coding does not mean they are without interest: by learning more about their evolution we will come to know more about the processes that have shaped the eukaryote genome as a whole. For example, it has been argued that some of these sequences spread solely for their own benefit, a tendency which has

Table 3.5 The different classes of non-coding repetitive DNA found in eukaryotes.

Class of repetitive DNA	Copy number	Organisation
Satellite DNA	Highly repetitive ($>10^4$)	Tandemly repeated
Minisatellites and microsatellites	Moderately repetitive	Tandemly repeated
Transposable elements	Moderately and highly repetitive	Dispersed

earned them the nickname of **selfish DNA**. It has even been suggested that some are 'ultra-selfish' because they can interfere with the function of other genes to increase their own copy number. An example of genes in this class are those involved in **meiotic drive**—the excess recovery of one allele in a pair in the gametes of a heterozygous parent. Furthermore, species-specific differences in the type and amount of non-coding repetitive DNA is a major reason why genome sizes differ between species. Finally, and perhaps rather unexpectedly, will see that some of these sequences are of great practical importance for those working in other areas of biology.

Tandemly repeated DNA

A great deal of the non-coding repetitive DNA in eukaryotes consists of short sequence motifs (i.e. strings of sequences) tandemly repeated many hundreds or thousands of times. One type of DNA in this class, termed **satellite DNA**, is located mainly in regions of heterochromatin and consists of motifs from 2 bp up to 40 Kb in length. For example, the α-satellite of primates is based on a 171 bp sequence. These motifs are tandemly repeated very many times such that the whole repeat unit can go on for hundreds of kilobases, and is often referred to as 'highly repetitive DNA'. DNA of this type can even form the bulk of genomes: for example, 60% of the genome of *Drosophila nasutoides* is made up of satellite DNA. Although it is usually assumed to be 'junk DNA', it is possible that satellite DNA is involved in the structure and function of centromeres.

Two similar types of repetitive DNA sequence are **minisatellites** and **microsatellites**. These are also composed of short sequence motifs (see below) that are tandemly repeated many times, although less frequently than satellite DNA (making them 'moderately repetitive DNA'). These short repetitive motifs are thought to be produced by mutation, unequal crossing-over and **DNA slippage**. This last process occurs when DNA strands mispair during replication and recombination so that short stretches of sequence slip against each other creating loops of DNA which, when repaired, result in the loss or gain of motifs.

Minisatellites, or VNTR loci ('variable number of tandem repeats'), are found in the euchromatic regions of vertebrates, fungi and plants. Each repeat unit contains a short G-rich 'core' sequence, ranging in size from 11 to 60 bp, although the complete repeat unit may be up to 200 bases long (Table 3.6).

Table 3.6 Comparison of selected minisatellite probe sequences. These probes are based on the core sequence and are used to detect variability in allele length at each minisatellite locus. Modified from Kendrew (1994).

Minisatellite probe	Sequence*
Core Jeffreys	GGAGGTGGGCAGGARG
Myoglobin	CTAAAGCTGGAGGTGGGCAGGAACGACCGARRT
33.15	AGAGGTGGGCAGGTGG
33.6	AGGGCTGGAGG
pλg3	AGAAAGGCGGGYGGTGTGGGCAGGGAGRGGCAGGAAT
λMS1	GTGGATAGG
YNH24	CAGCAGCAGTGGGAAGTACAGTGGGGTTGGTT
Insulin	ACAGGGGTGTGGGG
Harvey *ras* c1	GGGGGAGTGTGGCGTCCCCTGGAGAGAA
α-globin 3′HVR	AACAGCGACACGGGGGG
D14S1	GGYGGYGGYGGYGGYGGYGGY . . .
Core Nakamura	GGGNNGTGGGG

* R = A or G; Y = C or T; N = any base.

These units are then tandemly repeated along the chromosome, reaching sizes of 50 kb.

The most striking feature of minisatellites is their extraordinary variability: new variants arise on average at a frequency of 1–2% per gamete, per generation (although this can be as high as 15%) whereas most gene loci have a mutation rate of 10^{-5} to 10^{-6} per generation. This means that repeat copy number (i.e. allele size) can vary between individuals and that a large number of alleles are present even within a single population. It is this high mutation rate, along with a simple pattern of Mendelian inheritance (see Chapter 4), that gives minisatellites their practical worth: they are extremely powerful **molecular markers**. More precisely, because of variation in allele size, each minisatellite locus produces large numbers of DNA fragments following electrophoresis and these can be used to distinguish different individuals within a population. This is known as **DNA profiling** (or DNA fingerprinting).

Because it offers great accuracy of identification—the chance of an incorrect match in a DNA profile is extremely small (although there is some debate as to actually how small, depending on the structure of the population the samples are drawn from)—DNA profiling has been used with great success in both population biology and forensic science. In behavioural ecology, for example, minisatellites can be used to determine which male in a population is the father of a set of offspring, a task which is often difficult through observation alone.

Microsatellites, or STRs ('short tandem repeat polymorphisms'), are sequences composed of runs of repeat units 2–5 bp in length (repeats of the dinucleotide CA are especially common) which are found in the genomes of many eukaryotes, again mostly in regions of euchromatin and in the plant chloroplast genome.

Because of their simple structure, these repetitive motifs are also called **simple sequences**. In the human genome there are perhaps 35 000 microsatellite loci, occurring on average every 100 000 bp, and with allele lengths of usually between 2 and 50 repeats per locus. However, the differences in allele sizes between species are often much less than those expected given what is observed within species, suggesting that there is an upper limit on allele size, although the reasons for this are unknown. Such constraints also mean that microsatellites are generally poor indicators of interspecific relationships. It is also unclear exactly how allele sizes change between individuals and whether mutational events always involve the loss or gain of single repeats. The population genetic models describing microsatellite evolution are discussed in more detail in Box 4.2 in Chapter 4 (see p. 108).

Like minisatellites, the importance of microsatellites lies in their high mutation rates, between 10^{-2} and 10^{-5} per gamete, per generation, so that they also vary greatly in copy number between individuals. Such high levels of genetic diversity coupled with neutral evolution, codominance (so that heterozygotes can be distinguished from homozygotes) and simple Mendelian inheritance mean that they are also an ideal, and currently extremely popular, set of molecular markers. For example, an analysis of nine microsatellite loci from the worlds' most endangered species of canid, the Ethiopian wolf, showed that wild populations had very little genetic diversity and that females had hybridised with male domestic dogs. Information of this type is important for those involved in conserving such threatened species. Microsatellites have also been used in forensic cases: six different microsatellite loci taken from the femur of an unidentified female murder victim were compared with those from the victim's probable mother and father. The similarity of the banding patterns indicated, with great confidence, that the victim was indeed their daughter. How both minisatellites and microsatellites have helped shape our view of human origins is discussed in Chapter 4.

Finally, microsatellites may also be of medical importance as a number of human genetic diseases, such as fragile X syndrome, Huntingdon's disease, myotonic dystrophy and spino-bulbo-muscular dystrophy are associated with a dramatic increase in the copy number of trinucleotide microsatellite repeats. For example, fragile X syndrome, the most common form of inherited (X-linked) mental retardation, and recognised by a fragile (breakable) site in the X chromosome, appears to be caused by expansion of a CGG repeat in exon 1 of the *FMR1* gene. Normal alleles contain between 6 and 50 repeat units whereas clinically affected individuals have more than 200 repeats, and frequently more than 1000.

Transposable elements

Instead of relying on unequal crossing-over and DNA slippage, the second main class of non-coding repetitive DNA sequences—the **transposable elements (TEs)**—increase their copy number by jumping around the genome, making additional copies of themselves as they do so. If one group of DNA sequences deserve the title of 'selfish', it is these. Transposable elements make up a substantial portion of the genomes of many eukaryotes: more than 50% of the maize genome is made up of transposable elements and a similar figure may yet be uncovered in humans (10–20% of the genome of *Drosophila melanogaster* is already known to be composed of DNA of this type). Transposable elements are also found in bacteria where they are referred to as insertion sequences or transposons.

Transposable elements can be divided into three groups based on their mechanism of transposition (Fig. 3.21). **Class I** transposable elements, or **retroelements**, transpose through an intermediate RNA stage (i.e. DNA → RNA → DNA) using the enzyme reverse transcriptase. This process is called **retrotransposition**. Reverse transcriptase is also used by retroviruses such as HIV-1, which means that they can also be classed as similar to the group I elements. In contrast, **Class II** or **DNA elements**, transpose directly from DNA to DNA. Mutation rates are higher in retroelements than in DNA elements because reverse transcriptase lacks a proofreading ability and so is extremely prone to errors. Much less is known about the third group of transposable elements, the **miniature inverted-repeat transposable elements** or **MITEs**. These sequences, such as the *Stowaway* element from sorghum and the *Tourist* element in maize, are only 100–400 bp in length and transpose by as yet unknown means.

Retroelements can be split into two smaller groups. **Retrotransposons** possess long terminal repeat (LTR) sequences at both ends of their genome

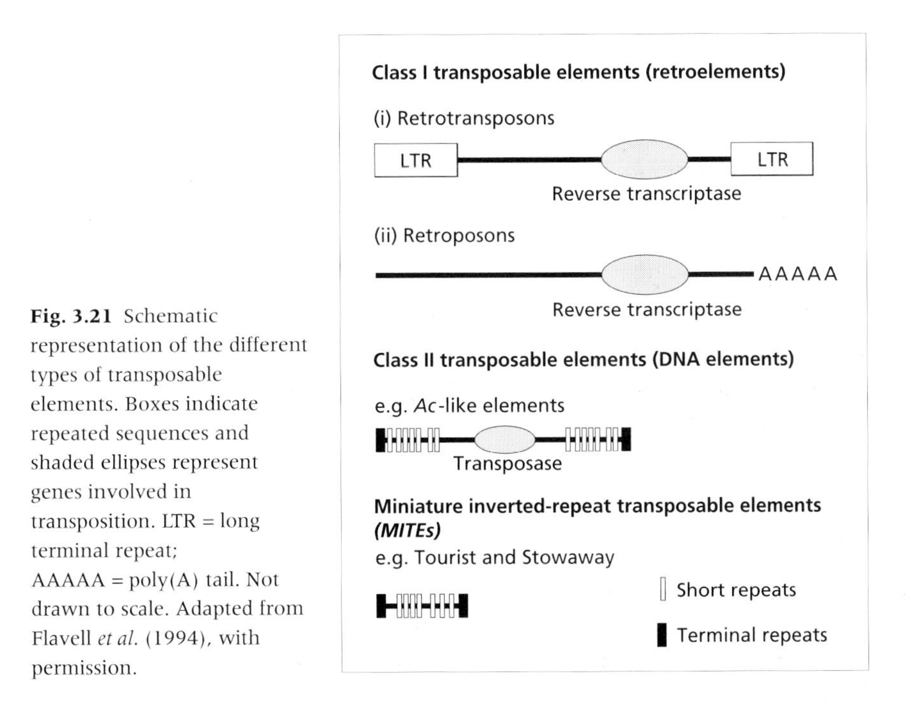

Fig. 3.21 Schematic representation of the different types of transposable elements. Boxes indicate repeated sequences and shaded ellipses represent genes involved in transposition. LTR = long terminal repeat; AAAAA = poly(A) tail. Not drawn to scale. Adapted from Flavell *et al.* (1994), with permission.

(and are also known as LTR retrotransposons). The *copia* element, for example, is 500 bp long with LTRs of 276 bp and is found 20–60 times in the genome of *D. melanogaster*, although *copia*-like elements make up nearly 1% of all DNA (and some 4.6% of the human genome is composed of retrotransposons). Other examples of retrotransposons are the *gypsy* elements of *Drosophila*, the *Ty* elements of yeast and the *stonor* elements of maize.

The second group of retroelements are the **retroposons**, such as the *I* elements of *Drosophila*, which have no LTR sequences and instead contain a poly(A) tail at their 3′ ends. These elements are also known as non-LTR retrotransposons or poly(A) retrotransposons. Furthermore, while retroposons use reverse transcriptase they do so in a different way from retrotransposons and retroviruses. One major type of retroposons are the **Long Interspersed Nuclear Elements (LINEs)** which are a major component of the G-banded regions of mammalian chromosomes (see Table 7.2). These elements are 6–8 kb in length and are present in many thousands of copies: for example, elements from the *L1* (Line 1) family have a consensus length of 6 kb (although most are truncated) and are present in a staggering 590 000 copies in the human genome, so that they make up almost 17% of all our genomic DNA—thereby occupying an order of magnitude more DNA than all the coding regions combined! LINEs can also have important mutagenic effects: insertion of a *L1* element into the gene which encodes the blood clotting protein factor VIII has led to cases of haemophilia.

Superficially similar to retroelements are the **Short Interspersed Nuclear Elements (SINEs)**, or *Alu*-like sequences, which are frequently found in the R-banded regions of mammalian chromosomes. Although these elements transpose through an RNA intermediate, they do not produce reverse transcriptase and so are not considered true retroelements. SINEs vary in size from 130 to 300 bp and have copy numbers ranging from 50 000 to over 1 000 000 per genome. The most studied SINEs are the *Alu* sequences, christened as such because they were first recognised by the restriction enzyme *Alu1*. *Alu* sequences are approximately 300 bp in length and are present in about 1 100 000 copies in the human genome (almost 12% of our total DNA content). Some SINEs have acquired important functions—an *Alu* sequence is used as a promoter sequence in the θ1 gene from the human α-globin gene family—although others cause deleterious effects, as in the case of the human genetic disorder neurofibromatosis which can be induced by the insertion of an *Alu* element into the *NF1* gene. Both LINEs and SINEs appear to be originally derived from RNA transcripts: LINEs from RNA polymerase II transcripts and SINEs from RNA polymerase III (tRNA) transcripts, although they now obviously lack their original cellular function.

Also related to retroelements are the **endogenous retroviruses** which make up about 1% of the mammalian genome. These are copies of retroviruses which have integrated as their DNA form (known as the provirus) into the germ-line of eukaryotes and which are now inherited along with the host genomic DNA. Although these elements were originally derived from infectious viruses, they have since acquired so many mutations that they are transcriptionally silent and so are non-infectious. The only endogenous retrovirus known to cause disease is MMTV which leads to mammary carcinoma in mice.

Class II (DNA) elements also possess terminal repeat sequences but unlike those of retrotransposons these are short—less than 100 bp in length—and frequently inverted. Class II elements also encode a special transposase protein which enables them to move around the genome, although the precise mechanisms by which this occurs are often unclear. Although some 1.6% of the human genome is composed of elements of this kind, the most studied Class II elements are the *P* and *hobo* elements of *Drosophila*, the *mariner* elements of animals, the *Tc1* elements of nematodes and the *Ac/Ds* ('Activator/Dissociation') elements of maize, which where the first transposable elements to be discovered by Barbara McClintock in the 1940s. Of these, most attention has been given to the ***P* elements** which are found in some species of *Drosophila*: for example, between 0 and 60 copies of this 2907bp element are found in the genome of *D. melanogaster*. *P* elements illustrate two of the most important aspects of the biology of transposable elements: that, as well as jumping around genomes, they are sometimes able to move between species and that they can affect the phenotype of their host organism.

The most notable phenotypic outcome of *P* elements in *D. melanogaster* is **hybrid dysgenesis**—an increased infertility due to chromosome breakage. However, *P* elements are only a recent introduction into wild populations of *D. melanogaster* and flies maintained in laboratory stocks established in the early part of this century do not carry them. This means that hybrid dysgenesis only occurs in the offspring of crosses between laboratory females which are free of *P* elements, and wild-caught males which carry them. Furthermore, *P* elements are not found in the fly species most closely related to *D. melanogaster*—*D. simulans, D. sechellia* and *D. mauritiana*—but are present in more distantly related species, such as those from the *D. willistoni* species group. This means that *P* elements in *D. melanogaster* must have been transferred from the *D. willistoni* group (perhaps by viruses or parasitic mites) after *D. melanogaster* split from its sibling species about two million years ago (Fig. 3.22).

Transposable elements can influence host biology in other ways. For example, elements inserted into host genes often inactivate them, or lead to other major mutations. In hybrid dysgenesis, *P* elements also create mutations, such as those seen in the *white* gene (in which flies lack the red eye pigment) and *Hobo*, another transposable element from *Drosophila*, leads to a number of chromosomal rearrangements. Recombination between elements of the same family that occupy different (non-homologous) sites on chromosomes—a process known as **ectopic exchange**—will also cause mutations. This has been widely demonstrated in the *P* elements of *Drosophila*, and causes both deletions and duplications of sections of chromosome. Mutations in host genomes can

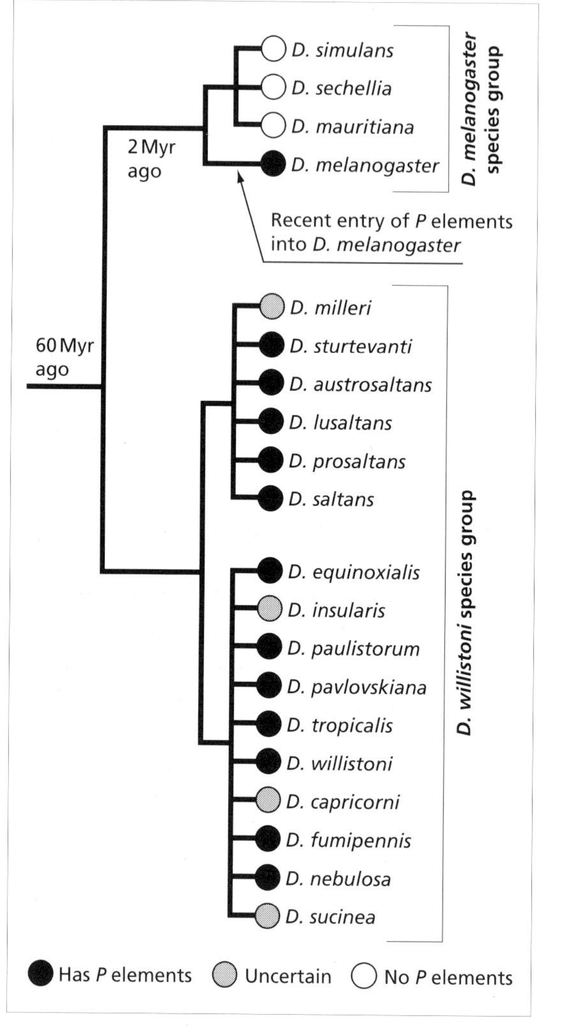

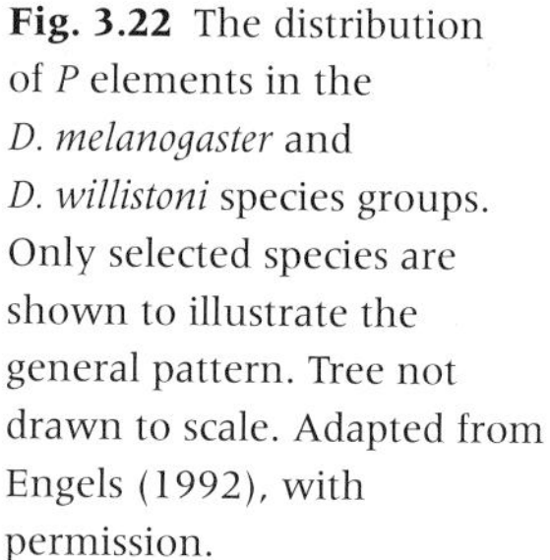

Fig. 3.22 The distribution of *P* elements in the *D. melanogaster* and *D. willistoni* species groups. Only selected species are shown to illustrate the general pattern. Tree not drawn to scale. Adapted from Engels (1992), with permission.

also occur when transposable elements are cut ('excised') from the genome, as this process can be quite inaccurate at the sequence level, with sections of indigenous DNA also being removed.

Although it has been claimed that transposable elements might have some beneficial effects, the great majority of the insertion events, chromosomal rearrangements and mutations induced by transposable elements will reduce the fitness of their host species and so be subject to selection pressure to remove them (see Chapter 4). Indeed, studies in *Drosophila* have shown that transposable elements are very rare in coding regions, where they are likely to be highly deleterious and so strongly selected against, yet occur at high frequency in regions of heterochromatin where fewer genes are found and where the rate of meiotic recombination, a mechanism by which they are removed from the population, is reduced. Because of their largely detrimental effects, both the host genome and transposable element are likely to evolve mechanisms that control their copy number and prevent runaway transposition. Indeed, insertion sequences in bacteria can regulate their own rate of transposition so that it decreases as their copy number increases. However, despite their deleterious effects, transposable elements may be used to beneficial effect in genetic engineering, where they can act as vectors for carrying genes to new locations.

Box 3.2 Techniques of molecular biology

Our knowledge of the structure and evolution of genes has been made possible by the dramatic advances in molecular biology. In this box we briefly describe some of the more important techniques in this ever-developing field.

One of the techniques central to the development of molecular biology was **gene cloning**. The DNA to be cloned, which will carry the gene of interest, may be derived directly from genomic DNA, or may be a **complementary DNA (cDNA)** copy of an mRNA produced in a particular tissue. A set of cloned DNA fragments from a single source, such as a particular cell type, is called a **DNA library**. During cloning fragments of DNA are isolated and purified. To do this they are first inserted into a special carrier DNA sequence, known as a **cloning vector**. The most common cloning vectors are: **plasmids**—non-chromosomal DNA elements found in bacteria which can replicate on their own accord (which can also transfer horizontally between bacteria species, sometimes bringing genes for antibiotic resistance with them); **bacteriophages**—viruses which infect bacteria; **cosmids**—plasmids which contain bacteriophage packaging sequences and **YACs**—yeast artificial chromosomes. The hybrid of cloning vector and inserted gene (also called a **recombinant DNA**) is then inserted into a bacteria where it is **amplified** to produce a large number of identical molecules.

As well as cloning, it is also possible to amplify genomic DNA directly using the **polymerase chain reaction (PCR)**. Starting with tiny amounts of a particular sequence (up to a length of about 5 kb), PCR generates millions of identical copies. PCR requires short sequence **primers**, usually about 20 bp long, which are specifically designed to bind to the ends of the sequence of interest. These primers are then mixed with the target DNA, a DNA polymerase, and free deoxyribonucleotides (A, C, G and T) and run through a thermal cycler (PCR) machine. This runs a cycle where the target DNA is first denaturated (the DNA strands are separated) at high temperatures (>91°C), the primers anneal (bind) to these strands at much cooler temperatures (~50°C), and then target DNA is synthesised by DNA polymerase at ~72°C (Fig. B3.2). This cycle is run 25–40 times, greatly increasing copy number each time. Most PCR reactions use a DNA polymerase called ***Taq* polymerase**, derived from a bacterium, *Thermophilus aquaticus*, which lives in hot springs and so is able to resist the high temperatures required to separate DNA strands. Each newly amplified

continued on p. 86

Box 3.2 *continued*

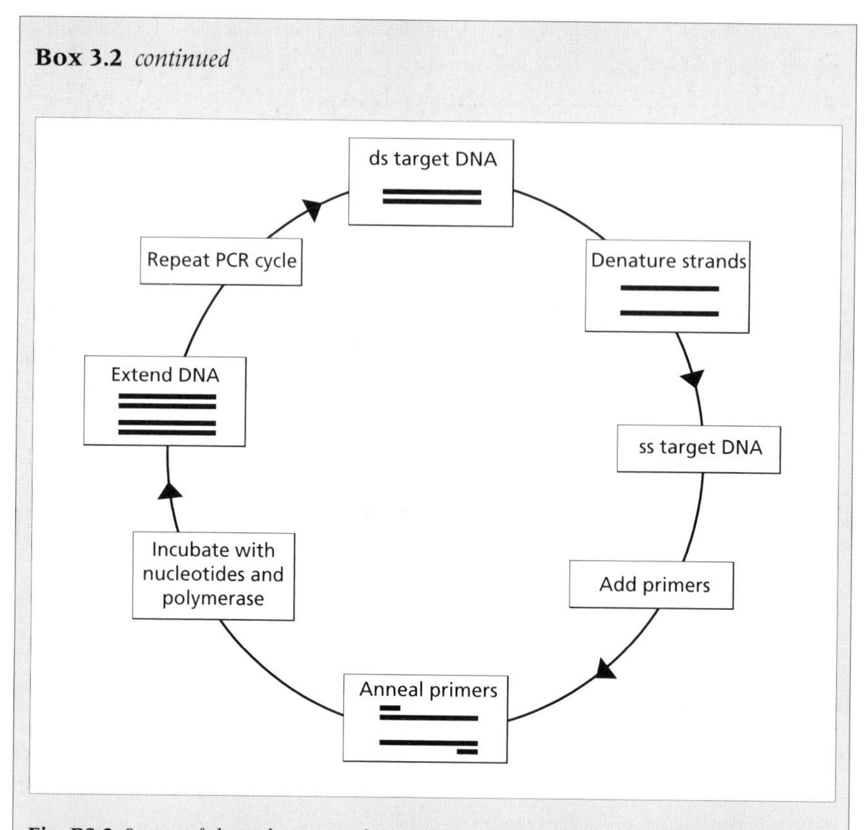

Fig. B3.2 Stages of the polymerase chain reaction (PCR). Adapted from Logan (1994).

single-stranded copy of the target DNA sequence can itself act as a template, so copy number doubles in each cycle (hence the term 'chain reaction').

Once large amounts of DNA have been isolated and purified it is possible to analyse it in more detail. One of the most important tools in this analysis are **restriction enzymes**—special bacterial enzymes that cut DNA at certain recognition sequences (or **restriction sites**). Restriction enzymes consist of a specific sequence, usually of 4–8 bp, and are often composed of small inverted repeat sequences called palindromes. The fragments of DNA produced by restriction analysis are then separated by size using **electrophoresis** (the separation of substances by charge, molecular weight and configuration by running an electric field through samples on gels made of agarose or polyacrylamide). Mutations produce fragments of different size, so that this technique is a useful way of surveying population variation (polymorphisms result from the loss or creation of a restriction site). This technique is called **Restriction Fragment Length Polymorphism (RFLP)** analysis. Another set of molecular markers used in population biology are **Randomly Amplified Polymorphic DNAs** or **RAPDs**. In this case, random primers of 8–10 bp in length amplify random

continued

Box 3.2 *continued*

DNA fragments during a PCR reaction, which can then be separated out by electrophoresis. Although simple to use, RAPDs markers can be dominant to each other so that it is difficult to identify heterozygotes.

Finally, it is possible to determine the precise running order of bases in a gene. There are two main methods for **DNA sequencing**, both of which were developed in the mid-1970s: the chemical sequencing method of Maxam and Gilbert and the more widely used Sanger dideoxy chain termination method. Both methods work on the same principle: four reactions, one specific for each base, generate a set of fragments which appear in electrophoresis as bands in one of four lanes. Much of gene sequencing technology is now automated through the use of DNA sequencing machines. In this case, instead of occupying different lines on a gel, the four bases are recognised by four different coloured dyes ('dye-terminators') and appear in a single lane. These can be read directly by a computer and displayed as peaks of different colours (see the back cover of this book).

3.4 Summary

1 Genetic information is stored in genomes at the chromosomal, protein and DNA sequence levels and various mechanisms exist by which this information is able to move between levels.

2 Genes themselves consist of a coding region flanked by various regulatory sequences which determine when and how the gene is expressed. Although genes from eukaryotes, Bacteria and Archaea are similar in this respect, coding regions in most eukaryotic genes also contain spliceosomal introns which are absent from Bacteria and Archaea.

3 Another difference between eukaryotes, Bacteria and Archaea is that most DNA in eukaryotes is non-coding and frequently composed of sequences that are repeated many times. In contrast, most DNA in Bacteria and Archaea encodes proteins.

4 The non-coding repetitive DNA of eukaryotes can be classified into: those composed of short motifs and which are tandemly-repeated—satellite DNA, minisatellites and microsatellites; and those where the repeat unit is longer and which occupy dispersed genomic locations because they are mobile—transposable elements.

5 Many coding DNA sequences are also repetitive, although in lower copy numbers, forming multigene families. Multigene families were an important evolutionary innovation, created by a variety of complex processes including gene duplication, unequal crossing-over and gene conversion. These processes also mean that multigene families often have mosaic phylogenetic histories.

3.5 Further reading

There are many books which describe the fundamental aspects of molecular genetics in more detail than presented here but Griffiths *et al.* (1993) and Kendrew (1994) are particularly good, whilst excellent overviews of the structure and evolution of eukaryote genomes and of genetic systems in general are provided by John and Miklos (1988), Li (1997) and Maynard Smith and Szathmáry (1995).

The extraordinary diversity of eukaryote genomes and the processes that have given rise to it are described by Henikoff *et al.* (1997), while Charlesworth *et al.* (1994) describe the processes responsible for the evolution of the repetitive part. With respect to the evolution of multigene families, that of the globins is described by Hardison (1991) and Dickerson and Geis (1983), whilst the *HOM/Hox* developmental genes are tackled by Bailey *et al.* (1997), Garcia-Fernàndez and Holland (1994) (who show the significance of *amphioxus*), Holland *et al.* (1994) and Krappen and Ruddle (1993), whilst Bird (1995) discusses the evolution of gene number in more general terms. The structure, evolution and application of another multigene family, the rDNA cluster, is well reviewed by Hillis and Dixon (1991). How unequal crossing-over produces the Lepore variant of human haemoglobin is explained by Weatherall (1991) (who also outlines the molecular basis to a number of human genetic disorders), whilst gene conversion in the γ-globins is tackled by Slightom *et al.* (1988).

Turning to some other components of the eukaryote genome, more detail on the structure and use of minisatellites can be found in Armour *et al.* (1996), Gill *et al.* (1985), Jeffreys *et al.* (1985), and of microsatellites in Gottelli *et al.* (1994) (who look at genetic diversity in the Ethiopian wolf), Hagelberg *et al.* (1991) (who discuss the murder case), Jarne and Lagoda (1996) and Tautz and Schlötterer (1994). Excellent reviews on the types and structure of transposable elements are provided by Flavell *et al.* (1994), Kidwell and Lisch (1997) and McDonald (1993), whilst the evolution of the *P* elements of *Drosophila* is examined in detail by Engels (1992).

With respect to some of the specific examples used, the story of AZT resistance in HIV-1 is initially told by Larder and Kemp (1989), mutation rates after Chernobyl by Dubrova *et al.* (1996) and penicillin resistance in *Neisseria* bacteria by Spratt *et al.* (1989 and 1992). Finally, the debate over the evolution of introns is given in Gilbert *et al.* (1997), Hurst (1994), Logsdon and Palmer (1994) and Palmer and Logsdon (1991).

Chapter 4
Genes in Populations

4.1 The fundamentals of population genetics

The peoples of Asia have high frequencies of a variant form of the enzyme aldehyde dehydrogenase (ALDH2) which helps in the breakdown of alcohol. Those with this mutant enzyme feel unwell and experience flushing when they drink alcohol and, discouraged by these ill-effects, are less likely to become alcoholics.

The genetic variation in aldehyde dehydrogenase is just one example of the many that exist within every species. The science of population genetics attempts to describe the structure of this variation and determine the evolutionary forces which have shaped it. This is an extremely important task because in one way evolution can be thought of as changes in the genetic composition of populations. Although this is not a textbook of population genetics, it is important that we understand the basics of this subject because, as we shall see more clearly in Chapter 7, molecular evolution is a process which occurs both within and between populations. Furthermore, although the molecular bases to many evolutionary processes are only just beginning to be uncovered, this is likely to change rapidly in the near future as gene sequence data accumulates with ever increasing speed.

4.1.1 Describing genetic variation

Traditionally, the most important pieces of data used by population geneticists have been the relative frequencies of different forms of genes within a population. These different forms, created by mutation, are called **alleles** and the coexistence of two or more alleles of a gene means that the population

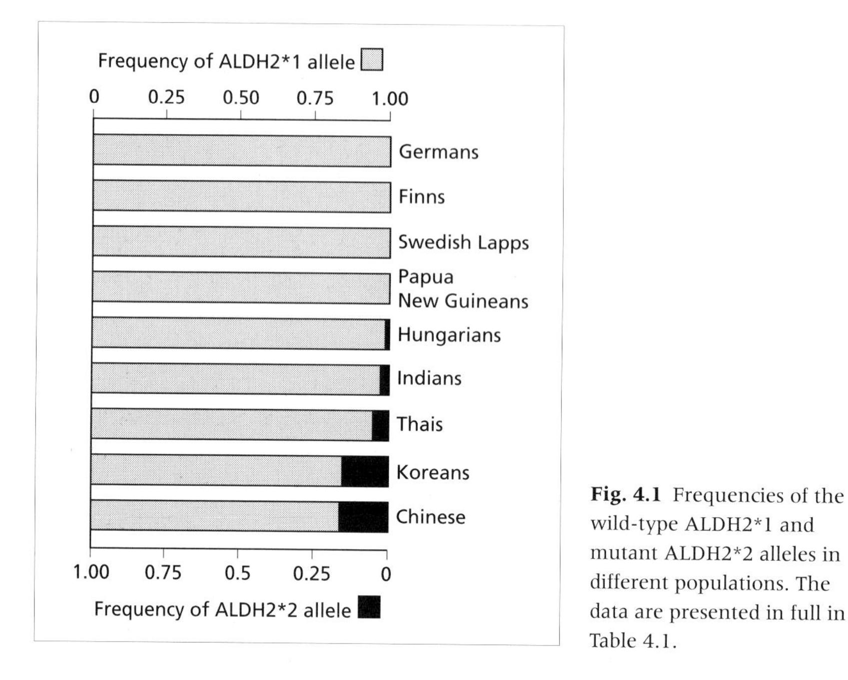

Fig. 4.1 Frequencies of the wild-type ALDH2*1 and mutant ALDH2*2 alleles in different populations. The data are presented in full in Table 4.1.

exhibits a **polymorphism**. The allele most frequently found is called the **wild-type**. In the case of ALDH2, for example, the wild-type allele worldwide, which produces a fully functional enzyme, is called ALDH2*1. A single base change creates the variant form, ALDH2*2, which can reach frequencies approaching 25% in some Asian populations, although it is only rarely found in Caucasians (Fig. 4.1).

To understand why allele frequencies look like they do, it is necessary to input the allele frequencies observed into theoretical (mathematical) models. Unfortunately, these models can become very complex and we will only present the most basic forms in this chapter, so that the fundamentals of the subject can be grasped. The simplest model has a single gene locus with two alleles, symbolised by A and a. In this case the frequencies of A and a in the population are given by the mathematical terms p and q, respectively. Although considering only a single gene with two alleles is highly simplistic, especially since studies at the DNA level have revealed enormous amounts of genetic variation, it is an informative first step.

If our study organism is diploid, so that it has two alleles for each gene (one on each homologous chromosome), then an individual may have one of three different combinations, or **genotypes**, of these alleles: AA, Aa or aa. Because AA and aa individuals carry two copies of the same allele they (or their genotypes) are said to be **homozygous**, while an individual with the Aa genotype is **heterozygous** at this locus because it possesses two different alleles. In some cases, one allele may be expressed preferentially over the other, in which case it is said to be the **dominant** allele. The other allele is then

recessive. Whether an allele is dominant or recessive becomes important when the genotype is heterozygous: if, in the *Aa* heterozygote, the *A* allele is dominant to the *a* allele, then the phenotype will be of the *AA* type. For example, the mutant ALDH2*2 allele is dominant to the wild-type ALDH2*1 allele, so that both homozygous and heterozygous individuals with this allele will lack the action of this enzyme in their liver. If both alleles are simultaneously expressed in the heterozygote, producing a third phenotype, they are said to be **codominant**. Most of the protein and DNA markers used in population genetics are codominant which is useful because it means that heterozygotes can be easily identified.

If we denote the frequencies of these three genotypes in a population by f_{AA}, f_{Aa} and f_{aa}, it becomes a simple matter to calculate the overall allele frequencies, p and q: homozygous individuals will carry only *A* or *a* alleles, whilst heterozygotes will have one-half *A* alleles and one-half *a* alleles, so that:

$$p = f_{AA} + \tfrac{1}{2}f_{Aa} \text{ and } q = f_{aa} + \tfrac{1}{2}f_{Aa} \qquad (4.1)$$

where $p + q = 1.0$. For example, the ALDH2 gene was studied in 218 Koreans (Table 4.1); 156 were found to be homozygous for the wild-type allele ALDH2*1 ($f_{AA} = 0.716$), 4 were homozygous for the mutant allele ALDH2*2 ($f_{aa} = 0.018$), whilst 58 were heterozygous ($f_{Aa} = 0.266$). Using equation (4.1) we find that $p = 0.716 + 0.133 = 0.849$ and $q = 0.018 + 0.133 = 0.151$. This procedure can be easily extended to cases where there are more than two alleles at a locus: the frequency of each allele is always calculated as the frequency of its homozygote form, plus half the frequency for all the heterozygotes in which it is present.

The allele frequencies p and q provide a very simple description of the amount of genetic variation in a population. However, because there are often

Table 4.1 Genotype and allele frequencies of alleles of ALDH2 from selected human populations. Adapted from Goedde *et al.* (1992).

	Genotype				Allele frequency	
Population	*n*	2*1/2*1	2*1/2*2	2*2/2*2	ALDH2*1	ALDH2*2
Germans	193	193	0	0	1.000	0.000
Finns	100	100	0	0	1.000	0.000
Swedish Lapps	100	100	0	0	1.000	0.000
Papua New Guineans	242	240	2	0	0.996	0.004
Hungarians	117	114	3	0	0.987	0.013
Indians	179	173	5	1	0.980	0.020
Thais	111	100	11	0	0.950	0.050
Koreans	218	156	58	4	0.849	0.151
Chinese	132	92	38	2	0.841	0.159

more than two alleles at a locus, dealing with their frequencies alone becomes cumbersome. Therefore, a more useful measure of genetic diversity is the total frequency of *heterozygotes* in the population otherwise known as the **heterozygosity** (h):

$$h = 1 - \sum_{i=1}^{m} x_i^2 \tag{4.2}$$

where m is the number of alleles and x_i the frequency of the ith allele at the locus. Heterozygosity will be greatest when there are many alleles, all at equal frequency. It is also possible to calculate the **average heterozygosity** (H) across all loci, which can also be thought of as the average fraction of heterozygotes per locus.

Until recently, heterozygosity was usually measured by analysing the frequencies of alleles in enzymes, detected using protein electrophoresis. This method involves running an electrical charge through proteins laid on a gel bed, usually made of starch. The proteins move down the gel at different speeds depending on their weight and charge, themselves governed by the underlying amino acid sequence. Enzymes with alleles that differ in electrophoretic mobility are called **allozymes**. Although studies of allozyme variation were the backbone of population genetics, today much more attention is given to DNA sequence variation where the level of resolution is much greater. Although heterozygosity can also be calculated from DNA sequences, where it is often called the **gene diversity**, measures of genetic variation which take into account the actual number of base changes between sequences, rather than just whether sequences are the same or different, are more informative. One of the most common of these measures is the **nucleotide diversity** (or π) which represents the average number of nucleotide differences per site between two sequences. Nucleotide diversity, as well as some other measures of DNA polymorphism, are described in Box 4.1.

Box 4.1 Measuring genetic variation within populations

One of the main tasks of population genetics is to measure the amount of genetic variation in populations. Although this can be done at a number of levels, from groups of allozymes to single bases, we will only consider variation at the nucleotide level, as this has the greatest degree of resolution and will doubtless be used with increasing frequency in years to come.

Perhaps the simplest measure which can be used is the number of variable nucleotide sites in a sample of sequences, which we can denote as S. This is also known as the number of **segregating sites**. However, S does not incorporate the length of the sequence analysed and so is not comparable across data sets. If we divide this number by the sequence length (L) we get the number

continued

Box 4.1 *continued*

of segregating sites per nucleotide site, which is a better measure of genetic diversity.

Another frequently used measure of molecular genetic variation in populations is the **average number of pairwise nucleotide differences** between sequences, Π. This can be estimated by;

$$\Pi = \frac{1}{[n(n-1)/2]} \sum_{i<j} \Pi_{ij}$$

where n is the number of sequences in the sample (so that $n(n-1)/2$ is the number of pairwise comparisons), and Π_{ij} is the difference between the ith and jth sequences. Once again, this number can be standardised by dividing it by the sequence length. This new measure, Π/L, is also known as the **nucleotide diversity** or π.

One of the most important measures of genetic diversity within populations is given by a parameter called θ, which is equal to $4N_e\mu$. θ is important because it describes the amount of variation expected at each nucleotide site if evolution is entirely neutral (and can actually be used to test the neutral theory—see Box 7.1, p. 242). There are various ways to estimate θ, including some which take into account the genealogical relationships between alleles, although the details of these methods are beyond the scope of this book. It is worth noting, however, that once θ has been estimated, and if μ, the mutation rate is known, then it is also possible to estimate the effective population size, N_e, directly from sequence data.

We can compare these measures using the sequence data obtained from a study of the evolution of HIV-1 in a single infected patient (p82), which we came across in Chapter 2. Sequences from the V3 region of the envelope glycoprotein gp120 were sampled over a number of years and it is possible to compute the amount of genetic variation in each sample, as has been done in Table B4.1.

Table B4.1 Various measures of molecular genetic variation in HIV-1 over five years of infection in a single patient. Adapted from Leigh Brown (1997), with permission. Copyright (1997) National Academy of Sciences, U.S.A.

Year	n	S	S/L	Π	$\Pi/L(\pi)$	θ
3	15	25	0.1082	7.37	0.0319	0.0964
4	11	21	0.0909	7.89	0.0342	0.0576
5	23	25	0.1082	6.27	0.0271	0.0505
6	15	33	0.1429	12.31	0.0533	0.1052
7	13	39	0.1688	10.12	0.0438	0.0915

The length of the sequence (L) is 231 bp and n is the number of sequences sampled in each year of the infection. θ is estimated using the maximum likelihood (and genealogy-based) method of Kuhner *et al.* (1995).

4.1.2 The Hardy–Weinberg theorem

Once allele frequencies and heterozygosities have been calculated, the next task facing population geneticists is to determine whether these frequencies are changing through generations and between populations. Such changes in allele frequency in time and space are the footprints of evolution. However, before we can examine whether allele frequencies are changing, it is necessary to decide why they might remain the *same*. Explaining how the variation observed in populations was maintained between generations was one of the most serious problems faced by early evolutionists. Charles Darwin, for example, who despite being a contemporary of Mendel was unaware of his work, had no mechanism for the inheritance of phenotypic characters between individuals. By the later editions of *On The Origin of Species* Darwin was advocating the 'blending' of characteristics, in which offspring are an intermediate mix of features inherited from both parents. Unfortunately, under blending variation will be halved in each generation and the population will become phenotypically homogeneous in a short time. Even after Mendel's work was rediscovered at the turn of the century, explaining the maintenance of genetic variation was a serious problem: if one allele was typically dominant over another, why was the recessive allele not lost completely? The first achievement of population genetics was to resolve this dilemma and show that genetic variation could be maintained intact through successive generations. This was done through what become known as the **Hardy–Weinberg theorem**, named after its co-discoverers, G.H. Hardy and W. Weinberg (in 1908). In many ways the Hardy–Weinberg theorem is the cornerstone of population genetics as it is a model of genetic inheritance *without* any evolutionary change, and only requires Mendel's simple laws of inheritance.

The Hardy–Weinberg theorem works as follows. Take a diploid population of males and females which are able to mate randomly with each other. We need to make a number of other assumptions about this population but they will be discussed later. As before, the frequency of allele A is p and the frequency of allele a is q. We can depict the outcomes of a set of random matings between these males and females, in terms of what genotype will be found in the offspring, using a **Punnett square** (Fig. 4.2), where each cell represents a particular genotype.

Crucially, the Punnett square shows how each offspring genotype relates to the parental allele frequencies, p and q. In other words, it is possible to calculate the frequency of each offspring genotype from the allele frequencies observed in the parents. This is one of the most important features of the Hardy–Weinberg theorem. For example, we can see from the Punnett square that the frequency of AA homozygotes in the offspring generation is simply the frequency of one A allele (p) multiplied by the frequency of another A allele: $p \times p$, or p^2 (the frequency of aa homozygotes is therefore $q \times q = q^2$).

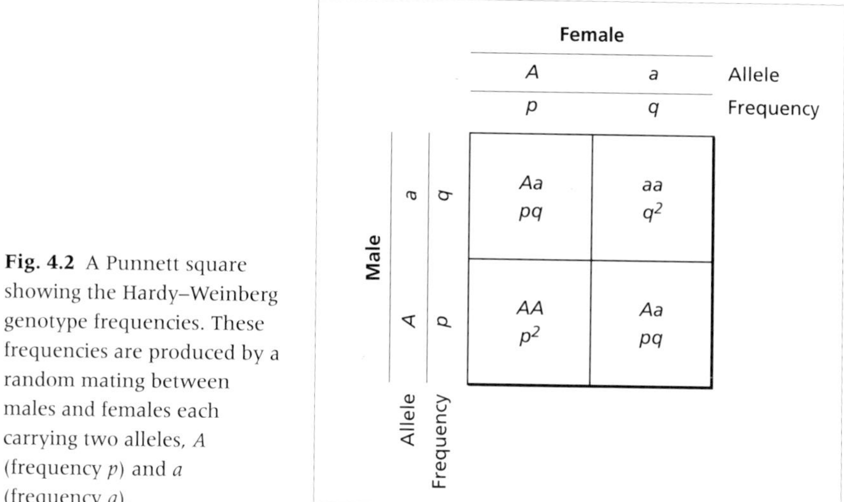

Fig. 4.2 A Punnett square showing the Hardy–Weinberg genotype frequencies. These frequencies are produced by a random mating between males and females each carrying two alleles, *A* (frequency p) and *a* (frequency q).

Likewise, the frequency of *Aa* heterozygotes is $(p \times q) + (q \times p)$ or $2pq$. The frequencies of the three genotypes after one generation of random mating are therefore given by:

$$AA = p^2;\ Aa = 2pq;\ aa = q^2 \tag{4.3}$$

We can illustrate this with a simple numerical example. As we have already seen, the frequencies of the ALDH2*1 (p) and ALDH2*2 (q) alleles in Koreans are 0.849 and 0.151, respectively. Now assume that males and females in this population mate randomly with each other. The genotype frequencies in the next generation will then be $p^2 = 0.721$, $2pq = 0.256$ and $q^2 = 0.023$. Multiplying these frequencies by the number in the population (218), we obtain $AA = 157$, $Aa = 56$, $aa = 5$, which are very similar to those observed in the previous generation ($AA = 156$, $Aa = 58$, $aa = 4$; Table 4.1). That the genotype frequencies are very similar to those in the previous generation suggests that the population has reached a stable condition, known as the **Hardy–Weinberg equilibrium**. Even more striking is that these equilibrium frequencies were arrived at after just one generation of random mating, regardless of what they were in the parental generation, and will remain the same in the generations that follow, providing all the assumptions still hold (see below). For example, if, in the first generation, p and q are also 0.849 and 0.151, respectively, but the genotype frequencies are very biased against heterozygotes, so that $AA = 185$, $Aa = 0$, $aa = 33$, our equilibrium genotype frequencies ($AA = 0.721$, $Aa = 0.256$, $aa = 0.023$) are still produced after a single generation of random mating (apply equation (4.3) to see for yourself). This is the most important part of the Hardy–Weinberg theorem because it means that a single generation of random mating establishes genotype (and allele) frequencies which will remain unchanged in the generations that follow. No genetic variation is lost. As we shall see later, equilibrium allele frequencies like those defined in the Hardy–Weinberg theorem are very important in population genetics.

The Hardy–Weinberg theorem can also be applied when there are more than two alleles at a locus, to the sex chromosomes, or to multiple loci. However, although the theorem is very simple, it only works if certain assumptions are met. These assumptions are that the organism: (i) is diploid; (ii) reproduces sexually; (iii) mates randomly with other individuals in the population; (iv) that this population is infinitely large; (v) suffers no mutation; (vi) no natural selection; and (vii) no migration with other populations. If these assumptions break down, as is often the case in real populations, then allele and/or genotype frequencies will change. Looking for deviations from what is predicted by Hardy–Weinberg is therefore a good starting point for studying the evolution of genes in populations. In the next section we discuss some of the forces which change allele frequencies through time—the mechanisms of evolutionary change.

4.2 Forces which change allele frequencies

4.2.1 Mutation and recombination

The only way in which new types of DNA sequence can arise is through mutation. Mutation therefore provides the fuel for evolution to run. Other forces which change allele frequencies work on what is ultimately provided by mutation. The different types of mutation are discussed in Chapter 3.

Despite its obvious importance in evolution, mutation is generally an infrequent occurrence at the molecular level (Table 4.2). Mutation rates are notoriously difficult to measure directly and so are shown here as average rates of synonymous substitution per site (which are good measures of the underlying mutation rate if changes are selectively neutral—see Chapter 7). For many eukaryotes, these substitution rates are only in the region of 10^{-9} per base, per year (or about 10^{-6} to 10^{-5} per gene, per year), and so low that mutation alone is a weak force in changing allele frequencies. Rates are higher in other organisms (such as *Drosophila*) and other genomes (such as mammalian mtDNA) and reach a peak in lytic RNA viruses, such as those that cause poliomyelitis and influenza, which lack any kind of repair mechanism. In these viruses, substitution rates may be in the order of 10^{-3} to 10^{-2} per base, per year, implying that a mutation occurs somewhere in the genome during every round of replication.

We can also show that mutation alone will change allele frequencies only very slowly with some simple mathematics. If we have no **back mutation**, so

Table 4.2 Average rates of synonymous substitution in various organisms and genomes.

Organism and genome	Rate of synonymous substitution (per site, per year)
Plant chloroplast DNA	$\sim 1 \times 10^{-9}$
Mammalian nuclear DNA	3.5×10^{-9}
Plant nuclear DNA[a]	$\sim 5 \times 10^{-9}$
E. coli and *Salmonella enterica* bacteria[b]	$\sim 5 \times 10^{-9}$
Drosophila nuclear DNA	1.5×10^{-8}
Mammalian mitochondrial DNA	5.7×10^{-8}
HIV-1[c]	6.6×10^{-3}
Influenza A virus	1.3×10^{-2}

If synonymous substitutions are selectively neutral, then these rates are a good measure of the background mutation rate, although they can be highly variable within each of the organisms and genomes shown (see Chapter 7) and so should only be taken as a rough guide. All data from Li (1997) except [a]Gaut *et al.* (1996) who studied the *Adh* gene of grasses and palms, [b]Lawrence and Ochman (1997) on bacteria and [c]Kasper *et al.* (1995) who examined the *gag* gene of HIV-1.

that *A* alleles can only change to *a* alleles but not vice versa, the change in allele frequencies due to mutation pressure can be calculated by:

$$p_t = p_0(1 - \mu)^t \qquad (4.4)$$

where p_t is the frequency of allele *A* after *t* generations, p_0 the starting frequency of *A* in the population and μ the mutation rate. Because mutation rates are so low it is actually possible to approximate $(1 - \mu)^t$ by $e^{-\mu t}$ (where e is the base of the natural logarithm, 2.718 ...) so that equation (4.4) can be rewritten as:

$$p_t = p_0 e^{-\mu t} \qquad (4.5)$$

Using these equations we can see that if the starting frequency of allele *A* (p_0) is 1 and the mutation rate from *A* to *a* is 10^{-5} per generation, then after 100 generations of mutation from allele *A* to allele *a*, the frequency of *A* in the population (p_t) will have hardly decreased at all, to only 0.999. It will take almost 70 000 generations for mutation to reduce the frequency of *A* in the population to 0.5.

The situation is a little more complicated if we have back mutation because it obviously means that we have two mutation terms in the equation: μ (*A* to *a*) and ν (*a* to *A*). In this case, the frequency of allele *A* after a period of mutation is given by:

$$p_t = \frac{\nu}{\mu + \nu} + \left(p_0 - \frac{\nu}{\mu + \nu}\right)(1 - \mu - \nu)^t \qquad (4.6)$$

Back mutation will also mean that neither the *A* nor *a* allele will end up as the only one in population, because mutants in both directions will continually arise. To use a slightly different language, neither allele will achieve a state of **fixation** in which $p = 1$ or $q = 1$. Rather, an equilibrium value (denoted $\hat{p}$ for allele A) will be reached, given by:

$$\hat{p} = \frac{\nu}{\mu + \nu} \tag{4.7}$$

Although these equations are very often shown in textbooks of population genetics, and represent a useful starting point, they are unrealistic for most genes in which there will be more than two alleles.

Other than in cases like RNA viruses, the only way in which mutation can greatly affect allele frequencies is when it occurs persistently at a single locus—a process known as **recurrent mutation**. Because the probability of the same mutation occurring at a single nucleotide in a sequence is very low, recurrent mutations usually involve large genetic changes, including whole genes or large segments of chromosome. A good example of recurrent mutations is the thalassaemia blood disorders, the commonest genetic diseases of humans, where a heterogeneous set of mutations often involving deletions of stretches of DNA, reduce the rate of synthesis of the α or β chains of adult haemoglobin. Mutations that cause these **deleterious** effects, so that they are removed from the population by natural selection (see section 4.2.2) but continually re-appear through mutation, are said to be in a state of **mutation–selection balance**—an equilibrium value set by the counteracting forces of mutation and selection.

Although mutation produces the genetic variation needed for adaptive evolution, it also entails a large cost because many mutations will be deleterious. In bacteria, for example, some strains have extremely high mutation rates ('**mutators**'), often due to a deficient mismatch repair system. Sometimes these mutator strains are able to out-compete those with normal mutation rates but do not take over in wild populations because they also produce more deleterious changes, which may in turn reduce their growth rates. It has also been suggested that some bacteria are able to make specific and adaptively useful mutations in response to external signals. Although there is little evidence for these highly controversial **directed mutations**, it is likely that some genes in bacteria are hypermutable, allowing greater genetic flexibility particularly in times of environmental stress.

The cost of an increased number of deleterious mutations is even more severe for multicellular eukaryotes because of their greater genome complexity and because the larger number of DNA replications required to make an organism with a long lifespan mean that there are more opportunities for mutation to act. Because of these costs, eukaryotes have generally reduced their mutation rates through the evolution of more efficient DNA repair systems, and instead generate much of their genetic variation through the process of **recombination** (see Chapter 3). Recombination leads to mutations being shuffled among chromosomes, often during meiosis, so that the progeny have different combinations of alleles from those of their parents. Although recombination does not change allele frequencies in itself, it greatly enhances the genetic variation produced by mutation.

The genetic diversity produced by recombination is one of the major plus points of sexual reproduction and gives sexual populations a number of evolutionary advantages over asexual communities, in particular a faster response to changing environments. In this case environments can be either physical or biological. For example, the extra genetic variation produced by sexual reproduction will lead to a better defence against parasites, because greater genetic diversity means that a wider spectrum of parasites can be recognised. Another advantage that recombination gives sexual populations is the easier removal of deleterious mutations. The steady build up of deleterious mutations in an asexual population may lead to a phenomenon known as **Muller's Ratchet** in which the fitness (a measure of evolutionary health—see section 4.2.2) declines through time as individuals without deleterious mutations become rarer and rarer. The loss of mutation-free individuals is irreversible and at each loss the ratchet clicks on one notch. The ratchet does not apply to sexual populations which are able to generate individuals without deleterious mutations through recombination and so maintain fitness.

A simple measure of the amount of recombination is the degree of **linkage disequilibrium**. Assume we have two gene loci each with a pair of alleles at equal frequency, *Aa* at the first locus and *Bb* at the second (take a look back at Fig. 3.14). There are four different combinations of these alleles: *AB, Ab, aB* and *ab*. If recombination occurs freely then, on average, there will be an equal proportion of each combination in the population. In this case we can say that our loci are in a state of linkage equilibrium. If, on the other hand, these alleles are not randomly assorted so that particular alleles are inherited together as a single unit, then there is a certain amount of linkage disequilibrium. This can occur either because there is a strong physical linkage between loci, perhaps because they work well in combination and so are favoured by natural selection, or because of population substructure (see section 4.2.5).

The degree of linkage disequilibrium (D) can simply be measured by:

$$D = P_{AB}P_{ab} - P_{aB}P_{Ab} \tag{4.8}$$

where P denotes the frequency of the particular allelic combination. A population is in complete linkage equilibrium when $D = 0$. Although this and other measures of linkage equilibrium are less than perfect when allele frequencies differ, values of D for allozymes in natural populations are usually very close to zero, indicating that recombination occurs frequently unless genes are very closely linked. However, it is also the case that some chromosomal regions are less liable to recombination than others, especially in areas of heterochromatin located at the centromeres and telomeres (see Chapter 3). In *Drosophila*, for example, genes located at the distal tip of the X chromosome and on the small fourth chromosome have very low recombination rates (see Chapter 7), while 95% of the mammalian Y chromosome does not recombine.

An important example of linkage disequilibrium at the molecular level occurs in the genes which make the major histocompatibility complex (MHC), a vital part of the vertebrate immune system (the evolution of this gene complex is discussed in detail in Chapter 7). Alleles at different MHC genes, called **haplotypes**, are often inherited in combination, with different haplotypes found in different populations. For example, the alleles A1, B8 and DR3 show significant linkage disequilibrium in Caucasian populations. Unfortunately, this haplotype has also been associated with a faster progression to AIDS in HIV-infected individuals.

4.2.2 Natural selection

Natural selection is at the heart of the evolutionary process. Because of the 'struggle for existence' that faces organisms as they compete for resources, those with genes that better adapt them to their environment have a greater probability of surviving this struggle, so that their favourable genes are preferentially passed on and will increase in frequency in the population. Although natural selection is synonymous with the name of Charles Darwin, it was R.A. Fisher who did much to show the power of this process at the genetic level.

Natural selection can act in a variety of ways and at different points during an organism's life-cycle. For example, genotypes may differ in their **viability** of producing adult organisms from zygotes. Later in the life-cycle, individuals may differ in their **fecundity**—the numbers of offspring they produce—which will also lead to natural selection. However, before this takes place there may also be **sexual selection** in which organisms differ in their mating success. Finally, gametes may have different probabilities of achieving fertilisation leading to a form of **gametic selection**.

The simplest way of thinking about whether one organism is better adapted than another is to describe its **fitness**. In population genetic terms, fitness can be defined as the capability that any particular genotype has to survive and reproduce. This is usually expressed in relative terms, such as whether the heterozygote *Aa* has a higher fitness than the homozygotes *AA* and *aa*. Fitness is also specific to each environment, because a genotype which is beneficial in one location might be deleterious in another (and vice versa).

In molecular evolution, fitnesses are most often expressed in terms of the **selection coefficient**, denoted s, which is a measure of the reduction in fitness compared with the best genotype in the population: for instance, an s of 0.01 means that a genotype has a 1% less chance of survival than the best genotype—it is 99% as fit. Despite this very simple notation, fitness is a complex entity which may change through time, and is difficult to measure in nature. Indeed, there are very few cases in which the differences between alleles at a single locus make enough difference to phenotype to have a measurable effect on fitness.

In most cases mutations will be deleterious and so lower fitness. These mutations will be removed fairly quickly from the population by **purifying** or **negative selection**. At other times a new mutation might have a higher fitness than all others in the population and so is able to multiply. This is usually called **positive selection** in molecular evolution, and will lead to the eventual fixation of the selectively favoured allele (see below). For example, it may be that allele *A* has a higher fitness than allele *a*. This means (assuming *A* is dominant to *a*) that selection will favour the *AA* and *Aa* genotypes over the *aa* genotype. In terms of selective coefficients, we can think of this as:

Genotype	*AA*	*Aa*	*aa*
Fitness	1	1	$1 - s$

On the other hand, if there is codominance, so that the *Aa* heterozygote has a fitness intermediate to those of the two homozygotes, then:

Genotype	*AA*	*Aa*	*aa*
Fitness	1	$1 - s$	$1 - 2s$

As well as looking at a single locus, it is also possible to think of natural selection acting on a phenotypic character which is under the control of many different loci, such as height or body weight. The genetics of these **quantitative characters** is discussed more fully in section 4.3. Typically, characters of this sort have a bell-shaped normal distribution of phenotypes, with most individuals somewhere near the middle (the mean) of the distribution, and decreasing numbers at more extreme values (represented by a variance). Positive selection on quantitative characters, more often referred to as **directional selection**, will move the mean of the distribution in one direction (Fig. 4.3a).

Some of the best examples of positive selection involve the evolution of resistance to antibiotics, drugs and pesticides, such as that built up by mosquitoes to the chemical DDT once used to control them. A spectacular example of this, and one that has been of great practical importance in recent years, is the resistance developed by HIV-1 to the drug AZT. AZT inhibits viral replication by terminating transcription of the enzyme reverse transcriptase. In the absence of AZT, the wild-type allele for reverse transcriptase was found to be between 0.4 and 2.3% more fit than alleles adapted for drug resistance ($s = 0.004 - 0.023$). Although the drug works extremely well for the first few months, and greatly reduces the amount of circulating virus, the benefit is only short-lived and within six months or so the virus is able to evolve resistance through a small number of mutations. The frequency of the wild-type allele then declines rapidly. Current approaches to HIV therapy therefore involve combinations of drugs because the virus has a much smaller chance of developing multiple resistance.

In other cases natural selection will favour the heterozygote over either of

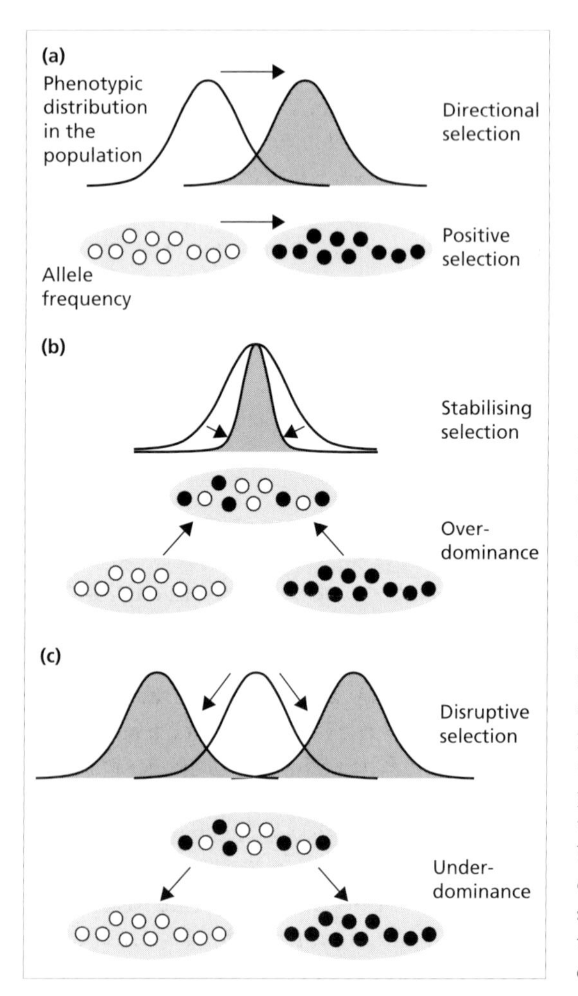

Fig. 4.3 Different types of natural selection and how they effect the phenotypic distribution of a character controlled by many genes (i.e. a quantitative character with a normal distribution) and the frequencies of alleles at a single locus where each shaded ellipse represents a population of individuals (circles) with an allele of a particular colour—black or white. The open and shaded bells represent the phenotypic distributions before and after selection, respectively, and the arrows signify the direction of selection.

the homozygotes. This is called **overdominance** (sometimes **heterozygous advantage**) and we can think of it in terms of fitnesses as follows:

Genotype	AA	Aa	aa
Fitness	$1 - s$	1	$1 - t$

where t is simply the selection coefficient for the aa genotype. Overdominant selection will produce high heterozygosity (i.e. over that expected under the Hardy-Weinberg equilibrium) and because it preserves both the p and q alleles will maintain genetic variation in populations, giving rise to a **balanced polymorphism**. The equilibrium allele frequencies (i.e. $\hat{p}$) which reflect this balance depend only on the relative fitnesses of the three genotypes, so that:

$$\hat{p} = \frac{t}{s+t} \qquad (4.9)$$

which takes exactly the same form as the equilibrium achieved under forward and backward mutation pressure (equation 4.7).

The most celebrated examples of overdominance are the sickle-cell polymorphism in human β-haemoglobin and the alcohol dehydrogenase (*Adh*) polymorphism in *Drosophila* (see Chapter 7). Overdominant selection is also likely to be operating on the MHC because a heterozygote with different MHC molecules will recognise more parasites than a homozygote with only one type of MHC molecule. The analogous process to overdominant selection for quantitative characters, in that genetic variation is actively maintained, is usually called **stabilising selection** and means that individuals in the mean of the distribution will benefit over those at the extremes (Fig. 4.3b).

The alternative to overdominance is where the heterozygote has a *lower* fitness than either homozygote:

Genotype	AA	Aa	aa
Fitness	1	$1 - s$	1

This is called **underdominance** and although homozygous genotypes are favoured, it still preserves both the *A* and *a* alleles although this polymorphism will be unstable, with a change in allele frequency away from the equilibrium leading to the fixation of *A* or *a*. For quantitative characters an analogous process is called **disruptive selection** whereby individuals at the extremes of the distribution have a higher fitness than those with mean values (Fig. 4.3c).

Underdominant selection appears to be controlling bill size in the African finch *Pyrenestes ostrinus* studied by Thomas Bates Smith. Bill sizes in this species are either large or small, but not of intermediate dimensions. The basis for this difference appears to be the hardness of the seeds eaten by the birds: during the reproductive season both the large- and small-billed morphs prefer soft seeds but during the dry season, when food abundance is low, the large-billed morph tends to feed on hard seeds while the small-billed morph expands its diet to include other foods. No seeds of intermediate hardness are found in the localities where these birds live so that selection acts against those with intermediate bill sizes.

Polymorphisms can also be maintained in populations through **frequency-dependent selection** (Fig. 4.4). This occurs when the fitness of a genotype depends on its frequency in the population. For example, the genotype at the *lowest* frequency may have the highest fitness (**negative frequency-dependent selection**). This may occur in host–parasite systems where a parasite at low frequency is not recognised by the host immune system and is therefore able to increase in frequency, whereas a strain at higher frequency may elicit a stronger immune response against it and so be more easily cleared. This will cause an oscillation in the frequency of different strains as they evade or are recognised by the immune system, leading to a polymorphism in the population. In other circumstances the genotype with the highest frequency may be favoured (**positive frequency-dependent selection**), which will

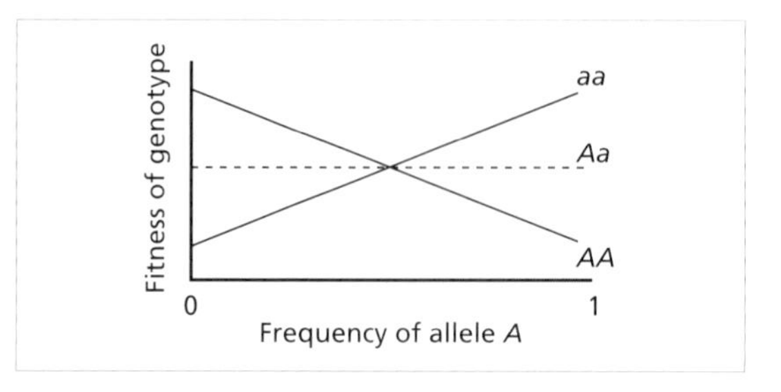

Fig. 4.4 Allele and genotype frequencies under frequency-dependent selection. Genotypes *AA* and *aa* are subject to negative frequency-dependent selection whereas the fitness of *Aa* is frequency-independent. To the right of the point where the lines cross, the *A* allele is favoured and increases in frequency although the fitness of the AA genotype decreases. To the left of this point, the frequency of the *A* allele declines and the *a* allele is favoured. Although the relationship between fitness and frequency is shown here as a straight line, in reality it can take on any shape. Adapted from Ridley (1996).

prevent polymorphisms from being established. Frequency-dependent selection may also occur if a population occupies a heterogeneous environment, so that different alleles are favoured in different patches (micro-environments) or at different times.

Unfortunately, detecting natural selection at the molecular level is a notoriously difficult thing to do because its effects can be subtle and because different evolutionary processes can leave similar signatures in sequences (see section 4.5 and Chapter 7). An added complication is that the agent of selection may no longer be present in the population. For instance, the chemokine receptor 5 (CCR5) gene is an important factor in allowing HIV-1 to infect cells called macrophages. About 10% of Caucasian people have a 32-bp deletion (called Δ32) which inactivates this gene and those homozygous for this deletion (about 1% of individuals) have a greatly reduced susceptibility to HIV infection, while those that are heterozygous may become HIV infected but seem to progress more slowly to AIDS. What is puzzling is that this mutation is at such high frequency in Caucasians (see Fig. 4.13) but is very rare in African and Asian populations. Because HIV is so recent an invader it cannot be the selective agent responsible for the high frequency of this mutation, which raises the possibility that a past infectious disease may have acted as a strong selective force favouring individuals with this deletion.

Despite this complexity, it is still possible to show how natural selection changes gene frequencies with some simple mathematics. Taking the simplest example, of positive selection favouring *AA* and *Aa* over *aa*, so that:

Genotype	*AA*	*Aa*	*aa*
Fitness	1	1	$1 - s$

it is easy to see that frequency of the *aa* genotype, or q^2 if we recall our Hardy–Weinberg formula, will be reduced by $1 - s$ in a given generation. The next thing we would like to do is calculate what p (the frequency of the favoured allele) will be after a generation of this selective process. We can denote this new frequency of p as p', and can calculate it by:

$$p' = \frac{p}{1 - sq^2} \tag{4.10}$$

The change in allele frequency (Δp) can then be simply computed as $p' - p$ or directly by:

$$\Delta p = \frac{spq^2}{1 - sq^2} \tag{4.11}$$

It is worth illustrating this with a numeral example. Let's assume p and q in the population are initially 0.25 and 0.75, respectively, and the selection coefficient against *aa* is at the top end of what we observed in the case of AZT resistance, say 0.025. Applying equation 4.11 to get Δp we discover that:

$$\frac{spq^2}{1 - sq^2} = \frac{0.025 \times 0.25 \times (0.75)^2}{1 - 0.025 \times (0.75)^2}$$
$$= \frac{0.0035}{0.9854}$$
$$= 0.0035$$

which means that p will have increased from 0.25 to 0.2535. Although this seems like a tiny increase it has happened in a *single* generation, and represents spectacular progress compared with what happens under mutation alone. If we run this process for long enough then the *A* allele will eventually achieve fixation. However, the rate at which this occurs is not constant, being fastest when there are many heterozygotes in the population and slowest when there are mostly homozygotes (plug in some different values for p and q and see for yourself). The same will also be true if there is codominance, in which case the change in allele frequencies takes on an S-shaped (symmetrical) form (Fig. 4.5). In sum, we can see that natural selection operates most efficiently when there are large amounts of genetic variation for it to work with. Furthermore, the efficiency of natural selection is also determined by the size of the population, working best when it is large. When population

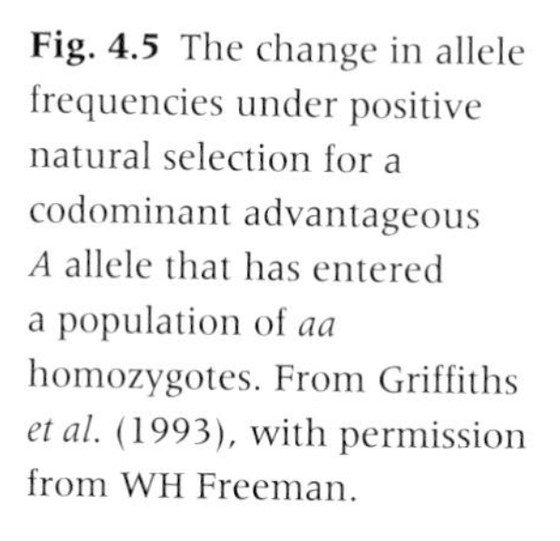

Fig. 4.5 The change in allele frequencies under positive natural selection for a codominant advantageous *A* allele that has entered a population of *aa* homozygotes. From Griffiths *et al.* (1993), with permission from WH Freeman.

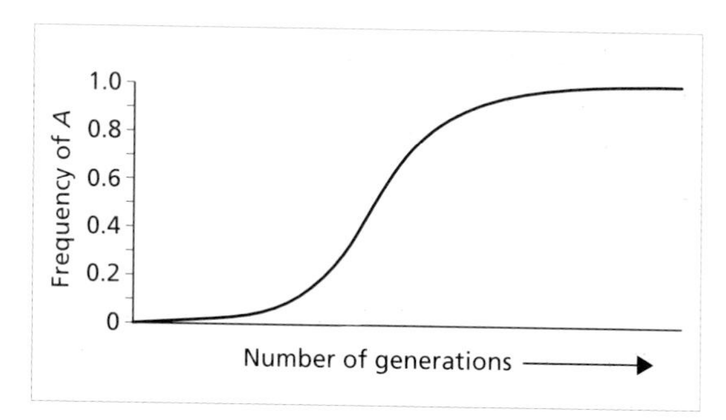

sizes are small, mutations are more under the control of chance processes. This is the subject of the next section.

4.2.3 Genetic drift

Once a mutation has arisen in a population it can experience one of two evolutionary fates: it can be fixed and so come to dominate the population, or it can be lost. Which outcome a new allele faces is not always down to how much better or worse it is compared to those alleles already present in the population. Instead it may simply be down to chance. For example, some alleles may reside in individuals who leave no offspring and even if an individual does produce offspring only a random sample of their genes, those present in the successful egg or sperm, will be inherited. Because of bad luck, most gametes will not make it to the next generation. This means that alleles are in effect randomly sampled in every generation, and this random sampling can change allele frequencies. This is called **genetic drift**.

To understand how allele frequencies change with genetic drift we need to develop a **stochastic model** of evolution, in which chance plays the leading role. This differs from our previous discussions of mutation and selection where we dealt with **deterministic models**, which ignored random effects. First imagine a diploid population of size N, so that there are a total of $2N$ allelic copies of each gene. Because of the random sampling of gametes there is a chance that some of these alleles will contribute no copies of themselves to the next generation, whilst others will contribute many. Some alleles will therefore be lost from the population in each generation because the gametes carrying them have not been passed on. This will cause allele frequencies to fluctuate slightly from generation to generation as some alleles are lost and others reproduce. If this stochastic variation in reproductive success continues for long enough there will eventually come a time when all $2N$ alleles will be descended from a single of our original $2N$ alleles because all others have failed to reproduce at some stage along the way (Fig. 4.6). This allele will then have been fixed by the process of genetic drift alone. The probability of fixation of an allele by random genetic drift is simply $1/(2N)$, which is its frequency in the population after it has arisen by mutation. This also means that the probability of fixation is greater when the population size is small, because $1/2N$ is larger.

Not only will genetic drift work better in small populations, but it will also be most important for those mutations which have no effect on phenotype, and so are neither advantageous nor deleterious compared to their predecessors. These mutations, which are free from the rigours of natural selection, are called **neutral mutations** and their evolution is discussed in detail in Chapter 7. Because an individual carrying a neutral mutation will not be at a selective advantage, the only way they can be fixed is through the chance action of genetic drift (the one exception—genetic hitchhiking—is discussed in Chapter 7). However, genetic drift will also affect new mutations with a selective

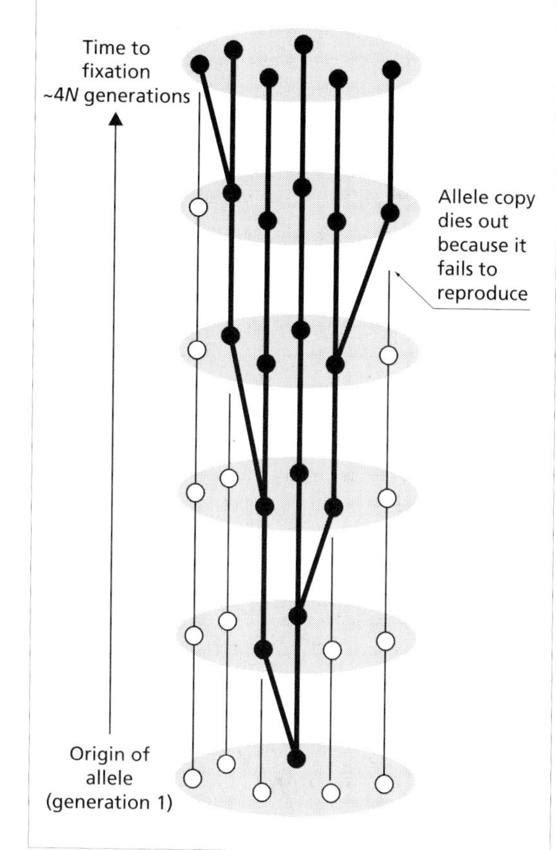

Fig. 4.6 Fixation of an allele (black circle) by genetic drift. The shaded ellipse represents a population observed at different time points. The other allele in the population (individuals with white circles) dies out because it fails to reproduce. On average it takes an allele $4N$ generations to get to fixation, although the probability of this happening is only $1/2N$.

advantage, most of which will also be lost by chance unless their benefit is substantial. For example, R.A. Fisher calculated that a mutant with a 1% selective advantage only has about a 2% chance of being fixed (so a 98% chance of being lost). Loss by genetic drift is especially likely soon after the mutation has appeared and so is still at low frequency.

Using a mathematical model it has been shown that it takes an average of $4N$ generations for a neutral allele to get to fixation through genetic drift, although the random nature of the process means that there is a large variance around this time. Mutations with a selective advantage will be fixed quicker than neutral alleles (that is, in less than $4N$ generations) because individuals with the better gene will produce more offspring than those with other alleles. Conversely, when natural selection is actively maintaining a polymorphism, alleles can coexist in the population without any going to fixation for as long as the selection pressure exists, so that they may survive for longer than $4N$ generations.

During the time it takes for an allele to be fixed by genetic drift, the population will exhibit polymorphism at the locus in question. Obviously, the larger the population size, the more time it will take for an allele to be fixed and the longer the polymorphism will last. However, if genetic drift continues for a long enough time and at enough loci, then eventually every locus will become homozygous because in each case a single allele will have drifted to fixation. Left to its own devices, genetic drift therefore tends to reduce genetic variation in populations. This situation is unlikely to be reached, however, because new mutations continue to arise and these will generate heterozygosity. Consequently, in the absence of natural selection and assuming that each mutation produces a new allele (this is the **infinite alleles model**—see Box 4.2), the level of heterozygosity reached can be thought of as a balance between the forces of genetic drift, which pushes the population towards homozygosity,

Box 4.2 Population genetic models of mutation

Understanding the process of mutation is vital if we are to make good inferences about the genetic structure of populations. A number of different models have therefore been devised which describe the accumulation of mutations in a population. Two of the most important are the **infinite alleles** and **infinite sites** models which relate the process of mutation to that of genetic drift. In the former is it assumed each mutation produces a new allele, not present anywhere else in the population (for example, in a gene of 1000 bp there will be 4^{1000} possible alleles). In the related infinite sites model it is assumed that a locus contains so many nucleotide sites that each new mutation occurs at a different site.

The infinite alleles model was very useful in the analysis of allozyme data because it led to predictions about how many alleles would be created through mutation and lost by drift. This also led to tests of the neutral theory of molecular evolution (see Chapter 7). For example, if more alleles were observed than expected then some force, such as natural selection, must be keeping them in the population. With the advent of DNA sequence data, in which mutations can be seen as discrete events, the infinite sites model has become more commonly used in population genetics and especially in coalescent approaches. This model has also led to tests of the neutral theory, for example by looking at the number of variable nucleotide positions—the segregating sites—between pairs of sequences (see Box 4.1 and Box 7.1, pp. 92 and 242).

Another important model of mutation used in population genetics is the **stepwise mutation model**, where mutations increase or decrease allele sizes by single units. This model was originally developed to look at the different patterns of mobility seen during protein electrophoresis but now is often used when analysing microsatellite data in which allele sizes often differ by single steps. In the case of microsatellites, adoption of the stepwise mutation model has led to a new measure of population subdivision, R_{ST}, which replaces the classical measure, F_{ST}, itself based on the infinite alleles model.

and mutation, which generates heterozygosity. In these circumstances, the equilibrium level of heterozygosity (H) is given by;

$$H = \frac{4N\mu}{4N\mu + 1} \qquad (4.12)$$

This is extremely important because it means that levels of genetic variation depend only on the mutation rate, μ, and the population size, N, so that heterozygosity will be lower in small populations and in those with low mutation rates (and vice versa).

4.2.4 Changes in population size

The case of genetic drift illustrates how differences in population size are an important part of molecular evolution because they alter the probabilities of random sampling. However, the measure of population size we have considered so far, the 'census', or total number of individuals in a population, N, is too simplistic in most cases because only a fraction of individuals will ever be able to produce progeny and so contribute genes to the next generation. How small this fraction is depends on factors such as the sex ratio, the mating system, variation in reproductive success and the extent of overlapping generations. It is therefore often more useful to think about how big a theoretically 'perfect' population (that is, one where every individual has the same probability of contributing genes to the next generation) would have to be for the random sampling of alleles to have the same affect as in the real population. This is known as the **effective population size**, denoted N_e, and it will almost always be smaller than N. Although it is commonly used in population genetics, N_e is very hard to measure in nature, although estimates are possible. For example, in a randomly mating population N_e roughly corresponds to the number of breeding individuals, whereas in species that go through large fluctuations in population size (see below), N_e can be estimated as the harmonic mean of N, which will give values nearer to the smaller than the larger sizes. More recently there have been attempts to estimate N_e using gene sequence data (see Box 4.1).

Another shorthand for population size frequently used in population genetics is the distinction between **finite** and **infinite** population sizes. Clearly, populations of infinite size do not exist so, once again, this is more of an idealised than a real condition. As we have already seen, genetic drift works best in populations with finite (small) sizes.

One of the most dramatic ways in which changes in population size influence molecular evolution occurs when a population experiences a rapid and severe decrease in the number of individuals, perhaps because of a natural disaster or a disease. This is known as a **population bottleneck** (Fig. 4.7). Bottlenecks greatly reduce the amount of genetic variation in populations because only a small number of alleles, contained in their lucky hosts, will

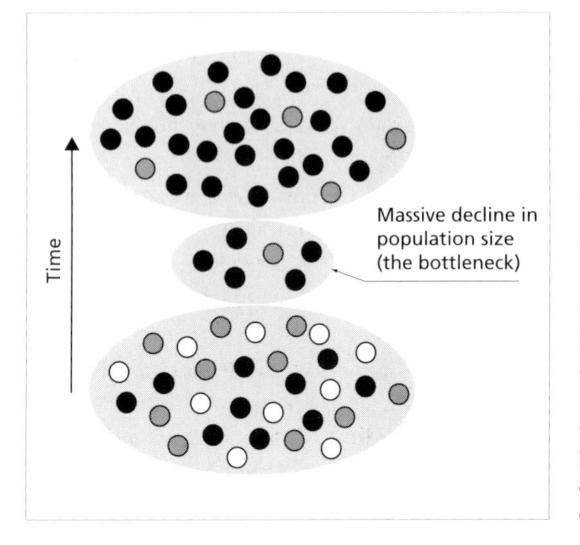

Fig. 4.7 How a population bottleneck can change allele frequencies. Three alleles, all at equal frequency, are found in the initial population. A bottleneck then occurs which drastically reduces the size of the population. By chance, larger numbers of the black allele survive than the others (the white allele is lost completely). When the population size increases again, the black alleles are overrepresented.

survive the drop in size. This reduction in size will also lead to an increase in the amount of inbreeding (which is more likely in small populations) and in doing so increase homozygosity. This effect is discussed in more detail in section 4.2.5.

Population sizes will also change if a small group of individuals becomes isolated from the main population, such as during the colonisation of an island. Because these isolated individuals will only carry a small proportion of the genetic variation from the larger population, genetic drift (and natural selection) can radically alter their genetic structure, sometimes leading to speciation. This is called the **founder effect**. More precisely, it is likely that the founding population will, by chance, have lost certain alleles present in the larger population (analogous to a bottleneck), or that alleles at low frequency in the large population will be overrepresented in the new population. Although many rare alleles will be lost during founder events, some will survive and may then reach high frequencies through a combination of genetic drift and inbreeding. This means that founder effects can produce high frequencies of otherwise rare, and perhaps even deleterious, alleles. This is highlighted by the case of the Ashkenazi Jews of Eastern Europe who have had a long history of population bottlenecks and who suffer a range of otherwise rare genetic abnormalities such as Tay–Sachs disease, familial dysautonomia and idiopathic torsion dystonia (ITD).

4.2.5 Population subdivision, non-random mating and gene flow

So far we have only thought about genes within a population of individuals that are able to interbreed freely. In nature, however, population structures are often much more complicated. In many cases populations contain smaller groups, known as **subpopulations** or **demes**, which may be partially isolated from each other. This means that complete interbreeding (or **panmixis**) is not always possible.

The most important genetic effect of population subdivision is that individuals are unable to mate randomly with each other—some will be out of reach. This is contrary to what is assumed in the Hardy–Weinberg theorem where all individuals are equally likely to mate with each other—random mating. The term 'random' in this context does not necessarily mean that individuals will mate with whoever they meet without any kind of choice. Rather, individuals mate randomly with respect to the genes we are interested in. For example, most of the allozymes traditionally used as markers in population genetics produce proteins which do not affect mate choice so that for them random mating is a fair assumption. In contrast, genes that affect body size and strength may directly influence which mate an individual chooses so that the assumption of random mating is harder to justify.

One form of non-random mating we have already come across is **inbreeding** in which individuals mate more frequently with their relatives than would be expected by chance. Inbreeding may take place if related organisms live in the same area, or because of specific reproductive and behavioural traits. The most extreme form of inbreeding is self-fertilisation which is common in crop plants such as barley and wheat.

A form of non-random mating similar to inbreeding is **positive assortative mating**. In this case individuals preferentially mate with each other because they share some sort of genetic similarity. For example, humans are more likely to mate with people of similar height and skin colour. In contrast, **negative assortative mating** occurs when individuals choose genetically different individuals. One possible, yet controversial, example of this may be the preference shown by mice for animals with particular body odours—odourtypes—which are present in urine. Odourtypes are under the control of H-2, the mouse MHC, and it seems that laboratory mice prefer mates with a different allele at this locus. This will have the effect of increasing genetic variation in a gene where diversity is required to recognise as many pathogens as possible. Whether this applies in a natural setting is uncertain and it is possible that the mating preference is the result of familial imprinting rather than genetic inheritance. It has also been suggested that mate choice in other animals, including humans, may be partly mediated by odours under the control of the MHC. Unlike inbreeding which will affect *all* the genes of an organism, assortative mating, either positive or negative, only applies to a specific trait.

Although neither inbreeding nor positive assortative mating change allele frequencies, both will lead to an increase in homozygosity above that predicted in the Hardy–Weinberg model. In contrast, because genetically different individuals are preferred, negative assortative mating generates more heterozygosity. Inbreeding increases homozygosity because individuals have a greater probability that they will inherit an identical allele from both parents

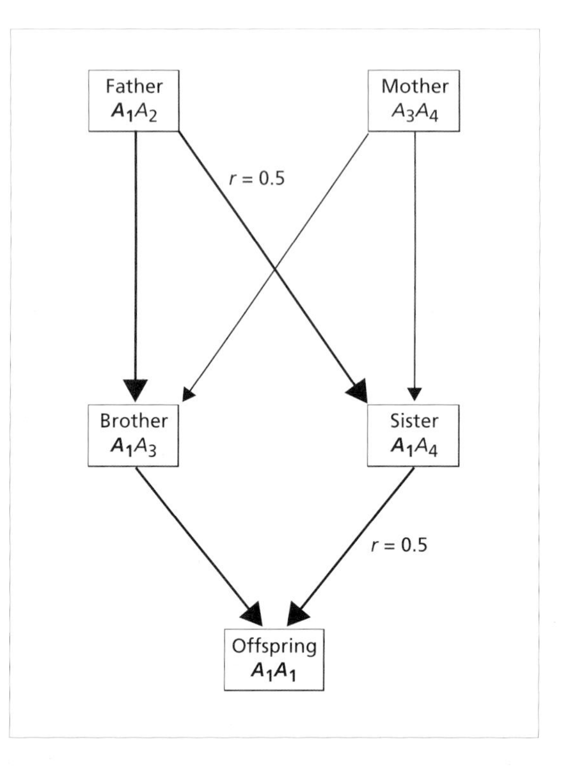

Fig. 4.8 Identity by descent (IBD) following a brother–sister (full sib) mating. The probability that two individuals share a copy of an allele by descent—called the **coefficient of relatedness**, r—is 0.5 for parents and offspring. In this case the father gives the A_1 allele to both his daughter and son, who then mate with each other. If they both pass on the A_1 allele on to their offspring (each with a probability 0.5, so an overall probability of 0.25) then the offspring will have become identical (homozygous) by descent. The descent of the identical alleles in the offspring is traced by the bold lines. r is also 0.5 for brothers and sisters, 0.25 for grandparents and grandchildren and 0.125 for cousins. Adapted from Griffiths *et al.* (1993), with permission from WH Freeman.

(who are genetically similar), so that their alleles become **identical by descent (IBD)** or **autozygous** (Fig. 4.8). This is an important concept in modern population genetics because it provides a link between genotype and phylogeny: homozygosity occurs because the alleles share an immediate common ancestor. The probability that two alleles are IBD is given by the **inbreeding coefficient** (F), where $F = 0$ represents a non-inbred population, and $F = 1$ signifies complete inbreeding (so that the population consists only of *AA* and *aa* homozygotes). The loss of genetic variation through inbreeding is similar to that which occurs under genetic drift: both result in an increased likelihood of bringing together alleles which share a common ancestor. Indeed, under inbreeding, heterozygosity is lost at the rate of $1/2N$ in each generation, which is the same as the probability that an allele will be fixed by genetic drift.

The increase in homozygosity due to inbreeding can also have severe phenotypic consequences. For instance, imagine that we have a deleterious recessive allele which causes a genetic disease when homozygous. Alleles of this sort are usually rare, because they are removed by selection, and exist only as heterozygotes. However, with inbreeding there is an increased probability that an individual could acquire two copies of this allele by descent (i.e. one from each parent), which may then drastically reduce fitness. In humans, more than 40% of offspring produced by full sib matings (between brother and sister) die young or develop serious disabilities. This phenomenon, known as **inbreeding depression**, is also quite common in captive animals and there is some evidence that the same is true of wild populations. For example, in an analysis of MHC and allozyme loci in wild cheetahs and lions, Steve O'Brien and colleagues discovered that these species have a great deal less genetic variation than other mammals, including the related domestic cat (Fig. 4.9). This is most apparent in a population of about 250 lions living in the Gir Forest sanctuary in India and which suffered a severe population bottleneck due to hunting at the turn of this century. No genetic variation in either allozymes or MHC RFLPs was seen in this population, and they are reported to show increased levels of abnormal sperm and lower amounts of testosterone.

Although population subdivision leads to local non-random mating, genes are sometimes able to move from one subpopulation to another when individuals migrate—a process known as **gene flow**. This represents another way, aside mutation, by which new genetic variation enters populations, although migration rates are usually much higher than mutation rates. Gene flow can be a powerful force determining allele frequencies and, because it means that genes are shared between populations, acts to homogenise allele frequencies among them, thereby preventing speciation (see section 4.4). Conversely, because new genetic variation is imported, there is an increase in heterozygosity *within* each subpopulation. The extent of genetic diversity within and between subpopulations can therefore be thought of as a balance between the opposing forces of genetic drift, which reduces diversity within populations but increases it between them, and gene flow which brings new genetic diversity into populations but reduces it between them (Fig. 4.10).

To understand how migration affects allele frequencies it is first necessary to devise a scheme for how subpopulations are arranged in space. The simplest way to do this is with the **island model**, where a large population is split into many smaller subpopulations of equal size which are dispersed over a geographical area and which exchange genes with equal probabilities. An alternative, although mathematically more complex, arrangement is the **stepping-stone model** where subpopulations have a more regular spatial ordering, and in which migration is only possible between adjacent subpopulations. The greater the geographical distance between the populations, the less the chance of gene flow, so that there is genetic 'isolation by distance'.

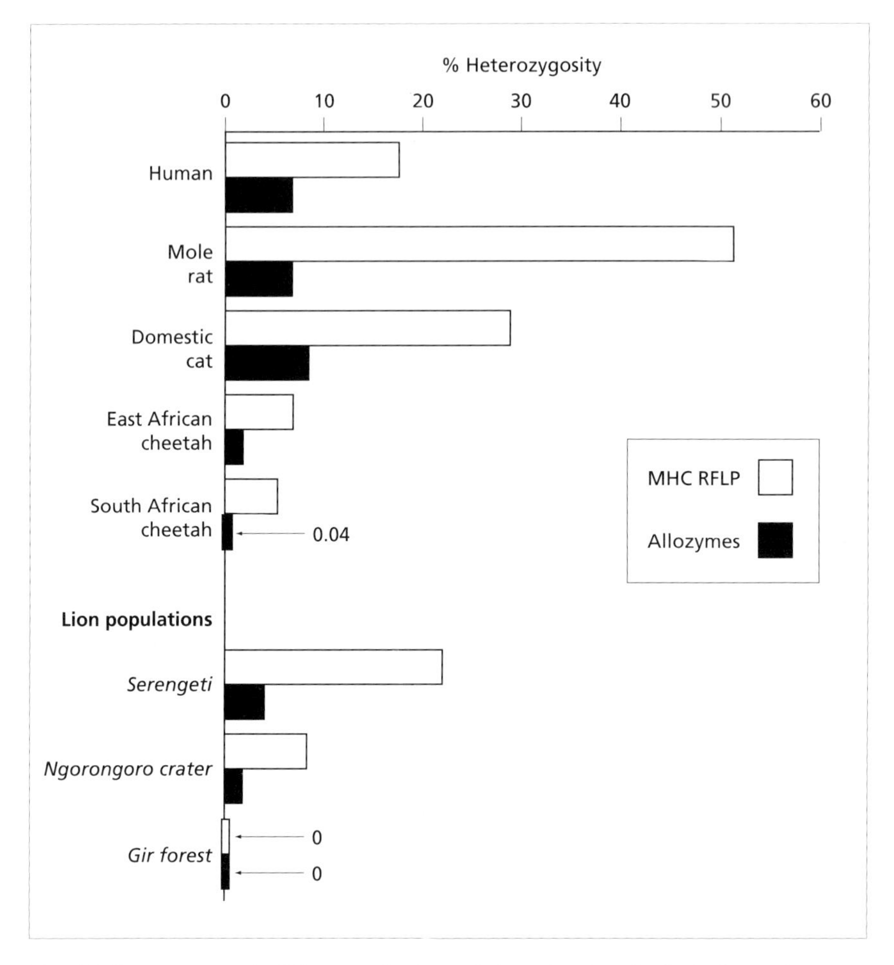

Fig. 4.9 Heterozygosity in different animal populations showing the limited genetic variation in cheetahs and lions. The Serengeti lion population numbers about 300, while just 100 live in the Ngorongoro crater (also in Tanzania) following a dramatic bottleneck in 1962. The Gir Forest (India) population today numbers about 250, although it also suffered a severe bottleneck at the turn of the century. Data from Yuhki and O'Brien (1990).

Given the island model, the rate of migration (gene flow) among subpopulations per generation is given by a quantity m, which is equivalent to the probability that an allele randomly chosen from a population comes from a migrant. Although this is often difficult to measure in nature, it can be done if the allele frequencies in the donor (p_1) and recipient (p_2) populations are known, as well as the change in allele frequency in the recipient population following migration (Δp):

$$m = \frac{\Delta p}{p_1 - p_2} \tag{4.13}$$

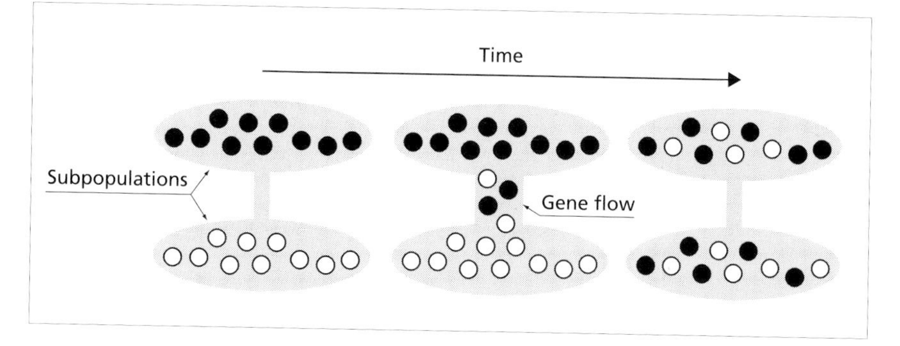

Fig. 4.10 How gene flow (migration) changes allele frequencies. In this example we have two subpopulations within a larger total population. We are able to observe these subpopulations at three points in time. Initially the subpopulations have very different allele frequencies but during the second time-point genes flow between them. This has the effect of homogenising the allele frequencies between the subpopulations.

Today the extent of gene flow between populations is often estimated using the geographical distribution of branches on molecular phylogenies: the more mixed the branches from different locations, the greater the amount of gene flow between them.

If we know the migration rate it then becomes possible to work out how this changes allele frequencies in the recipient population over time. This can be done with:

$$p_t = \bar{p} + (p_0 - \bar{p})(1 - m)^t \tag{4.14}$$

where p_t is the frequency of allele A after t generations, p_0 is the initial frequency of allele A in the subpopulation, $\bar{p}$ is the probability this allele has in fact come from another subpopulation (i.e. that it is found in a migrant) and m is the rate of migration per generation. From this equation we can infer that the change in allele frequency following migration is proportional to the difference in frequencies between the donor and recipient populations and that even low migration rates are able to keep subpopulations genetically homogeneous. Roughly speaking, genetic homogeneity is maintained if the number of migrants per generation, Nm, is one or more, irrespective of population size. Migration rates of this magnitude are not unreasonable, although they vary extensively between species. For example, an Nm of 9.9 has been reported in *Drosophila willistoni*, although the rate is only 1.0 in *D. pseudoobscura*, and a paltry 0.22 in the salamander *Plethodon cinereus* (see Hartl and Clark, 1989). In humans, values of Nm of around 2.0 have been estimated (Table 4.3) implying that gene flow has been a fairly regular occurrence in our species (although it is also possible that human populations are not at equilibrium, thereby complicating estimations of Nm, and that the similarity between them is simply caused by a very recent origin—see section 4.6).

Data	Nm	F_{ST}
84 protein loci	2.0	0.11
33 blood group loci	1.3	0.16
8 HLA and immunoglobulin loci	2.0	0.11
61 DNA markers	2.0	0.11
mtDNA: Papua New Guinea	1.1	0.31
mtDNA: worldwide	0.6	0.46

For nuclear loci Nm is calculated here as $(1 - F_{ST})/4F_{ST}$ whilst for mtDNA, where the genetic material is maternally inherited, Nm refers to the number of migrant females (or $N_f m$) which is estimated as $(1 - F_{ST})/2F_{ST}$.

Table 4.3 Estimates of Nm (number of migrants per generation) and F_{ST} (fixation index) among human populations. Adapted from Takahata (1993), with permission.

Because population subdivision, inbreeding and gene flow are some the most important factors shaping the genetic structure of populations, it is important to measure how frequently they occur. This can be done by assessing how the observed levels of heterozygosity differ from those expected under Hardy–Weinberg (inbreeding, for example, will lead to an excess of homozygotes). This is most commonly done using the ***F*-statistics** devised by Sewell Wright.

F-statistics work as follows. Let us assume that we have three levels of population structure: individuals (I), subpopulations (S) and the total population (T). We can think about heterozygosity at each of these three levels: H_I is the heterozygosity we *observe* in an individual in a subpopulation; H_S is the heterozygosity we would *expect* that individual to show in an equivalent subpopulation which is mating randomly; H_T is the heterozygosity expected to be shown by that individual in an equivalent randomly-mating total population. By combining these measures and expectations of heterozygosity together in various ways—the *F*-statistics—we can come to a detailed description of population structure. The most important of these is F_{ST}, or the **fixation index**, which can be used to measure the extent of population subdivision, the reduction in heterozygosity caused by genetic drift, or the amount of gene flow between subpopulations (high levels of gene flow will mean low values of F_{ST}). F_{ST} can be estimated in a number of different ways, a common method being:

$$F_{ST} = \frac{H_T - \bar{H}_S}{H_T} \tag{4.15}$$

where $\bar{H}_S$ is the mean heterozygosity across all subpopulations. If the number of migrants per generation is known, and assuming an island model, then F_{ST} can also be estimated by $1/(1 + 4Nm)$. Values of F_{ST} for humans range from 0.11 in nuclear loci to 0.46 in mitochondrial genes (Table 4.3), indicating that there is relatively little genetic differentiation among populations (the values

As well as time and place of origin, there is also controversy as to whether a population bottleneck accompanied the emergence of modern humans. The best way of testing this idea at the molecular level is to calculate the distribution of coalescence times of alleles: we would only expect a few alleles (within their lucky hosts) to make it through a severe bottleneck, so that any polymorphisms observed would have arisen since the bottleneck, making most coalescence times very short (take a look at Fig. 4.17 again). An analysis of this kind has been performed on the MHC class II locus *DRB1* and showed that *all* 58 human alleles have been present for the last 500 000 years, with some as old as 30 million years (so that they are found in other primate species), probably because they have been maintained by balancing selection. To retain such a high level of polymorphism population sizes would have to have been consistently high, with a mean of perhaps 100 000 individuals over many years, and at no time could drop below 2000–3000. Similar conclusions have been drawn from analyses of polymorphisms in β-globin genes, on the Y chromosome and in mtDNA. Although these calculations are heavily dependent on the assumptions made, together they suggest that while population sizes are likely to have undergone large fluctuations, severe bottlenecks, such as those proposed in the 'Noah's Ark' model, are unlikely to have played a major role in our evolutionary history.

To conclude, it seems most likely that anatomically modern humans evolved in Africa at around 200 000 years ago, and then spread around the world. However, the amount of sequence information available to date is very limited: for example, an analysis of 2.6 kb of the Y chromosome from 16 humans of different geographical origins (a total of 41 600 bases) uncovered just three nucleotide polymorphisms. There is clearly a need for more sequence data to fully understand human origins.

4.7 Summary

1 Population genetics is concerned with understanding the factors responsible for the origin and maintenance of genetic variation in populations. Traditionally this was based on the analysis of allele frequencies in allozymes but today more attention is given to gene genealogies of DNA sequences because different evolutionary processes and rates of population growth leave different signatures in these genealogies.
2 The simplest model in population genetics is one in which allele frequencies remain the same in time and space. This is the Hardy–Weinberg theorem.
3 Evolution at the genetic level results in a change in allele frequencies. A number of different processes can lead to such changes, including natural selection, genetic drift, changes in population size and gene flow. Crucially, all these processes will also change the shape of gene genealogies.
4 Many phenotypic characteristics are produced by the concerted action of multiple gene loci. The study of the molecular evolution of these quantitative

trait loci is only just beginning. The same is true of the molecular population genetics of speciation, although phylogenies are being increasingly used as a way of choosing between different evolutionary scenarios.

5 Gene sequence data have been particularly informative with respect to understanding the origin of modern *H. sapiens*, giving support for the model in which human populations emerged in Africa only within the last 200 000 years, followed by a later migration to other parts of the world.

4.8 Further reading

A number of textbooks exist which deal with the practicalities of population genetics with Hartl and Clark (1989) being particularly complete, although good summaries are found in Griffiths *et al.* (1993) and Ridley (1996). Those interested in coalescent theory should also refer to Donnelly and Tavaré (1995), Harding (1996) and Hudson (1990), while descriptions of how to measure DNA sequence variation within populations can be found in Nei (1987) and Li (1997). A particularly elegant demonstration of how molecular phylogenies can provide important information about speciation and other ecological processes is provided by Avise (1994).

With respect to the more specific examples used, details of the ALDH2 polymorphism are given in Goedde *et al.* (1992) and Ferguson and Goldberg (1997), while a beautiful example of the Hardy–Weinberg equilibrium not mentioned in the text is provided by Shuster and Wade (1991). The population genetics of AZT resistance is given in Goudsmit *et al.* (1996) (but also see Leigh Brown (1997)), while the story of the ΔF508 mutation of cystic fibrosis is told by Tsui (1992) and the Δ32 mutation of the CCR5 gene by Dean *et al.* (1996) and O'Brien and Dean (1997). Those more interested in selection on morphological characters, in this case on bill sizes in *Pyrenestes ostrinus*, should look at Smith (1993) and the accompanying article by Holmes and Harvey (1993), while the strange case of mice odour types is summarised by Potts and Wakeland (1993). The circumstances surrounding the low levels of genetic variation seen in cheetahs and lions are described in Yuhki and O'Brien (1990). The quantitative genetics of human cognition is told in McClearn *et al.* (1997) and Gottesman (1997) and of *Drosophila* bristles by Lai *et al.* (1994). In regard to speciation, the hybrid zone in fire ants is described by Shoemaker *et al.* (1996) and the evidence for sympatric speciation in the cichlids by Schliewen *et al.* (1994), while the search for speciation genes in *Drosophila* is documented in Coyne *et al.* (1994) and put into context by Coyne (1992). Finally, the molecular population genetics of human origins is well summarised by Takahata (1993) and Hey (1997), while the extraordinary story of Neanderthal mtDNA is told by Krings *et al.* (1997).

Chapter 5
Measuring Genetic Change

Given two or more sequences the instinctive reaction of most molecular biologists is to calculate a genetic distance. Just what do these distances measure? How might we choose between the many different kinds available? Why are there so many alternatives? This chapter focuses on the notion of distance as a measure of amount of evolution. It also addresses the problems of relating distances between genes to distances between populations. We stress that distances themselves are only relevant when coupled with an underlying phylogenetic structure.

5.1 Sequence alignment and homology

5.1.1 Homology

In this book we use homology in the evolutionary sense of the word: two nucleotides in different sequences are **homologous** if and only if the two sequences both acquired that state directly from their common ancestor (Fig. 5.1). A key point is that determining whether a feature is homologous requires knowledge of the evolutionary relationships among the species having that feature. In Fig. 5.1 the character state 'black' may be either homologous or homoplasious depending on the relationships of the species a and b to species c and d. If a and b are each other's closest relatives then the simplest interpretation is that black is homologous; however, if a and b are distantly related then we would attribute a and b sharing the same feature to independent evolution.

Whether a feature is homologous depends in part on what aspect of the feature we are considering. To use a classic example from comparative anatomy,

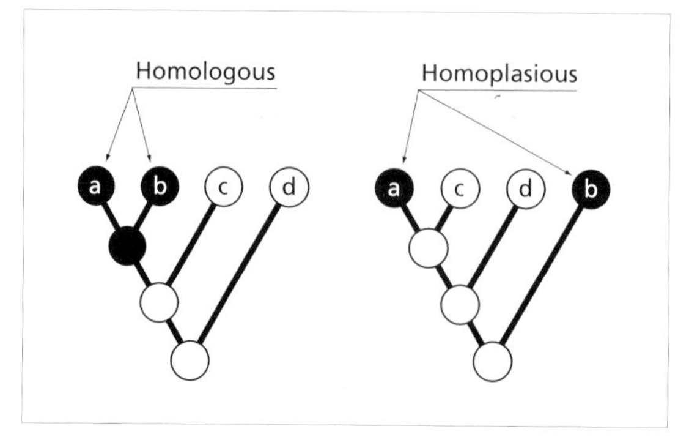

Fig. 5.1 The difference between homology and homoplasy. In the tree on the left the character state 'black' is homologous in the two species possessing it (a and b) because black is inherited directly from an ancestor which was also black. In the tree on the right, the two species sharing black are not closely related and the occurrences of black are not homologous because they have evolved independently from ancestors that were white.

the wings of a bird and a bat are homologous as *forearms*, but not as *wings*. Both birds and bats have inherited their forearms from their last common ancestor; forearms have not evolved independently *de novo*. However, in both groups the forearm has independently undergone modification for flight (the common ancestor of birds and bats could not fly), hence when considered as wings the same structures are not homologous but **homoplasious**.

In the same way, at the molecular level our judgement of homology may vary depending on the properties in which we are interested. Two proteins in two different organisms may be encoded by the same gene (i.e. the two genes are direct descendants of a gene present in the ancestor of those two species). The two genes may share many amino acids in common, and have similar function. However, if that functionality has been acquired independently then that functionality is not homologous.

The classic molecular example is the parallel evolution of amino acid sequences in the lysozyme enzyme in leaf-eating langur monkeys and in cows (Fig. 5.2). Both animals have independently evolved foregut fermentation using bacteria, and in both cases lysozyme has been recruited to degrade these bacteria. Therefore, langur and cow lysozymes are homologous as genes; however, as digestive enzymes they are not homologous because this functionality was not present in the ancestral lysozyme.

Homology can sometimes be difficult to distinguish from homoplasy, especially at the molecular level. At the organismal level complex structures that superficially appear similar, such as bird and bat wings, on closer inspection may prove quite different in detailed structure, mechanics and development, which may cause us to question whether they are really the same thing (i.e. homologous). However, at the molecular level the features we observe may be literally identical. The cow and langur lysozyme sequences share the identical

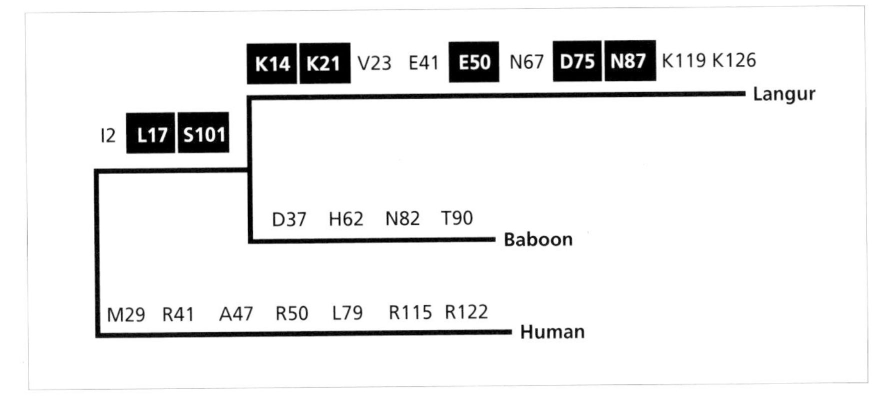

Fig. 5.2 Independent evolution of amino acid replacements in cows and langur monkeys. Although langur monkey lysozyme is phylogenetically closely related to other primate lysozymes it has independently acquired several amino acid substitutions in common with cow lysozyme (these are indicated by the black squares). Redrawn from Li and Graur (1991).

amino acid (aspartic acid) at position 14. Although chemically the same entity, from an evolutionary perspective the aspartic acid in the cow is not the same as the aspartic acid in the langur. As a consequence of the redundancy of the genetic code the same amino acid may be encoded by different codons in the two taxa, which may provide evidence that the amino acid is not homologous in the two genes. However, at the level of nucleotides we reach the limits of any attempt to distinguish homology from homoplasy by simply looking for differences in the two features being compared; a guanine in one species is the same as a guanine in any other species. Deciding whether the two instances of this nucleotide are homologous requires knowledge of the evolutionary relationships among the sequences being compared, a point we discuss in section 5.2.1.

5.1.2 Homology among genes

Many genes are members of gene families—suites of genes that are the descendants of an ancestral gene. Thus in some sense, different genes may be homologous because they are descendants of the same gene. As mentioned in Chapter 2 (see section 2.4.2) we can distinguish two basic types of homology among genes: **orthology** and **paralogy** (a third relationship among genes is **xenology**, where one gene owes its presence in an organism to horizontal gene transfer from another organism). Recall that paralogous genes are descendants of an ancestral gene that has undergone one or more gene duplications. If there is no duplication in the gene tree then the sequences are orthologous. Staying with the example of lysozyme, the lysozyme genes expressed in the foregut of cows and leaf-eating monkeys are orthologous; they are both

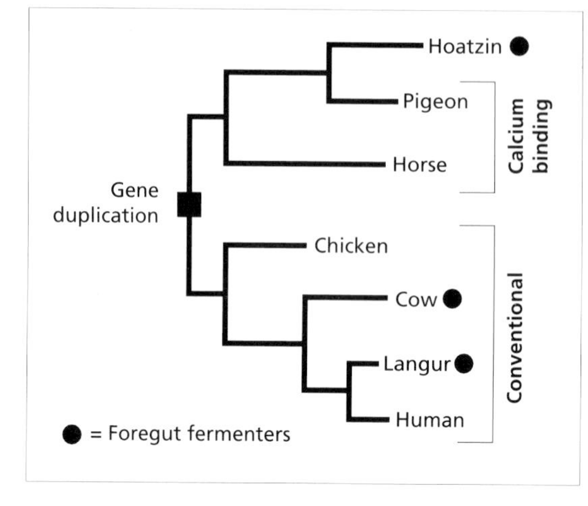

Fig. 5.3 Phylogeny of some bird and mammal lysozymes. The lysozyme in cows, langur monkeys and the hoatzin bird have all independently evolved similar digestive properties. The two mammalian genes are orthologous, and are paralogous with respect to the hoatzin gene, which is related to calcium binding lysozymes found in birds and mammals. After Kornegay *et al.* (1994).

descendants of conventional lysozyme (Fig. 5.3). The bizarre hoatzin bird of South America, apparently a relative of the cuckoos, is the only bird known to be a foregut fermenter, and is hence the 'cow' of the bird world. In this species, a lysozyme is also recruited as a digestive enzyme and shows striking parallel evolution with mammalian lysozymes. However, although hoatzin lysozyme is related to cow and monkey lysozymes, it belongs to a different gene lineage and is paralogous to the gene in the mammalian foregut fermenters (Fig. 5.3).

Hence, when using the term 'homologous' we need to be careful about what kind of homology we are talking about; does it relate to the genes, their sequences, or their function? In the example above, all the lysozyme genes are homologous (they are all ultimately descendants of the same gene). The homology between cow and langur genes is due to their being descendants of the same copy of lysozyme, hence they are orthologous. The calcium-binding and conventional lysozymes trace their ancestry back to different copies of the ancestral lysozyme, and are hence paralogous. The functionality of the calcium-binding lysozymes of horse and pigeon are homologous because the ancestral function of those genes was calcium binding. Because the digestive function of lysozymes in ruminant mammals, leaf-eating monkeys, and the hoatzin has arisen independently in each group this functionality is not homologous.

5.1.3 Homology among sequences

So far we have emphasised the need to compare homologous structures or features of organisms. In the case of nucleotide or protein sequences it may not be immediately obvious which features should be compared. This is because the sequences comprise only a few symbols (e.g. the four nucleotides) in varying order along the sequence. Given two sequences that are similar but different the first task is to establish which regions of the two sequences are homologous, that is, correspond to the same region of the gene and hence may be compared. Consider these two nucleotide sequences:

Sequence 1 A T G C G T C G T T

Sequence 2 A T G C G T C G T

These two sequences are clearly very similar, but are not identical. A useful graphical tool for comparing two sequences is the **dot plot** (Fig. 5.4). Dot plots are constructed by making the two sequences the axes of a graph and plotting a dot where the two sequences agree. If the sequences are identical an unbroken diagonal line or 'path' can be drawn from the bottom left to the top right connecting dots in adjacent cells. Sequence 2 is a subsequence of sequence 1, differing only in lacking the last T in sequence 1, so there is an unbroken path through the diagonal of the plot (path 1 in Fig. 5.4). This path corresponds to the following **alignment** of the two sequences:

```
A T G C G T C G T T
| | | | | | | | |
A T G C G T C G T
```

The alignment specifies the relationship between each nucleotide site in the two sequences. In this case the alignment seems trivial; however there are other possible alignments we could create. The dot plot (Fig. 5.4) shows a number of other diagonal paths that are offset from the main diagonal, such as path 2. This path corresponds to the alignment

```
A T G - - - C G T C G T T
| | |       | | |
A T G C G T C G T
```

In this case the first occurrence of the string of bases 'CGT' in sequence 1 is aligned with the second occurrence of the CGT string in sequence 2, rather than the first. Doing so requires us to postulate a **gap** in sequence 1 of three

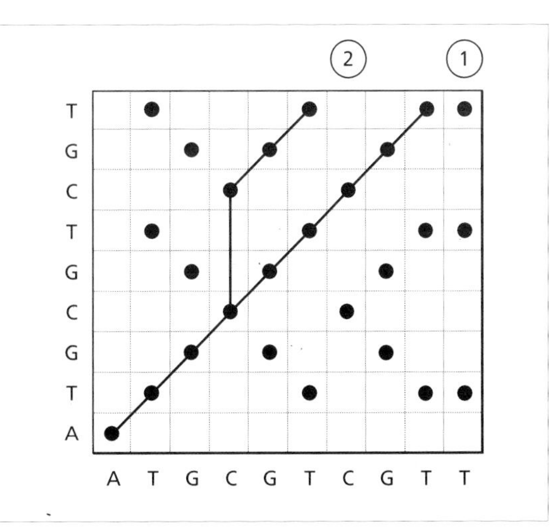

Fig. 5.4 A dot plot for two sequences. In each cell in the plot a dot is placed if the two sequences have the same nucleotide at that pair of sites. Two possible paths are shown through the plot, each corresponding to two separate alignments of the two sequences. Redrawn from Li and Graur (1991).

bases. This gap corresponds to the vertical 'kink' in path 2 in Fig. 5.4, and is the result of either an insertion in sequence 2 or a deletion in sequence 1 (i.e. an insertion/deletion or **indel**). This alignment is possible, but intuitively seems worse than the first alignment because it requires us to hypothesise at least one extra event, namely the insertion of three bases in sequence 1 (or a corresponding deletion in sequence 2).

Now consider these two sequences and their dot plot (Fig. 5.5).

Sequence 1 A T G C G T C G T T

Sequence 3 A T C C G T C A T

In this instance the diagonal path crosses cells that do not contain dots. These cells represent sites where the two sequences have different nucleotides at the same position, so that the alignment is

```
A T G C G T C G T T
| |   | | | |   |
A T C C G T C A T
```

In this case the two sequences differ at the third and eighth positions. It is possible to align any two sequences by postulating some combination of gaps and substitutions. By counting the number of these events, or some function of them, we can compute the 'cost' of a particular alignment. For example, if we have two aligned sequences the cost of that alignment may be given by

$$D = s + wg \tag{5.1}$$

where s is the number of substitutions, g is the total length of any gaps, and w is the **gap penalty**. The gap penalty specifies the cost of a gap relative to a substitution. For example if $w = 1$ then a gap of a single nucleotide has the same cost as a single substitution; if $w = 2$ then gaps are twice as costly as base pair mismatches. Note that equation (5.1) ignores the lengths of different gaps;

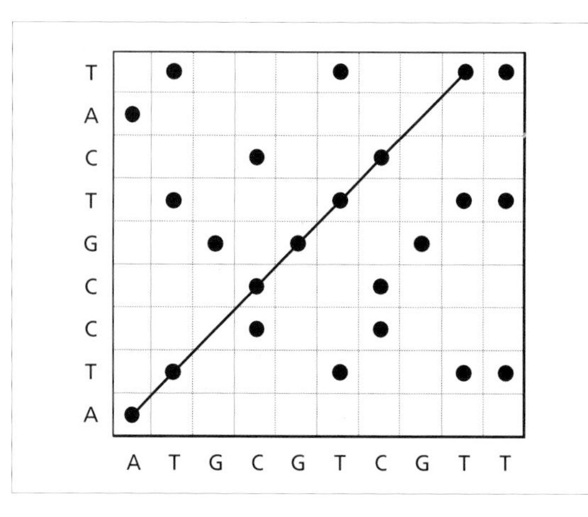

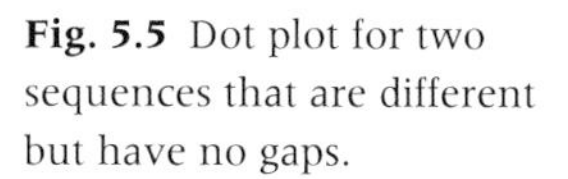
Fig. 5.5 Dot plot for two sequences that are different but have no gaps.

two single nucleotide gaps have the same cost as one gap two nucleotides long. There are other formulae that take into account the length of the gap. Choosing a gap penalty makes implicit assumptions about how the sequences have evolved. If indels are thought to be rare then w should be large; conversely, if indels are frequent then low values of w are more appropriate.

The basic task of sequence alignment is to find the alignment with the lowest cost, for which several algorithms are available. However, note that which alignment is deemed best can depend greatly on the value of the gap penalty. If the gap penalty is low then alignments with large numbers of gaps may be favoured; in the extreme case where $w = 0$ (and hence gaps have no cost) we can always introduce enough gaps to align the sequences. Conversely, high values of w will favour alignments that introduce substitutions to explain sequence mismatches.

Figure 5.6 shows a dot plot for two sequences and two alternative paths corresponding to the following alignments:

```
        A T - - G C G T C G T T
Path 1  | |     | | | | |
        A T C C G C G T C

        A T G C G T C G T T
Path 2  | |   | |   | | |
        A T C C G - C G T C
```

The first alignment postulates a single gap of two nucleotides; the second postulates a one nucleotide gap and two substitutions. If gaps have low cost (e.g. $w = 1$) then the first alignment would be favoured as its cost is $D = 0 + 1 \times 2 = 2$, whereas alignment 2 has cost $2 + 1 \times 1 = 3$. However, if gaps are thought to be less likely then the second alignment would be favoured as it has the smaller gap; for example, if $w = 3$ then the costs of the two alignments

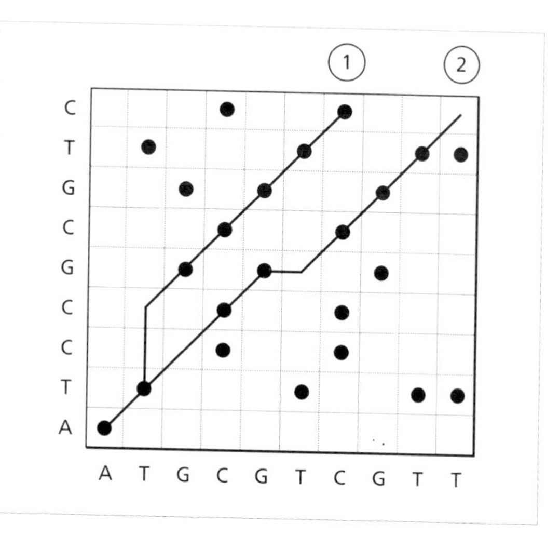

Fig. 5.6 Dot plot for two sequences and two alternative alignments (paths 1 and 2). Path 1 requires a two nucleotide insertion/deletion; path 2 requires a single nucleotide insertion/deletion and two nucleotide substitutions. The choice between the two alignments depends on the cost assigned to postulating indel events (i.e. the gap penalty). After Li and Graur (1991).

are 6 and 5, respectively. In some cases we may have additional information that may allow us to decide whether the gaps introduced by an alignment are reasonable. If the two sequences being aligned are for functional protein coding genes then any gaps would be expected to have lengths that were in multiples of three (i.e. 3, 6, 9 etc.) to preserve the reading frame of the gene, so that each gap corresponded to the insertion or deletion of one or more codons corresponding to one or more amino acids. For ribosomal genes there may be aspects of the secondary structure that can be used to evaluate the plausibility of various gaps introduced in alignment.

5.1.4 Alignment of protein sequences

Alignment of protein sequences differs from alignment of nucleotide sequences in two important respects. Firstly, there are more symbols (typically 20) in the sequence because there are more amino acids than nucleotide bases. Secondly, alignment is not simply a matter of aligning the symbols so that the greatest number match.

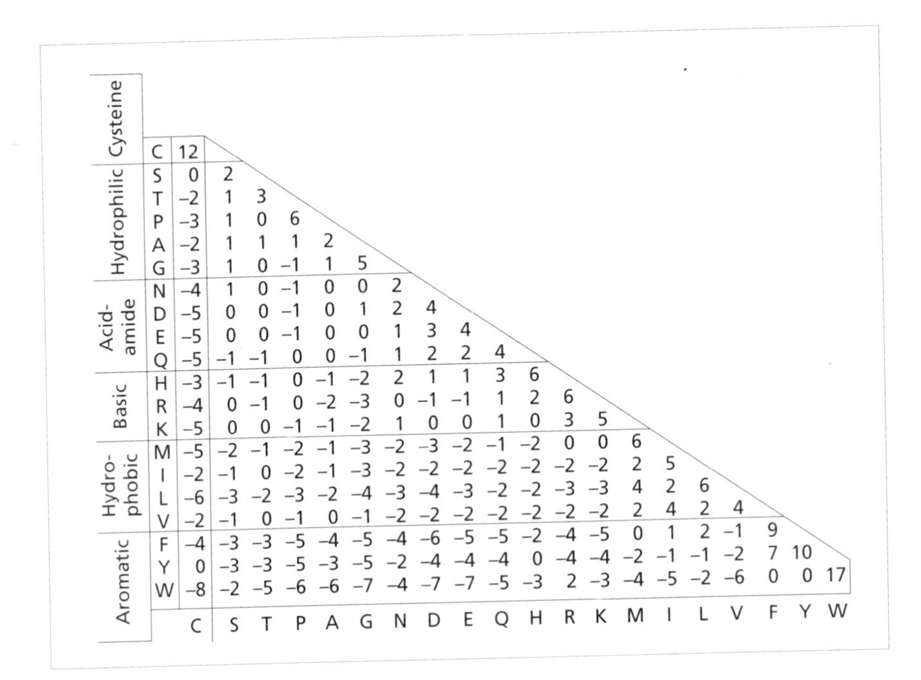

		C	S	T	P	A	G	N	D	E	Q	H	R	K	M	I	L	V	F	Y	W
Cysteine	C	12																			
Hydrophilic	S	0	2																		
	T	-2	1	3																	
	P	-3	1	0	6																
	A	-2	1	1	1	2															
	G	-3	1	0	-1	1	5														
Acid-amide	N	-4	1	0	-1	0	0	2													
	D	-5	0	0	-1	0	1	2	4												
	E	-5	0	0	-1	0	0	1	3	4											
	Q	-5	-1	-1	0	0	-1	1	2	2	4										
Basic	H	-3	-1	-1	0	-1	-2	2	1	1	3	6									
	R	-4	0	-1	0	-2	-3	0	-1	-1	1	2	6								
	K	-5	0	0	-1	-1	-2	1	0	0	1	0	3	5							
Hydro-phobic	M	-5	-2	-1	-2	-1	-3	-2	-3	-2	-1	-2	0	0	6						
	I	-2	-1	0	-2	-1	-3	-2	-2	-2	-2	-2	-2	-2	2	5					
	L	-6	-3	-2	-3	-2	-4	-3	-4	-3	-2	-2	-3	-3	4	2	6				
	V	-2	-1	0	-1	0	-1	-2	-2	-2	-2	-2	-2	-2	2	4	2	4			
Aromatic	F	-4	-3	-3	-5	-4	-5	-4	-6	-5	-5	-2	-4	-5	0	1	2	-1	9		
	Y	0	-3	-3	-5	-3	-5	-2	-4	-4	-4	0	-4	-4	-2	-1	-1	-2	7	10	
	W	-8	-2	-5	-6	-6	-7	-4	-7	-7	-5	-3	2	-3	-4	-5	-2	-6	0	0	17

Fig. 5.7 The PAM 250 matrix. For each pair of amino acids (see Table 3.1, p. 41, for key to the one-letter codes for amino acids) the matrix gives the ratio of the frequency at which the pair is observed in pairwise comparisons of proteins to that are expected due to chance alone, expressed as a 'log odd'. Amino acids that regularly replace each other have a positive score, amino acids that rarely replace each other have negative scores. Note that replacements more often occur among chemically related amino acids (indicated on the left). From Dayhoff (1978: Fig. 84).

For nucleotide sequences the cost of a mismatch between two nucleotides is often scored simply as '1', whereas a match has zero cost. Hence, nucleotides are either the same or they are different. Some methods are more sophisticated and have different costs for transitions and transversions, as these two classes of substitution often differ in frequency (see pp. 150–151). However, for amino acids we need to take into account the possible pathways in which one amino acid might be replaced by another. For example, a cysteine encoded by the triplet UGU can be replaced by a tyrosine (UAU) by a single nucleotide substitution at the second codon position, whereas replacing cysteine by methionine (AUG) requires three nucleotide substitutions, one at each codon position. Hence aligning cysteine with tyrosine is less costly than aligning cysteine with methionine. The cost for every pair of possible amino acid replacements defines a cost matrix that can be used to score the alignment. Protein sequence alignment programmes typically use matrices derived from empirical comparisons of protein sequences. The best known are the Dayhoff matrices, one of which is shown in Fig. 5.7, and are often used when estimating distances from protein sequences (see section 5.2.4).

5.1.5 Multiple alignment

So far we have considered aligning pairs of sequences. However, in many if not most cases we will have more than two sequences to align. There are various ways of performing such a **multiple alignment** (Fig. 5.8). The first is to find the alignment that minimises the total cost of all the pairwise alignments. For the five sequences in Fig. 5.8 there are ten possible pairings

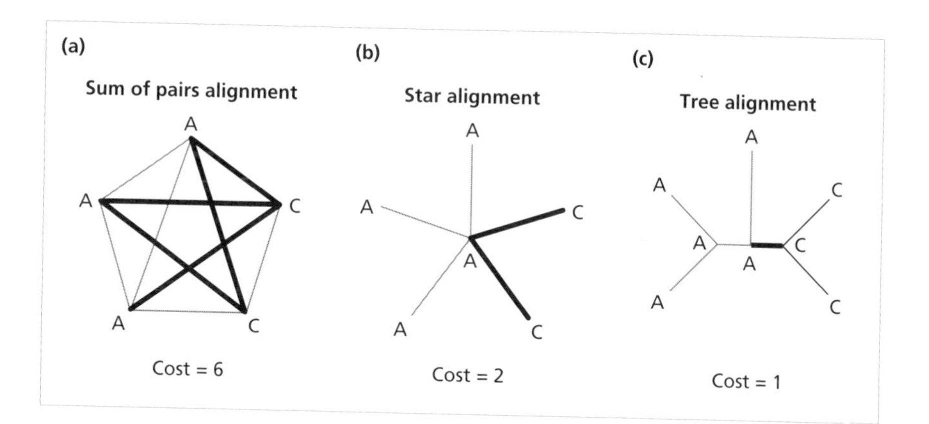

Fig. 5.8 Three different kinds of alignments for the same position in five nucleotide sequences. The sum-of-pairs alignment (a) counts the cost of aligning every pair of sequences. A star alignment (b) aligns the sequences on a star tree, which is a special case of a tree alignment (c). Pairwise alignments between different nucleotides (cost = 1) are indicated by solid lines, and pairwise alignments between identical nucleotides (cost = 0) by grey lines. After Altschul and Lipman (1989).

sequences, of which six connect different nucleotides hence the cost of the alignment is 6. A difficulty with this approach is that this cost is very difficult to interpret biologically; how many actual evolutionary events does this cost represent? From Fig. 5.8(a) it is clear that a single substitution in one sequence may contribute several times to the total cost. An alternative to considering all possible pairs is to align the sequences using a tree. If all sequences are equally related to each other then we can depict this using a star tree (Fig. 5.8b). If we knew that some sequences were more closely related to some sequences than to others then we could align those sequences on that tree to obtain a **tree alignment** (Fig. 5.8c). For the sequences shown in Fig. 5.8 the cost of the tree alignment is 1; there is a single substitution on an internal branch of the tree. (The method used to work out the minimum number of substitutions on a given tree for a single site is described below in section 5.3.)

Tree alignments have the advantage of being readily interpretable in terms of actual biological events. A complication, however, is that the alignment may change depending upon the tree on which the sequences are aligned. This has important implications, because most molecular phylogenetic studies align the sequences first, then compute a phylogeny based on that alignment. However, if the details of the alignment depend on the phylogeny of the sequences, then a particular alignment obtained, say by sum-of-pairs multiple alignment, may bias the result of a tree building method. One solution to this dilemma is to infer both the alignment and the phylogeny at the same time, so that the optimal alignment and tree are obtained together. This approach is becoming increasingly commonly used, and most current alignment packages make some use of phylogeny, either as a guide to the order in which pairs of sequences should be aligned, or by aligning the sequences and estimating the phylogeny simultaneously.

5.2 Genetic distance

In one sense, DNA sequences are not very informative about their evolutionary history. When comparing homologous sites in two DNA sequences, we simply observe that the sequences are the same or not. Hence, for any given site the maximum number of differences we can observe is one. Furthermore, there are just four possible states each site can have: A, C, G or T. This means that if more than one substitution occurs at a site, we lose any record of the previous substitution. When we look at, say, a bird's wing and our arm we can at least imagine that some considerable evolutionary change has taken place along the phylogenetic path between the bird and ourselves, but at a given nucleotide site, a bird might have A and ourselves G. In itself, this tell us little about how much change has taken place at that site. At least one substitution must have occurred, but many more might have taken place.

5.2.1 Kinds of substitution

Given two nucleotide sequences, we can ask how their similarities and differences arose. Figure 5.9 shows the six possible kinds of substitution that can occur. The first is a **single substitution**, resulting from the replacement of a nucleotide by another, different nucleotide just once in the history of one lineage. As a result, the two descendants have different nucleotides. If the number of nucleotide substitutions that have occurred since the two sequences last had a common ancestor is small, then most substitutions are likely to be single substitutions simply because the probability of the same site mutating more than once is fairly small. However, as the number of substitutions increases the probability that the same site may undergo more than one substitution becomes higher. **Multiple substitutions** or **multiple 'hits'** can greatly obscure the actual evolutionary history of a pair of sequences. Figure 5.9(b) shows a simple case where one lineage has accumulated two substitutions (A → C, and C → T), whereas in Fig. 5.9(c) each lineage has had a single, different substitution. In both these two cases the two descendant sequences show only a single difference (they either have the same nucleotide or they do not), yet there have been two substitutions. Simply counting the number of differences between the two sequences underestimates the real amount of evolutionary change.

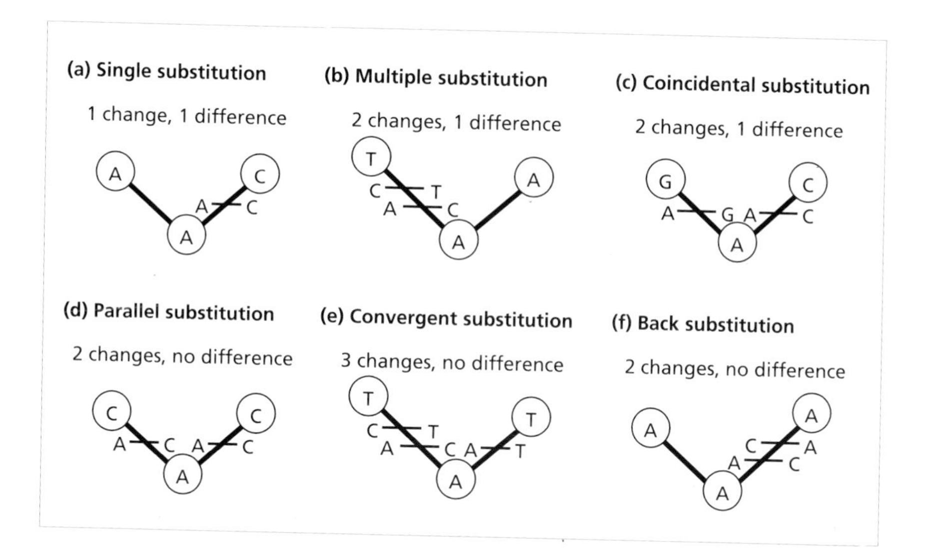

Fig. 5.9 Six kinds of nucleotide substitution. In each case the ancestral nucleotide was A. In all except the case of a single substitution, the number of substitutions that actually occurred is greater than would be counted if we just compared the two descendant sequences. In the lower three cases the nucleotides are identical in both descendant sequences, but this similarity has not been directly inherited from the ancestral sequence. Such similarity is termed 'homoplasious'.

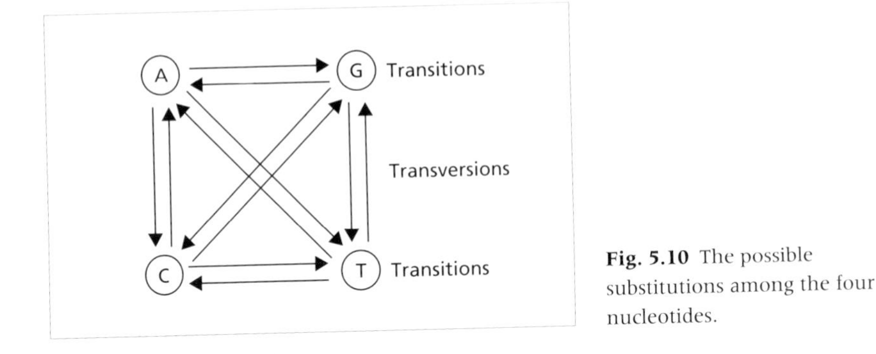

Fig. 5.10 The possible substitutions among the four nucleotides.

The last three kinds of multiple substitution shown in Fig. 5.9 have potentially more serious consequences. In each case the two descendant sequences are identical, yet in no case is that similarity inherited directly from the ancestral sequence. Similarity that is inherited from the ancestor is **homologous** similarity, whereas independently acquired similarity is **homoplasious** similarity (Fig. 5.1). The occurrence of homoplasy can obscure the actual number of evolutionary events: in each case shown in Fig. 5.9 the two descendant sequences are identical, even though between two and three substitutions have occurred. Furthermore, homoplasy can mislead our attempts to infer the evolutionary relationships among sequences. All tree-building methods rely on there being sufficient homologous similarity among sequences for us to recover the evolutionary tree linking those sequences.

We can also classify substitutions in other ways. Nucleotides are either purines or pyrimidines. Substitutions that exchange a purine for another purine, or a pyrimidine for another pyrimidine are called **transitions**, and in some genes are more common than the remaining substitutions purine → pyrimidine or pyrimidine → purine (**transversions**) (Fig. 5.10).

We can also classify substitutions in protein coding genes by the consequences they have for the protein the gene encodes. Because of the degeneracy of the genetic code (Chapter 3), not all substitutions in a codon need result in a different amino acid. A substitution that does not change the amino acid is a **synonymous** substitution, whereas one that does change the amino acid is **non-synonymous**. More detail about types of mutation is given in Chapter 3.

5.2.2 Distance measures for nucleotide sequences

Given two sequences one obvious question to ask is 'how much evolutionary change has occurred between these two sequences?' Answering this seemingly straightforward question has spawned a veritable industry of measures of sequence difference.

Observed differences

The simplest measure of the distance between two nucleotide sequences is to count the number of nucleotide sites at which the two sequences differ. We have already encountered this measure in our discussion of pairwise sequence alignment (section 5.1.3). However, for all but very similar sequences this is a poor measure of the actual number of evolutionary changes. If change has been relatively common (or if not all sites are equally likely to change) then the same site may undergo repeated substitutions, each successive substitution obliterating information the site conveys about evolutionary change. As time goes by, the number of differences between two sequences becomes less and less of an accurate estimator of the actual number of substitutions that occurred since two sequences diverged from their common ancestor.

Figure 5.11 shows the relationship between number of observed nucleotide differences and time since common ancestry for some mitochondrial DNA sequences from bovid mammals (cows, buffaloes, and their relatives). The youngest pairs of sequences show approximately 17 substitutions per million years (Myr). Extrapolating this back in time, we would expect the species that diverged 20 Myr ago to show about 340 substitution differences. In fact, only some 94–103 changes are observed. Indeed, sequences separated by 20 Myr are no more different than sequences that diverged 15 Myr ago.

The relationship between sequence difference and the time elapsed since divergence is not linear but is instead deflected downwards due to multiple hits at the same site. As more substitutions are accumulated between two sequences they become progressively more **saturated**; most of the sites changing have already changed before. As a result, more recent substitutions make little or no impact on the total number of observed differences between the sequences. Hence in Fig. 5.11, sequences that diverged 20 Myr, such as the

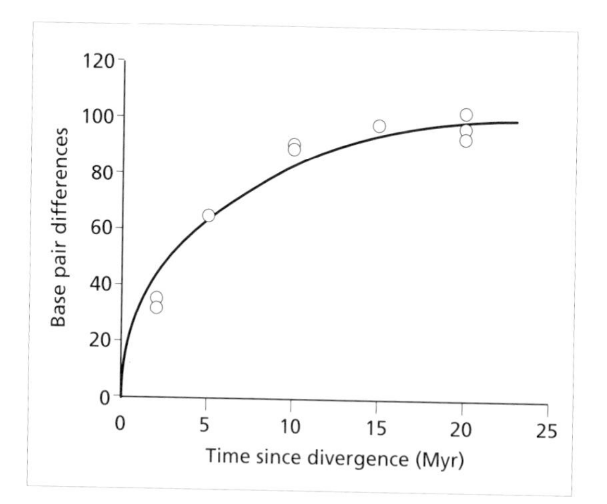

Fig. 5.11 Number of nucleotide substitutions between pairs of bovid mammal mitochondrial sequences (684 basepairs from the *COII* gene) against estimated time of divergence. Notice that the observed number of substitutions is not linear with time but curvilinear. Data from Janecek *et al.* (1996).

goat and the cow, show no more total sequence difference than that observed between the cow and the lesser kudu, which diverged some 15 Myr ago.

Models of sequence evolution

Given that observed distances may underestimate the actual amount of evolutionary change, there has been a considerable amount of research on developing methods of converting observed distances into measures of actual evolutionary distance. These techniques are often termed **distance correction** methods; their goal is to 'correct' the observed distances by estimating the amount of evolutionary changes that has been overprinted (Fig. 5.12).

A bewildering array of methods has been proposed, all with various assumptions about the nature of the molecular evolutionary process. Rather than review every single method, we will cover some of the more commonly used measures. Most of the methods for 'correcting' distances are interrelated, differing only in how many parameters they attempt to include. For example, some methods may allow for variation in nucleotide frequencies; more complex methods may allow different kinds of substitution to occur with different probabilities, while others may take into account variation in the rate of substitution between sites.

We can use a single general framework to show how these models are interrelated. Within this framework, the probability of a given nucleotide

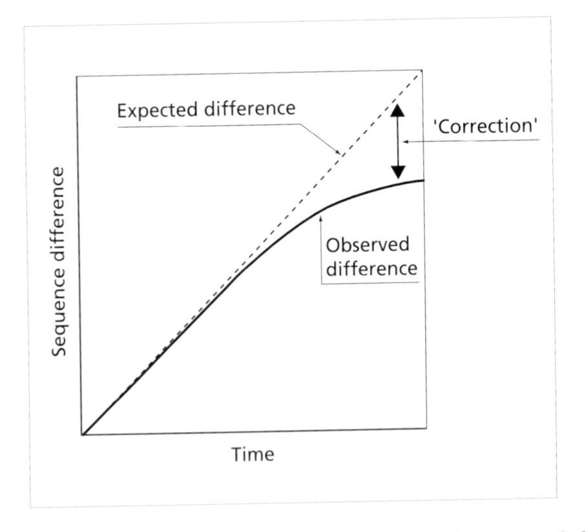

Fig. 5.12 The need to correct observed sequence differences. The extent of observed differences between two sequences is not linear with time (as we would expect if the rate of molecular evolution is approximately constant) but curvilinear due to multiple hits. The goal of distance correction methods is to recover the amount of evolutionary change that the multiple hits have overprinted and to 'correct' the distances for unobserved hits. In effect, the methods seek to 'straighten out' the line representing observed differences.

substitution remains constant over time, and the base composition of the sequences is in equilibrium. Given these assumptions (which are discussed further in section 5.2.3), the substitution probability matrix is given by

$$\mathbf{P}_t = \begin{bmatrix} p_{AA} & p_{AC} & p_{AG} & p_{AT} \\ p_{CA} & p_{CC} & p_{CG} & p_{CT} \\ p_{GA} & p_{GC} & p_{GG} & p_{GT} \\ p_{TA} & p_{TC} & p_{TG} & p_{TT} \end{bmatrix} \quad (5.2)$$

where p_{AC} is the probability that a site that started with A had nucleotide C at the end of time interval t, and so on. In most models the matrix is symmetric so that, for example, $p_{AC} = p_{CA}$. The diagonal elements p_{AA}, p_{CC}, p_{GG} and p_{TT} correspond to the case where there has been no (apparent) change at the site; the site has the same nucleotide at the end of time interval t (note that this does not necessarily mean that there has been no change, only that the end result of any substitutions has been the same nucleotide we started with). If we assign the numbers 1–4 to each nucleotide (e.g. A = 1, C = 2, etc.) then the value of the diagonal elements is given by

$$p_{ii} = 1 - \sum_{j \neq i} p_{ij} \quad (5.3)$$

In other words, the probability of observing an A at a given site at time 0 and again at time t is 1 minus the probability of observing the substitution of A by any of C, G, or T.

The base composition of the sequences can be represented by a vector

$$\mathbf{f} = [f_A \; f_C \; f_G \; f_T] \quad (5.4)$$

where f_A is the equilibrium frequency of A, and so on. In some models $f_A = f_C = f_G = f_T$ so that all four bases are present in equal proportion, whereas in other models the bases may vary in their relative frequency.

By specifying the probabilities of a given nucleotide substitution and the expected base frequencies a plethora of models can be, and have been, generated. We consider four of the best known below.

Jukes–Cantor (JC)

The Jukes–Cantor (JC) model was one of the first proposed and is perhaps the simplest model of sequence evolution. It assumes that the four bases have equal frequencies, and that all substitutions are equally likely. Under this model the distance between two sequences is given by

$$d = -\tfrac{3}{4}\ln(1 - \tfrac{4}{3}p) \quad (5.5)$$

where p is the proportion of nucleotides that are different in the two sequences, and ln is the natural log function. Expressed using equations 5.2 and 5.4 above, the JC model can be represented using the following substitution probability matrix and base composition vector:

$$\mathbf{P}_t = \begin{bmatrix} . & \alpha & \alpha & \alpha \\ \alpha & . & \alpha & \alpha \\ \alpha & \alpha & . & \alpha \\ \alpha & \alpha & \alpha & . \end{bmatrix}, \qquad \mathbf{f} = [\tfrac{1}{4}\ \tfrac{1}{4}\ \tfrac{1}{4}\ \tfrac{1}{4}]$$

where α is the probability of a substitution and is the same for all possible substitutions (the values for the diagonal elements (p_{AA}, p_{CC}, p_{GG} and p_{TT}) are given by equation (5.3) above).

Kimura's 2 parameter model (K2P)

From Fig. 5.10 one might expect transversions to be more common than transitions as only one of the three possible substitutions for any nucleotide is a transition. However, the reverse is typically the case: transitions are generally more frequent than transversions. This is especially true of mitochondrial DNA (Table 5.1).

If we consider separately the two classes of substitution in the bovid mtDNA sequences (Fig. 5.13) we can see that transitions accumulate rapidly and begin to reach saturation, whereas transversions are much rarer and appear to accumulate approximately linearly with time. Kimura's 2 parameter model (K2P) incorporates this observation that the rate of transitions per site (α) may differ from the rate of transversions (β), giving a total rate of substitution per site of ($\alpha + 2\beta$) (remember that for any nucleotide there are three possible changes, one of which is a transition, the remaining two being transversions). The transition : transversion ratio α/β is often represented by the letter kappa (κ). In the K2P model the number of nucleotide substitutions per site is given by

$$d = \tfrac{1}{2}\ln[1/(1 - 2P - Q)] + \tfrac{1}{4}\ln[1/(1 - 2Q)] \qquad (5.6)$$

where P and Q are the proportional differences between the two sequences

Table 5.1 Transition bias in four types of DNA sequence (from Wakeley, 1996).

Type of sequences	Transition/transversion ratio (κ)
mtDNA	9.0
12S rRNA	1.75
α- and β-globins	0.66
Pseudo η-globin	2.70

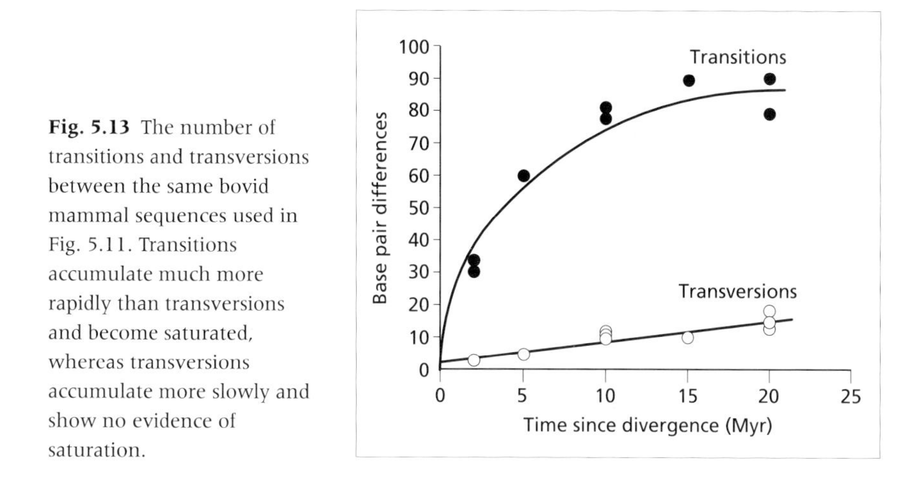

Fig. 5.13 The number of transitions and transversions between the same bovid mammal sequences used in Fig. 5.11. Transitions accumulate much more rapidly than transversions and become saturated, whereas transversions accumulate more slowly and show no evidence of saturation.

due to transitions and transversions, respectively. This model has the following substitution probability matrix and base frequency vector:

$$\mathbf{P}_t = \begin{bmatrix} \cdot & \beta & \alpha & \beta \\ \beta & \cdot & \beta & \alpha \\ \alpha & \beta & \cdot & \beta \\ \beta & \alpha & \beta & \cdot \end{bmatrix}, \qquad \mathbf{f} = [\tfrac{1}{4}\ \tfrac{1}{4}\ \tfrac{1}{4}\ \tfrac{1}{4}].$$

Note how the substitution probability matrix differs from the Jukes–Cantor model in having different probabilities for transitions and transversions. Of course, if $\alpha = \beta$, then the K2P model becomes the JC model.

Felsenstein (1981)

Another reason that some nucleotide substitutions might be more frequent than others is variation in base composition. For example, insect mitochondrial DNA tends to be rich in adenine and thymine compared to that of vertebrates; among eubacteria the total percentage of guanine and cytosine (G + C) ranges from 25% to 75%. If some bases are more common than others then we might expect some substitutions to be more frequent than others; if a sequence contains very few Gs then we are unlikely to see many substitutions involving that nucleotide. Felsenstein's (1981) model (F81) addresses this concern by allowing the frequencies of the four nucleotides to be different. Note, however, that the method assumes that the frequency of each nucleotide is approximately the same over all the sequences being considered. If different sequences have different base compositions then this assumption is violated (see pp. 157–158). Base composition can vary among genes, and among the same gene in different species.

The F81 model has the following form:

$$\mathbf{P}_t = \begin{bmatrix} . & \pi_C\alpha & \pi_G\alpha & \pi_T\alpha \\ \pi_A\alpha & . & \pi_G\alpha & \pi_T\alpha \\ \pi_A\alpha & \pi_C\alpha & . & \pi_T\alpha \\ \pi_A\alpha & \pi_C\alpha & \pi_G\alpha & . \end{bmatrix}, \qquad \mathbf{f} = [\pi_A\ \pi_C\ \pi_G\ \pi_T]$$

where π_i is the frequency of the ith base averaged over the sequences being compared. Note that if $\pi_A = \pi_C = \pi_G = \pi_T = \frac{1}{4}$ then the F81 model is the same as the JC model.

Hasegawa, Kishino and Yano (1985)

The HKY85 model proposed by Hasegawa, Kishino and Yano (1985) essentially merges the K2P and F81 models by allowing transitions and transversions to occur at different rates, and allowing base frequencies to vary as well. It has the following form:

$$\mathbf{P}_t = \begin{bmatrix} . & \pi_C\beta & \pi_G\alpha & \pi_T\beta \\ \pi_A\beta & . & \pi_G\beta & \pi_T\alpha \\ \pi_A\alpha & \pi_C\beta & . & \pi_T\beta \\ \pi_A\beta & \pi_C\alpha & \pi_G\beta & . \end{bmatrix}, \qquad \mathbf{f} = [\pi_A\ \pi_C\ \pi_G\ \pi_T]$$

General reversible model (REV)

The REV, or general reversible model (Rodríguez *et al.*, 1990; Yang *et al.*, 1994) is, as its name implies, more general than the previous models. Its probability matrix has six parameters, such that each possible substitution has its own probability:

$$\mathbf{P}_t = \begin{bmatrix} . & \pi_C a & \pi_G b & \pi_T c \\ \pi_A a & . & \pi_G d & \pi_T e \\ \pi_A b & \pi_C d & . & \pi_T f \\ \pi_A c & \pi_C e & \pi_G f & . \end{bmatrix}, \qquad \mathbf{f} = [\pi_A\ \pi_C\ \pi_G\ \pi_T]$$

By suitably constraining the six parameters a–f and the base frequencies we can generate any of the previous models. For example, if $a = c = d = f$ and $b = e$ then we have the HKY model; further constraining $a = b = c = d = e = f$ gives the F81 model, and so on.

Comparing the models

The models we have discussed above are clearly interrelated (Fig. 5.14). The K2P and F81 models each extend the JC model by adding extra parameters to

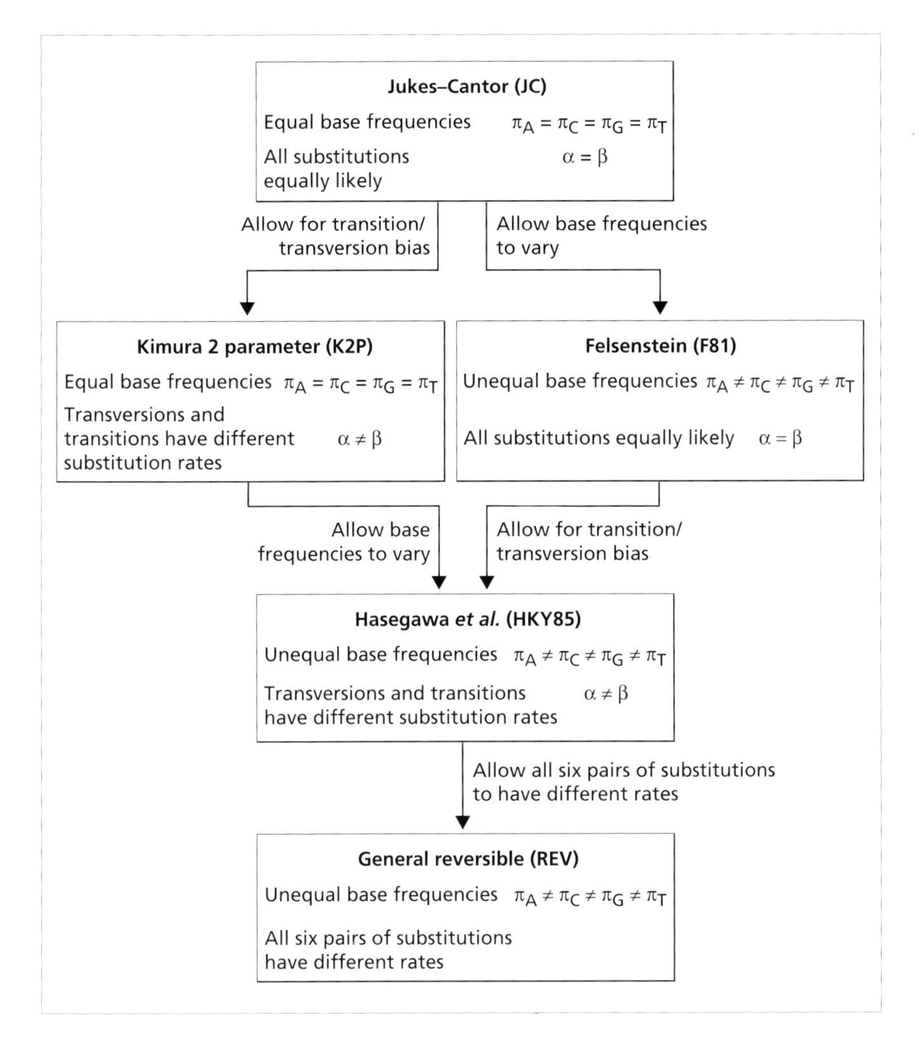

Fig. 5.14 Interrelationships among five models for estimating the number of nucleotide substitutions among a pair of DNA sequences. The JC, K2P, F81 and HKY85 models can all be generated by constraining various parameters of the REV model.

allow transitions and transversions to occur at different rates (K2P) and base frequencies to vary (F81). The HKY85 model combines these two models. The REV model encompasses all four models, and hence each can be regarded as a special case of the REV model.

These models are merely some of the best known and most commonly applied models in use. There are many other methods that have been proposed. Given the range of models proposed, how do we choose which model to use when computing a distance measure? One criterion would be to choose the most realistic model available. However, realism comes at a cost. For a given data set, each time we add a parameter to add realism (say, allowing transitions and transversions to occur at different rates) we must estimate that parameter

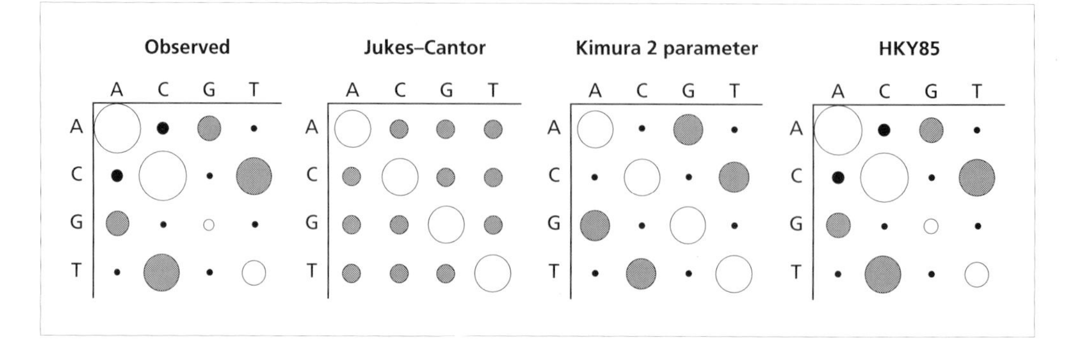

Fig. 5.15 Observed and expected numbers of nucleotide pairs between human and chimpanzee mtDNA sequences for three different models. As the models add parameters they more closely approximate the observed pattern. Data from Tamura (1994).

using our data. The more parameters we add, the greater the uncertainty in our estimates as, in effect, we are spreading our data more and more thinly. Hence we might choose a more realistic model but end up with large sampling errors, and hence loose statistical power. Clearly it would be desirable to use only the minimum number of parameters that yield a reasonable estimate. Fewer parameters might give an inaccurate estimate; more parameters may decrease the precision of our estimate.

Figure 5.15 shows the observed pattern of nucleotide substitutions between mtDNA sequences from humans and chimpanzees. It is apparent that the frequencies of the bases are not equal, and that transitions are more frequent than transversions. Given the observed nucleotide frequencies the different models predict different patterns of nucleotide substitution, based on their different substitution probability matrices. Recall that the Juke–Cantor model (pp. 149–150) assumes that each nucleotide is present in equal proportions, and that all substitutions are equally likely. For the data in Fig. 5.15, 307 of the 1333 sites differ among human and chimps (i.e. $p = 307/1333 = 0.230$), so JC distributes these changes evenly among all pairs of different nucleotides resulting in an expected pattern of nucleotide substitution that does not resemble the observed pattern. One source of the disparity is the transition/transversion bias; for these data $P = 0.219$ and $Q = 0.011$ (see pp. 150–151). The Kimura 2 parameter model adds the transition/transversion ratio as a parameter and yields a pattern that looks more like that observed than does JC. However, K2P still assumes that each nucleotide is equally common and hence greatly overestimates the number of GG and TT sites. It also underestimates the number of A $\leftrightarrow$ C transversions. A better fit to the observed data is achieved by incorporating the different nucleotide frequencies, which for these data are $\pi_A = 0.371$, $\pi_T = 0.176$, $\pi_C = 0.405$, and $\pi_G = 0.048$. The expected pattern under the HKY85 model (p. 152) matches the observed data rather well (Fig. 5.15). Further refinements, such as allowing the transition rate to be different between purines and between pyrimidines, provide an even better fit.

The relative degree of fit between each model and the data can be evaluated by computing for each model the likelihood (see Box 5.1) of obtaining the observed data. For the data shown in Fig. 5.15 the theoretical best value for the log likelihood (ln L) is −2064.80. Obtaining the observed data under the

Box 5.1 Likelihood

Given some observed data and an hypothesis of how that data came about, how can we decide whether our hypothesis is an adequate explanation of the data? One approach to this problem is the concept of **likelihood**. Likelihood is the probability of observing the data given a particular model. Different models may make the observed data more or less probable. For instance, if you are flipping a coin and get a 'heads' only one time in 100 then you would suspect that the coin you are flipping is biased. If the coin was fair then you would expect to have seen 'heads' about 50 times, not once. The chances of getting just one head out of 100 flips with perfectly fair coin seems very small; in other words, given a fair coin the likelihood of the result of 1 head and 99 tails is very low.

It is important to distinguish between the probability of getting the observed data, and the probability of the underlying model being correct. Likelihood says nothing about the probability of the model itself. The philosopher of biology Elliot Sober has used the following example to make the distinction clear. Suppose you hear a loud noise in the room above you. You could suggest that it was caused by gremlins playing ten-pin bowling in the attic. Given this hypothesis your *observation* (a loud noise above you) has a high likelihood; if gremlins were indeed bowling above you would almost certainly expect to hear it! However, the probability that your hypothesis is true (namely, that it is *gremlins* making the noise, as opposed to any other possible cause) is something else again; it is almost certainly not the case that there are gremlins in the attic. Hence in this case your hypothesis confers a high likelihood upon the data, but is itself highly improbable.

Given a model that specifies probabilities of observing various events we can compute the likelihood L of obtaining the observed data. This is often written as

$$L = \Pr(D|H)$$

where $\Pr(D|H)$ is the probability of getting the data D given hypothesis H. As L is often a very small number, likelihoods are often expressed as natural logarithms and referred to as **log-likelihoods**.

Likelihoods for different models can be compared if those models are nested, that is, one model is a special case of the other. Twice the difference between

continued on p. 156

Box 5.1 *continued*

the log-likelihoods of two models is χ^2 distributed, hence it is possible to test whether one model is significantly better than the other. This makes likelihood a powerful tool for hypothesis testing. A common use is to compare models in which one parameter (such as base composition, transition/transversion ratio, or rate of evolution) is fixed in one model and allowed to vary in the other.

JC model is much less likely ($\ln L = -2691.76$). Adding the transition/transversion ratio parameter (the K2P model) significantly increases the likelihood of obtaining the observed data ($\ln L = -2424.79$). Allowing nucleotide frequencies to vary (HKY85) results in an even greater increase in likelihood ($\ln L = -2075.41$). The likelihood values confirm the visual impression given in Fig. 5.15 regarding the relative merits of the models. In order of fit to the data the models can be ranked HKY85 > K2P > JC, with HKY85 most closely resembling the observed patterns.

5.2.3 Assumptions

All the methods we have discussed share these assumptions:

1 All nucleotide sites change independently.

2 The substitution rate is constant over time and in different lineages.

3 The base composition is at equilibrium.

4 The conditional probabilities of nucleotide substitutions are the same for all sites and do not change over time.

While these assumptions make the methods tractable they are in many cases unrealistic. We consider some of these assumptions below.

Independence

The assumption of independence means that change at one site has no effect on change at any other site. There are a number of clear exceptions to this assumption. For example, once transcribed, ribosomal RNA assumes a complicated secondary structure comprising 'stem' and 'loops' where the stems are held together by Watson–Crick base pairing. A substitution in a stem may result in a pair of nucleotides that cannot pair correctly (Fig. 5.16), reducing the stability of the structure. Often a **compensatory change** may occur, where a substitution occurs that restores Watson–Crick base pairing. This tendency for sites in stems to covary means that these changes are not truly independent. Because of the complicated secondary structure of the RNA molecule, parts of the sequence that are far apart when the sequence is thought of as just a string of letters may be folded into close proximity with each other. Hence, correlated changes may occur between sites that are separated by many intervening nucleotides.

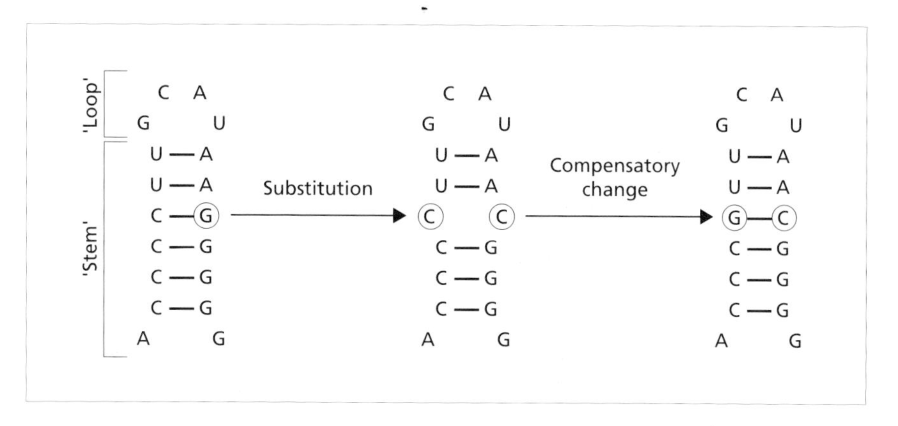

Fig. 5.16 RNA molecules have a secondary structure comprising 'stems' of Watson–Crick paired nucleotides and 'loops' of unpaired nucleotides. A substitution in a stem can destroy the Watson–Crick bond and reduce the stability of the molecule. A substitution that restores the stem is a compensatory change. After Hickson *et al.* (1996).

This non-independence of substitutions in RNA has led to some discussion about 'down-weighting' sites in stems to compensate for the correlated changes because a single substitution may ultimately result in two substitutions: the original substitution and the subsequent compensation (Fig. 5.16). The issue of weighting characters is discussed further in Chapter 6. However, compensatory changes need not occur very quickly; basepair mismatches in stems may persist for millions of years.

Base composition

The assumption that base composition is at equilibrium means that over the collection sequences being studied the base composition is roughly the same. Deviations from this assumption do occur and can lead to problems inferring the correct evolutionary tree. For example, the tree in Fig. 5.17 shows two

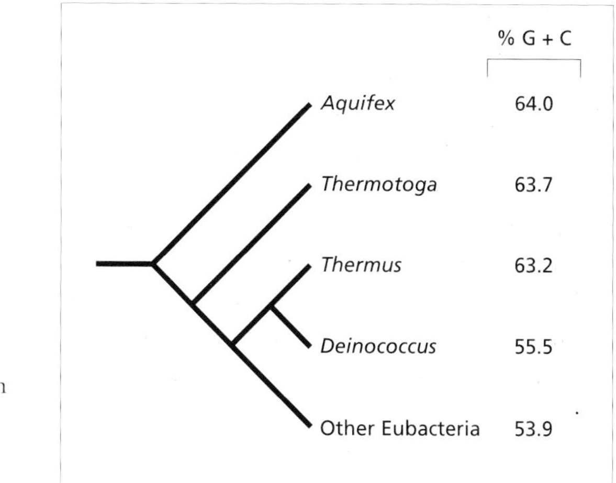

Fig. 5.17 Part of the phylogeny of bacteria showing the variation in percentage G + C content in small subunit rRNA. After Galtier and Gouy (1995: Fig. 4).

closely related bacteria, *Thermus* and *Deinococcus*, that differ in their base composition, *Deinococcus* having a composition more typical of other bacteria. Trees inferred from distances computed using most methods will not group these two bacteria together, but instead will group *Deinococcus* with bacteria with which it shares little other than a similar base composition. One method developed specifically for this kind of problem is the LogDet measure (see Box 5.2).

Box 5.2 LogDet

The principal goal of the distance measures described in this chapter is to measure the number of nucleotide substitutions between a pair of DNA sequences. However, one recently proposed method for comparing sequences is not intended to estimate sequence divergence as such, although under some circumstances it can be so used. This method, called **LogDet**, is designed to circumvent the problem of variable base composition among sequences. While some methods, such as the HKY85 model, allow base frequencies of the four nucleotides to vary, they assume that these frequencies are approximately constant across the sequences being compared. If different sequences have different base compositions then these methods may be a poor choice. This has particular consequences for phylogenetic tree construction (Chapter 6), where some distance measures can lead to sequences being grouped together on the basis of similarity in base composition rather than actual evolutionary relationship.

The LogDet transformation is designed to recover an additive distance (see Chapter 2) between sequences, even when base composition is variable. For each pair of DNA sequences, x and y, a 4×4 matrix is constructed with the entries being the proportion of sites which have each possible pair of nucleotides. For example, this table shows the number of times a nucleotide in a 900 basepair stretch of 16S rRNA sequence from photosynthetic organelles from *Euglena* matched each nucleotide in the same gene in the organelle from *Olithodiscus*.

		Olithodiscus			
		a	c	g	t
Euglena	a	224	5	24	8
	c	3	149	1	16
	g	24	5	230	4
	t	5	19	8	175

From this table we construct a 4×4 matrix **F** with rows and cells corresponding to each nucleotide, and the cells being the proportion of sites with each possible pair of nucleotides. In this example,

$$\mathbf{F}_{xy} = \begin{bmatrix} 0.249 & 0.006 & 0.027 & 0.009 \\ 0.003 & 0.166 & 0.001 & 0.018 \\ 0.027 & 0.006 & 0.256 & 0.004 \\ 0.006 & 0.021 & 0.009 & 0.194 \end{bmatrix}$$

continued

Box 5.2 *continued*

and the LogDet distance is

$d_{xy} = -\ln[\det \mathbf{F}_{xy}] = 6.216$

where det is the determinant of the matrix $\mathbf{F}_{xy}$. For more details see Lockhart *et al.* (1994), from which this example is taken.

Variation in rates of substitution among sites

The methods for calculating evolutionary distances between sequences described above make various assumptions about the probability of substitutions between various nucleotides, but one assumption that they all share is that each nucleotide site in a sequence is equally likely to undergo a substitution. This assumption simplifies the mathematics but at the cost of biological realism. Different regions of DNA sequence may have quite different probabilities of change. Figure 5.18 shows the relative rates of substitution in different parts of a sample of mammalian genes and pseudogenes. Pseudogenes, which have lost all functionality evolve most rapidly, with fourfold degenerate sites close behind. As we might expect, non-degenerate sites (at which any nucleotide substitution results in an amino acid replacement) evolve relatively slowly. The reasons why different regions of DNA sequence evolve at different rates are discussed in more detail in Chapter 7.

Variable rates of substitution can have considerable impact on sequence divergence. Consider firstly the simple case where some sites are free to vary,

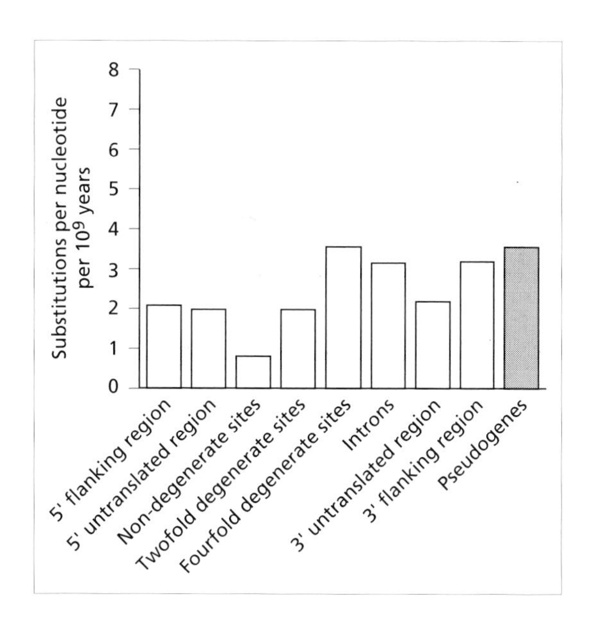

Fig. 5.18 Average rates of substitution in different parts of mammalian genes and pseudogenes. From Li and Graur (1991).

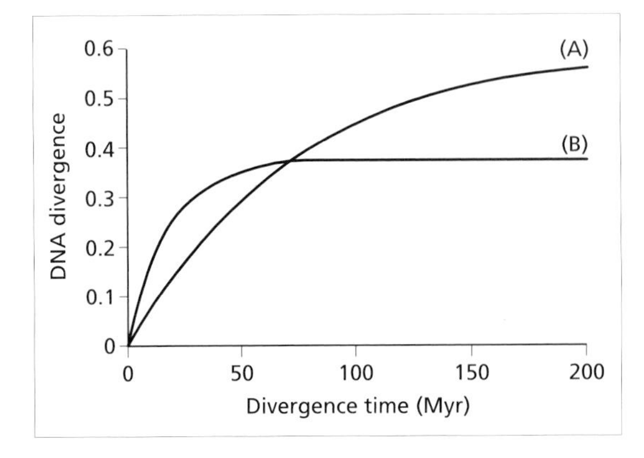

Fig. 5.19 DNA sequence divergence plotted against time since divergence. For the upper curve (A) the rate of substitution is 0.5%/Myr and 80% of sites are free to vary, whereas for the lower curve (B) the rate of substitution is higher (2%/Myr) but only half the sites are free to vary. After Palumbi (1989).

and others are constrained to be invariant. This can be thought of as crudely modelling the case where substitutions are either neutral, or are strongly deleterious and hence selected against. If some sites are not free to vary then sequences that evolve at a fast rate can, over evolutionary time, paradoxically show less divergence than more slowly evolving sequences that have fewer constraints. Figure 5.19 shows the expected divergence of two sequences, A and B, that differ in rate of substitution and fraction of sites free to vary. In sequence A, substitutions occur at a rate of 0.5% per million years (Myr) and 80% of sites can undergo a substitution, whereas in sequence B the rate of substitution is higher, 2% per Myr, but half the sites are constrained. The more rapidly evolving sequence (B) initially shows more sequence divergence but then begins to saturate such that by 80 Myr the slower sequence (A) starts to show more divergence. In this hypothetical example, were we to compare the divergence between the two sequences in a pair of species that diverged from their common ancestor 100 Myr ago we might mistakenly conclude that sequence A was evolving more rapidly because the sequences are more divergent.

Distribution of rates

In reality, sites show a range of probabilities of substitution, rather than simply the two categories of zero and non zero used in Fig. 5.19. The challenge is to develop tractable models of this rate variation. The most widely used approach makes use of the gamma distribution. This distribution has 'shape' parameter α which specifies the range of rate variation among sites. Small values of α result in an L-shaped distribution with extreme variation of rates; most sites are invariable but a few have very high rates of substitution. Conversely, the larger α the smaller the range of rates; when $\alpha > 1$ the distribution is bell-shaped (Fig. 5.20).

Estimates of α obtained from nuclear and mitochondrial genes range from 0.16 to 1.37 (Table 5.2), in other words the distribution of rates is generally L-shaped. These values are for all codon positions combined; when codon

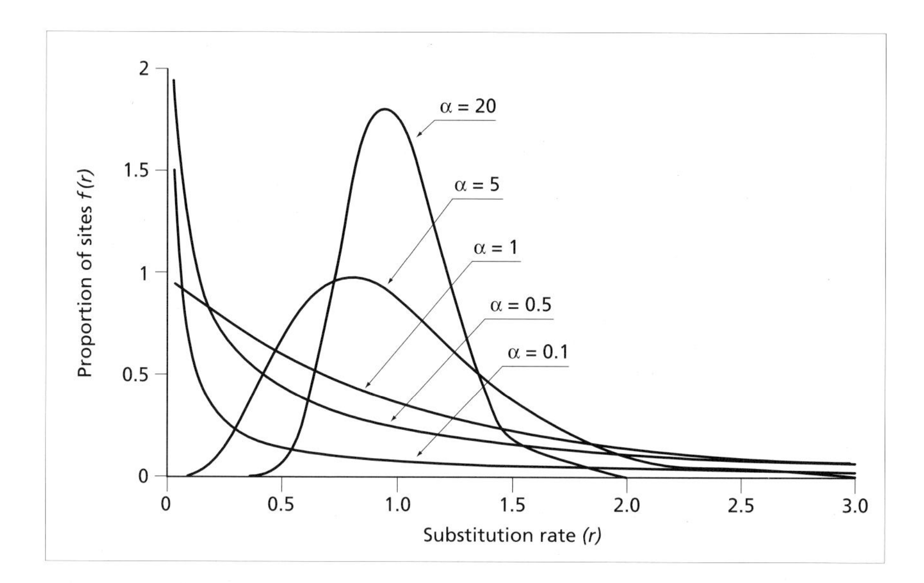

Fig. 5.20 The distribution of relative substitution rate r corresponding to different values of the gamma shape parameter α. Low α corresponds to large rate variation. As α gets larger the range of variation diminishes, until as α approaches ∞ all sites have the same substitution rate. After Yang (1996: Fig. 1).

Table 5.2 Estimates of α shape parameter of the distribution of rate variation for various genes (after Yang, 1996, Table 1).

Type of sequences	α
Nuclear genes	
Albumin genes	1.05
Insulin genes	0.40
c-myc genes	0.47
Prolactin genes	1.37
16S-like rRNAs, stem region	0.29
16S-like rRNAs, loop region	0.58
ψη-globin pseudogenes	0.66
Viral genes	
Hepatitis B virus genomes	0.26
Mitochondrial genes	
12S rRNAs	0.16
Position 1 of four genes	0.18
Position 2 of four genes	0.08
Position 3 of four genes	1.58
D-loop region	0.17
Cytochrome *b*	0.44

positions are analysed separately, the value of α for first and second positions is much smaller than that for third codon positions, the latter tending to have a more even distribution of rates.

The models of evolutionary change discussed on pp. 148–152 above can be modified to include the gamma distribution, typically represented by the symbol Γ. For example, adding a gamma distribution to the HKY85 model yields the 'HKY85 + Γ' model.

5.2.4 Distance measures for protein sequences

In contrast to the plethora of methods for computing distances between nucleotide sequences, there are relatively few methods for computing distances for protein sequences. This is largely due to the greater complexity of the problem; instead of the four states required for nucleotide data, protein sequences comprise some 20 different amino acids. A simple measure of dissimilarity between protein sequences due to Kimura is given by

$$d = -\ln(1 - p - 0.2p^2) \quad (5.7)$$

where p is the proportion of amino acids differing in the two protein sequences. This measure does not take into account the differing probabilities of replacement among amino acids (see Fig. 5.7), and hence lacks realism. The development of better measures is an active area of research.

5.3 Measuring evolutionary change on a tree

So far in this chapter we have concentrated on measuring evolutionary change between two or more sequences by comparing those sequences directly. We can also estimate evolutionary change on a tree. For any pair of sequences the total amount of evolutionary change that occurred since the time they diverged from their common ancestor is the sum of the evolutionary change along each edge of the path in the tree between the two sequences (Fig. 5.21).

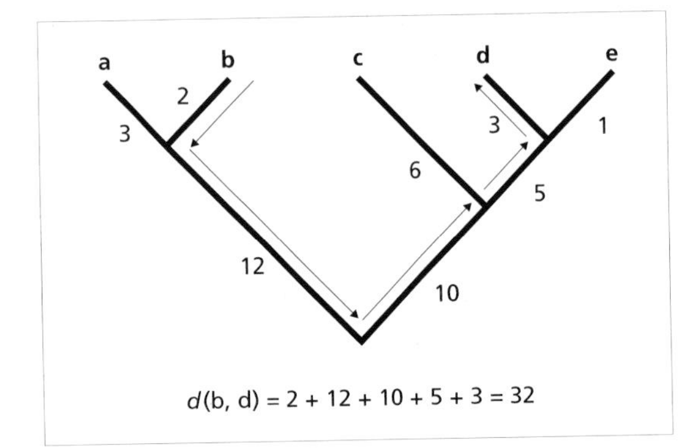

Fig. 5.21 The evolutionary distance between b and d is the sum of the edge lengths along the path in the tree between the two sequences.

There are a range of methods for inferring branch lengths, and these are directly related to the methods of tree construction discussed in Chapter 6. In this chapter we will focus on parsimony, which is one of the few methods that explicitly seeks to reconstruct the ancestral sequences themselves, rather than just the edge lengths.

Given that ancestors are almost never actually observed (although see the HIV example in Chapter 2), we need some rules for inferring what they looked like. In this context, we need rules to construct an ancestral sequence from the sequences and relationships of its descendants. One approach might be to simply 'average' the descendant sequences and use that as the ancestor. Indeed, such an approach is often used to construct a consensus sequence which summarises the common features of a set of sequences (we encountered the notion of consensus in Chapter 2 in relation to trees). However, a consensus sequence is not necessarily the same as an ancestral sequence (see Box 5.3). Using the parsimony criterion we can construct the ancestral sequences that can give rise to the observed sequences with the fewest possible evolutionary changes by following a simple method. For each internal node in the tree we use the sequences of its immediate descendants to construct a hypothetical ancestral sequence.

Under parsimony we assign to each node in the tree a **character state set**. For observed DNA sequences this is simply the observed nucleotide at a

Box 5.3 Consensus sequences

Suppose we have the following six sequences for a homologous codon in six different species.

1 CGA
2 CGA
3 ATT
4 TTT
5 TGT
6 TGG

There is sequence variation among these sequences. If we want to summarise these sequences we could compute a **consensus sequence** which might, for example, comprise the base found in a majority of sequences at each site. For these six sequences the consensus sequence is:

consensus TGT

While this approach does summarise the variation among the sequences in terms of an 'average' sequence, it should not be confused with an ancestral sequence. This is because sequence consensus methods implicitly assume that all the sequences are equally related, that is by a star tree (shown below left).

continued on p. 164

Box 5.3 *continued*

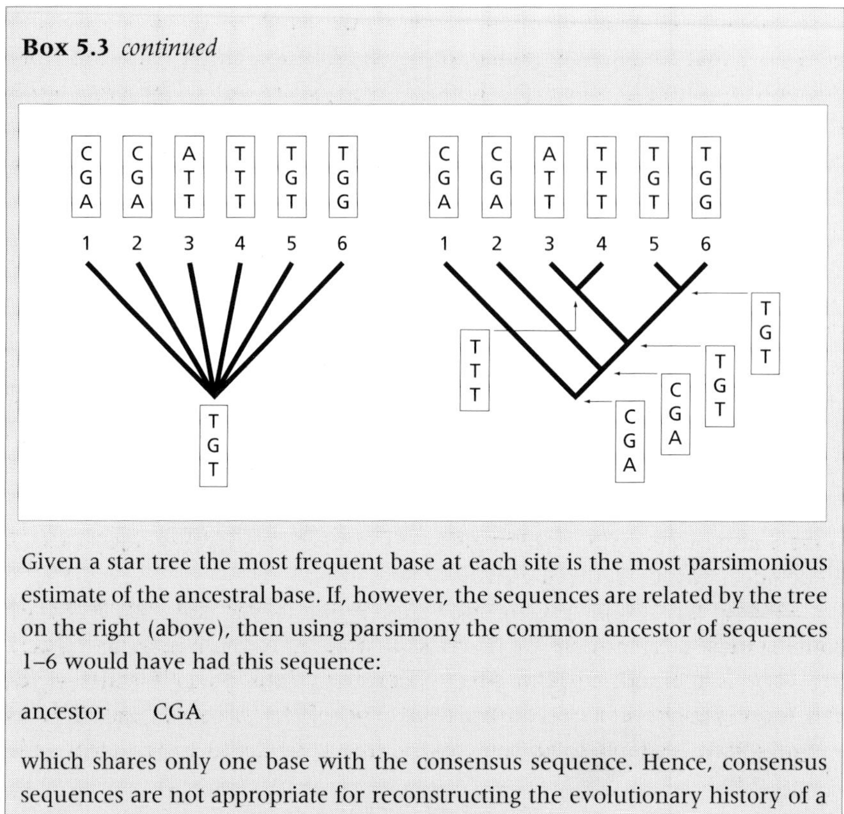

Given a star tree the most frequent base at each site is the most parsimonious estimate of the ancestral base. If, however, the sequences are related by the tree on the right (above), then using parsimony the common ancestor of sequences 1–6 would have had this sequence:

ancestor CGA

which shares only one base with the consensus sequence. Hence, consensus sequences are not appropriate for reconstructing the evolutionary history of a set of sequences (unless those sequences are actually related by a star tree); they are simply tools for summarising sequence variation.

given site. For example, if a sequence x has A then its state set is $s_x = \{A\}$. Suppose we have two sequences a and b and we want to infer the state possessed by their ancestor x. The parsimony rules are as follows.

1 If the state sets of a and b are identical then x has the same state set, i.e.

$s_x = s_a = s_b$.

2 If a and b have different state sets then if the state sets s_a and s_b have any states in common the ancestor has those states, otherwise the ancestor has the combined state sets of a and b, i.e. if $s_a \cap s_b \neq \emptyset$, $s_x = s_a \cup s_b$, otherwise $s_x = s_a \cap s_b$.

The first rule seems obvious; if both descendants have A then the simplest interpretation is that the ancestor also had A. The second rule is a little more complex. If the two sequences have different states then the state set of the ancestor has both states. For example, given descendants with state sets {A} and {T} we would assign the state set {A, T} to the ancestor. This does not mean that the ancestor literally had both nucleotides (i.e. was polymorphic), rather it means that we lack the information to decide at this point which one of the alternative states it possessed. The second part of rule 2 comes in to play when we are looking at ancestors whose descendants include at least one node with a state set that contains more than one state, for example {A, T}. If the other descendant of this node has {T} then the most parsimonious interpretation is that the ancestor has state set {T}.

We can see these rules in action using the following example. Suppose we have the tree shown in Fig. 5.22. Using parsimony we walk along the tree making sure we visit the descendants of a node before visiting the node itself (technically, this is known as a **post-order traversal**). Hence, the first node we visit is node 1 (we could also have started with node 3, as it makes no difference to the result). The descendants of this node have states A and T, hence by rule 2 above, node 1 has the union of these two state sets, i.e. {A, T}. Having visited node 1 we now know the state sets for the two descendants of node 2, {A, T} and {T}. These two sets share {T}, which is (by rule 2) the state set for this node. Before visiting node 3 we must first get the state set for node 3. Both the descendants of node 3 have the same state set {G} so by rule 1, node 3 also has {G}. Continuing down the tree, node 4 has {T} ∪ {G} = {G, T} and node 5 has {G, T} ∩ {G} = {G}.

We now have state sets for all the internal nodes in the tree. However, some state sets have more than one state and hence are ambiguous. There are various methods for choosing which one of the possible states yields the most parsimonious reconstruction. One of the simplest (called Farris optimisation) is simply to go back up the tree assigning to any internal node

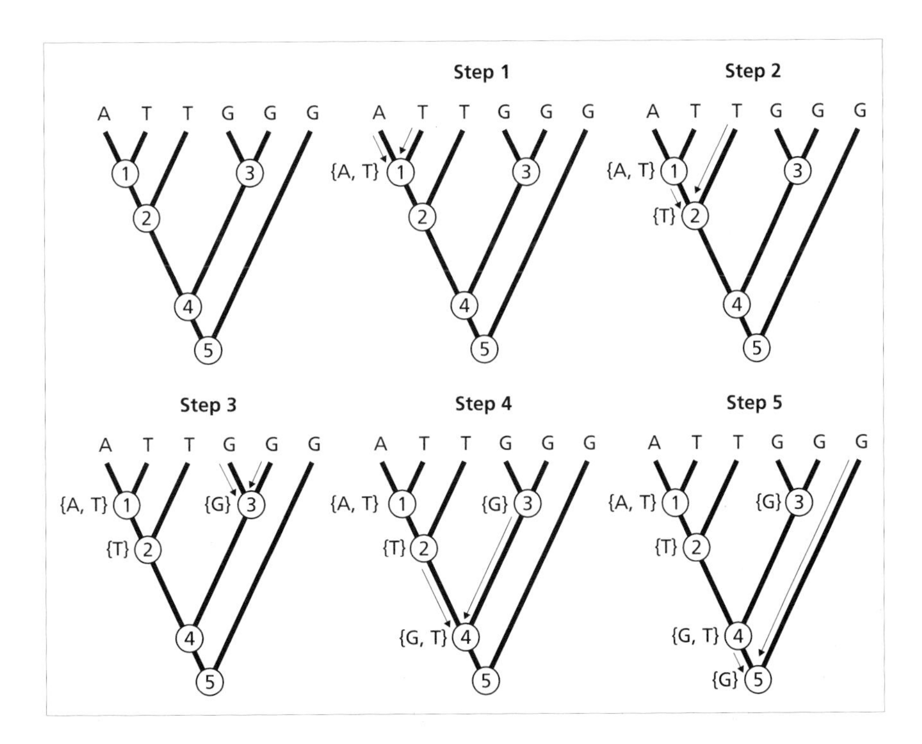

Fig. 5.22 Assigning state sets to internal nodes on a tree (see text).

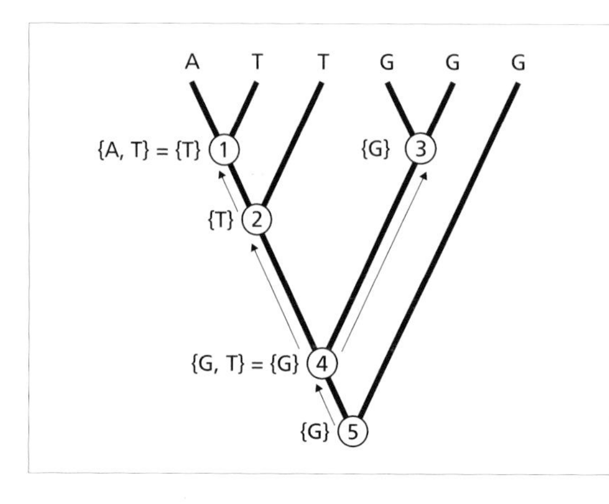

Fig. 5.23 Resolving the ambiguity in ancestral reconstructions by going back up the tree and assigning to ambiguous nodes the intersection of its state set with that of its ancestor.

that is ambiguous the intersection of its state set with that of its immediate ancestor.

Having obtained states for the ancestors we can infer where on the tree evolutionary changes took place, and the nature of those changes. In the tree shown in Fig. 5.23 we can see that a total of two substitutions occurred (between nodes 1 and its left descendant, and between nodes 2 and 4), and that one change was T → A and the other was G → T.

In the example given above, there is only one most parsimonious solution for the ancestral states. However, this need not always be the case, there may be a number of alternative reconstructions that postulate the same number of events. However, the different reconstructions may imply rather different sequences of events and hence different biological consequences. We should also stress that the amount of change recovered by parsimony is, by definition, the smallest possible consistent with the data. The actual amount of evolutionary change may have been somewhat larger. Furthermore, our estimates of the amount of change can depend on how many sequences we have examined.

5.3.1 Estimating branch lengths

Just because two sequences may have the same nucleotide at the same position, this does not mean that there has been no evolutionary change between those two sequences. As we noted above (Fig. 5.9) the same nucleotide may arise independently in different lineages. Imagine that we have a few sequences for a group of species and of those sequences, 1 and 2 are the most closely related (Fig. 5.24a). If those two sequences share the same nucleotide (say, A) at the same site then the simplest interpretation is that the A is homologous, that is, 1 and 2 inherited it from some common ancestor. If, however, we discovered more sequences that were more closely related to 2 than to 1, and these sequences have either T or G at this position, then the occurrence of A in sequences 1 and 2 is more likely to be of independent origin (Fig. 5.24b). In

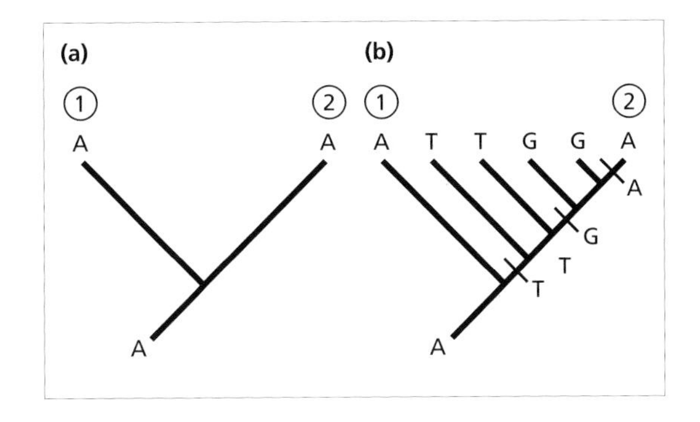

Fig. 5.24 The effects of sampling on estimates of the amount of sequence divergence. (a) Two sequences that appear to be each others closest relative. If the ancestral state is A then apparently there has been no evolution along the path between sequences 1 and 2. If, however, more sequences are found that are more closely related to sequence 2 than to sequence 1, and these sequences have different nucleotides at the same site then we now have evidence for at least two substitutions between 1 and 2.

this instance, obtaining more sequences to flesh out the evolutionary tree has uncovered more evolutionary change. Furthermore, use of an evolutionary tree detects more substitutions (3 versus none) than would have been counted just by directly comparing the two sequences.

This phenomenon of the dependence of the amount of inferred evolutionary change on how densely we sample sequences can be illustrated by an oft observed pattern in molecular trees. If trees are plotted with branch lengths proportional to change (i.e. as additive trees, see Chapter 2) then in many cases lineages that show fewer cladogenetic events also show the least amount of evolutionary change. Figure 5.25(a) shows this pattern in a phylogeny for mammalian ribonuclease genes. If we add up the total amount of evolutionary change along the path from each sequence to the root of the tree, there is a clear relationship between the length of that path and the number of internal nodes we encounter along that path. The tips of the tree that are separated from the root by the most cladogenetic events show the most evolutionary change (Fig. 5.25b).

5.3.2 Testing ancestral reconstructions

Ancestors, as much as trees themselves, are hypotheses. The sequences we assign to them are hypotheses that we can rarely test directly because the ancestors themselves are long extinct. There are few exceptions to this, such as the experimental phylogenies discussed in Chapter 6. However, in the absence of such direct evidence we must use other criteria.

Of course the real ancestral sequences were not merely strings of nucleotides, they were functioning entities (if we exclude pseudogenes and other non-functional sequences). Hence one check on the reliability of our inferred

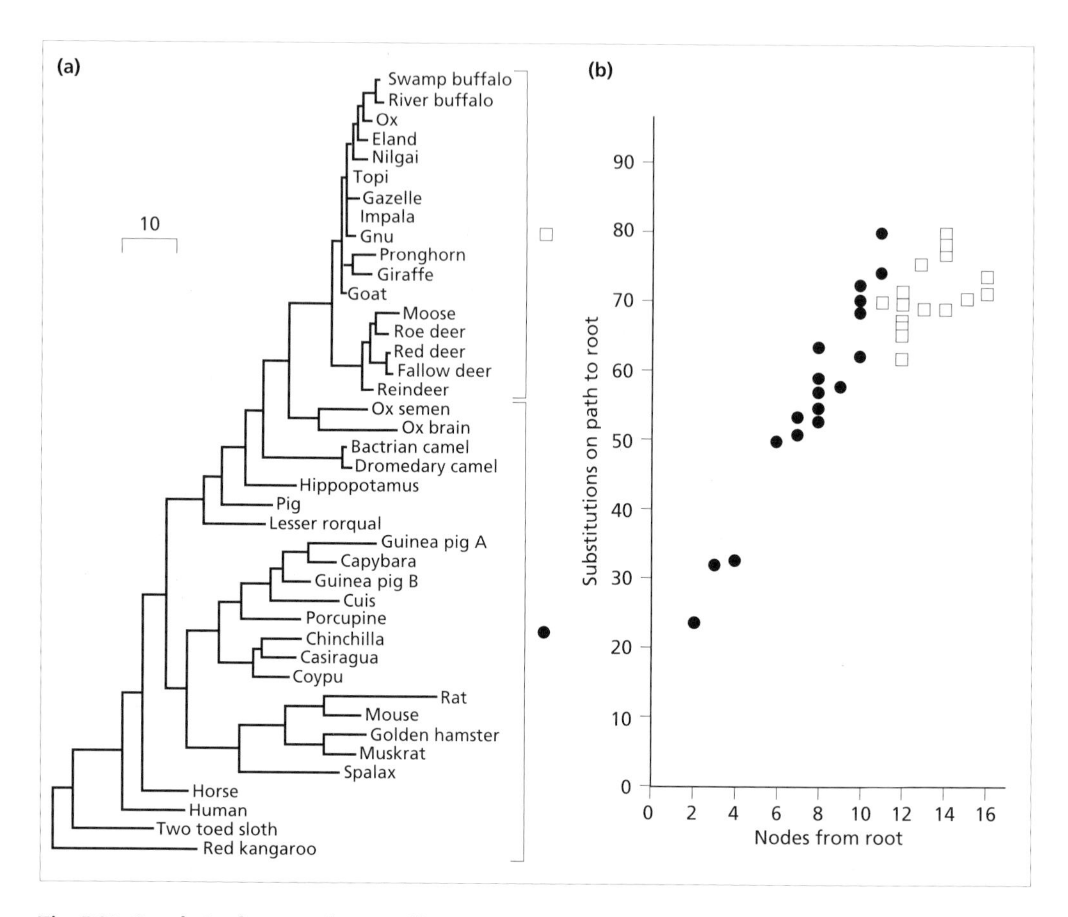

Fig. 5.25 Correlation between density of lineage sampling and total amount of evolutionary change inferred. (a) Phylogeny of ribonucleases with branch lengths proportional to the minimum number of nucleotide substitutions required by parsimony. (b) Plot of number of substitutions along the path from tip to the root, against the number of nodes along that path. Note the correlation between the two values for sequences < 12 nodes from the root. A square (□) denotes the clade reindeer to swamp buffalo; the remaining taxa are indicated by a circle (●). Adapted from Fitch and Beintema (1990).

ancestral sequences is whether they are, or could be, functional. For an example of this approach we can return to the ribonucleases (RNAse) discussed above. In a pioneering study by Jermann and colleagues the sequences for 13 ancestral RNAses were inferred for artiodactyl mammals (this order includes the pig, camel, deer, sheep and ox) using parsimony. These hypothetical ancient proteins were then synthesised in the laboratory and their properties compared with present day RNAses (Fig. 5.26). All 13 RNAses had catalytic activity consistent with being functional enzymes, suggesting that the hypothetical ancestral sequences inferred using parsimony were indeed plausible.

What is particularly interesting is that the inferred properties of the RNAses at the base of the artiodactyl phylogeny differ from those of their more recent

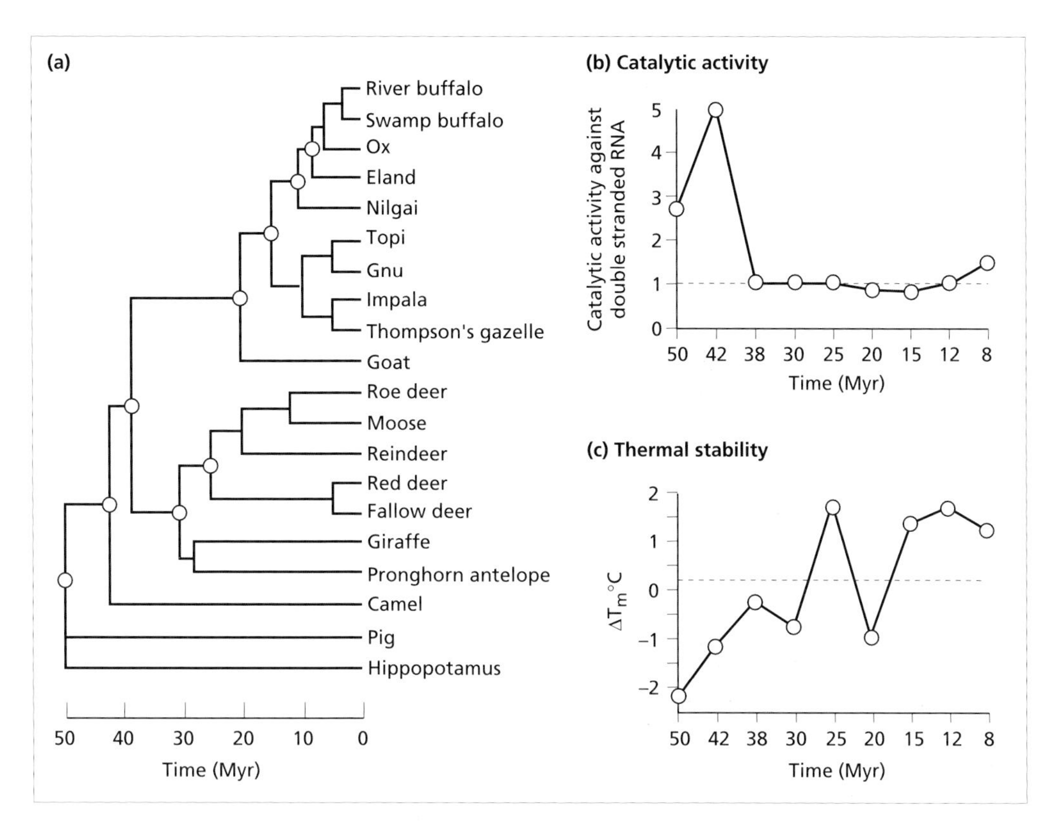

Fig. 5.26 Recreating ancient protein sequences. Left is a phylogeny of artiodactyl ribonucleases. The ancestral sequences marked with (○) were inferred using parsimony, then created in the laboratory and their catalytic and thermal properties measured with respect to present day ribonuclease (panels at right: values for present day RNAse indicated by dotted lines). After Jermann *et al.* (1995).

descendants. Specifically, the earliest artiodactyls had much higher levels of activity against double-stranded RNA and were slightly less thermally stable. These changes occurred at the same time that these mammals evolved foregut fermentation, which requires digestion of large amounts of RNA generated by fermenting bacteria in the foregut of these animals. Hence the reconstructed ancestral sequences suggest new insights into the evolution of both the gene and the physiology of the organisms themselves.

Box 5.4 Using parsimony reconstructions to test for saturation

Parsimony reconstructions by definition provide only minimum estimates of evolutionary change. Although this can be seen as a limitation, it does provide us with a useful lower bound—the actual amount of evolutionary change must

continued on p. 170

Box 5.4 *continued*

have been at least as much as inferred using parsimony. For example, we can use this knowledge to look for evidence of saturation of pairwise distances (both corrected and uncorrected) between sequences. Philippe *et al.* (1994) compared branch lengths inferred using parsimony with observed pairwise distances between sequences of the Cu–Zn superoxide dismutase (SOD) gene and 28S rRNA genes for a range of eukaryotes.

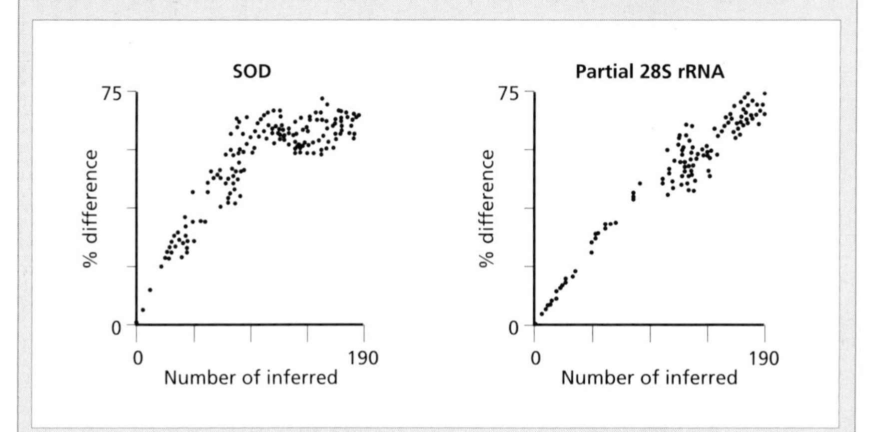

For the SOD sequences there is a noticeable levelling off in the relationship between sequence difference and parsimony branch lengths: a 65% difference can correspond to anything from 70 to 190 substitutions. By contrast the 28S rRNA sequences show only the beginnings of saturation. This approach can be used to detect saturation, and to evaluate the effectiveness of different methods of correcting for multiple hits.

5.4 Summary

1 Homology is the sharing of a feature due to direct inheritance from an ancestor. Independent evolution of the same feature is termed homoplasy.

2 Different genes may be descendants of the same copy of an ancestral gene (orthologous), or descendants of different copies (paralogous).

3 Sequences must be aligned to establish homology among nucleotide positions. The alignment can be sensitive to the relative weights assigned to different events, such as insertion and deletion events versus substitutions.

4 Multiple 'hits' (multiple substitutions at the same site) result in the observed number of sequences differences underestimating the actual number of substitutions that have taken place. A variety of methods has been proposed to correct for these unobserved substitutions.

5 Standard assumptions that most distance correction methods make, such as independence of substitutions at different sites, and that each site is equally likely to undergo substitution, are generally not true.

6 Evolutionary change can be measured on a phylogeny. The quality of these estimates can depend on how many sequences have been sampled.

7 Ancestral sequences can be reconstructed on an evolutionary tree.

5.5 Further reading

There is a huge literature on sequence alignment. William Day has made a comprehensive bibliography available on the Internet from the Classification Society of North America's home page (http://www.pitt.edu/~csna/). For a discussion of tree alignment see Hein (1990) and Wheeler (1996). Zharkikh (1994) gives a detailed (and demanding) mathematical account of various models of sequence evolution. See Goldman and Yang (1994) for more sophisticated models for analysing protein sequences. Tamura (1994) discusses using maximum likelihood to evaluate different models of substitution. Yang (1996) gives an excellent overview of methods for handling variation in rate of substitution across sites. A detailed discussion of reconstructing ancestral character states is given in Maddison and Maddison's (1992) book that accompanies the MacClade program.

Chapter 6
Inferring Molecular Phylogeny

6.1 Introduction

The task of molecular phylogenetics is to convert information in sequences into an evolutionary tree for those sequences. A great (and ever increasing) number of methods have been described for doing this, which raises the inevitable question of how to come to grips with this plethora of possibilities. Two ways which seem useful to us are either to divide the methods by how they handle data, or to divide them by the approach taken when building trees. Both these divisions can help us appreciate the differences among the various tree building techniques. Given the rapid development of methods in this field, we cannot hope to cover completely every method that has been proposed, nor would this necessarily be helpful. Our goal is to cover the major methods, and to show how they interrelate.

6.1.1 Kinds of data: distances versus discrete characters

This division is based on how the data are treated; distance methods first convert

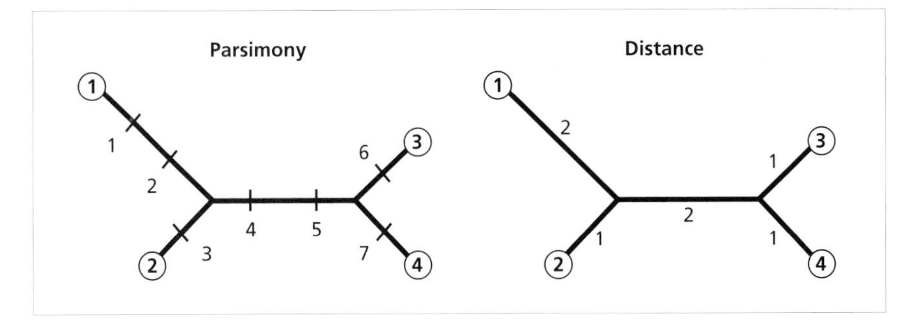

Fig. 6.1 A parsimony tree and a distance tree for the same sequence data. Note that both trees have the same topology and branch lengths, but that the parsimony tree identifies which site contributes to the length of each branch.

aligned sequences into a pairwise distance matrix, then input that matrix into a tree building method, whereas discrete methods consider each nucleotide site (or some function of each site) directly. As an example, consider the following sequences and corresponding (uncorrected) distance matrix:

sequences

sites

sequences	1	2	3	4	5	6	7
1	T	T	A	T	T	A	A
2	A	A	T	T	T	A	A
3	A	A	A	A	A	T	A
4	A	A	A	A	A	A	T

distances

sequences			
2	3		
3	5	4	
4	5	4	2
	1	**2**	**3**

sequences

The trees obtained by parsimony (a discrete method) and minimum evolution (a distance method) are identical in topology and branch lengths (Fig. 6.1). The parsimony analysis identifies seven substitutions and places them on the five branches of the tree. The distance tree apportions the observed distances between the sequences over the branches of the tree, and you can see that both methods arrive at the same estimates of the lengths of each branch. Under parsimony each of the seven sites requires one change, for a total of seven changes; if we sum the branch lengths on the distance tree we obtain the same value: $2 + 1 + 2 + 1 + 1 = 7$. Note, however, that the parsimony tree gives us the additional information of which site contributes to the length of each branch. Once we convert sequences into distances we lose this information. Furthermore, discrete methods allow us to infer the attributes of extinct ancestors, in this case extinct nucleotide sequences. These reconstructed ancestors can offer insights about molecular evolution (see Chapter 5).

6.1.2 Clustering methods versus search methods

Another way of dividing tree building methods is by the way they construct

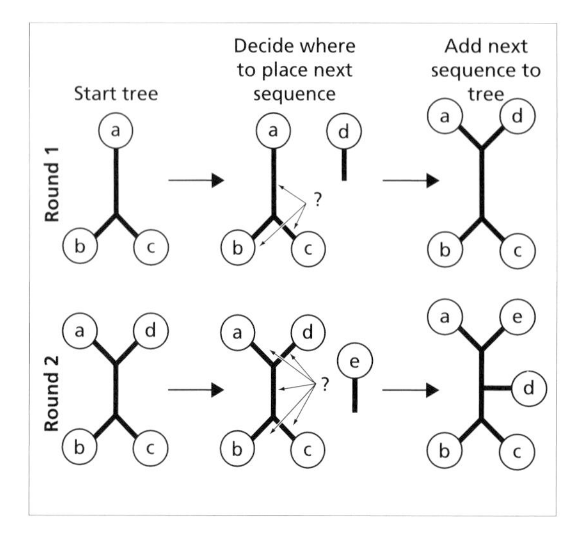

Fig. 6.2 An example of how a clustering method builds a tree. The tree is constructed by starting with the tree for three sequences, then adding each remaining sequence in turn until finally all sequences have been added.

trees. **Cluster methods** follow a set of steps (an algorithm) and arrive at a tree. For example, if we have five sequences we might start with three of them (remember that there is only one possible unrooted tree for three sequences) and decide where to place the fourth sequence. Given the resulting tree for four sequences, we then decide where to add the fifth and last sequence to our tree (Fig. 6.2).

Clustering methods have the advantage of being easy to implement, resulting in very fast computer programs. Furthermore they almost always produce a single tree. This combination of speed and an apparently unambiguous answer is naturally very appealing, and accounts for much of the sustained popularity of clustering methods. However, they have some severe limitations as analytical tools. The result obtained from simple clustering algorithms often depends on the order in which we add the sequences to the growing tree. For example, had we started with sequences b, d and e rather than a, b and c in Fig. 6.2 we might have arrived at a different tree. But the biggest limitation is that cluster methods do not allow us to evaluate competing hypotheses, they merely produce a tree. It may be that two different trees could explain our data equally, or nearly equally, well. Unless we have some way to measure the fit between tree and data, we will not be aware of this.

Tree-building methods in the second class use **optimality criteria** to choose among the set of all possible trees (Fig. 6.3). This criterion is used to assign to each tree a 'score' or rank which is a function of the relationship between tree and data (examples include maximum parsimony and maximum likelihood). Optimality methods have the great advantage of requiring an explicit function that relates data and tree (for example, a model of how sequences evolve). These methods also allow us to evaluate the quality of any tree, hence we can compare how well competing hypotheses of evolutionary relationship fit the data. The Achilles' heel of optimality methods is that they are computationally

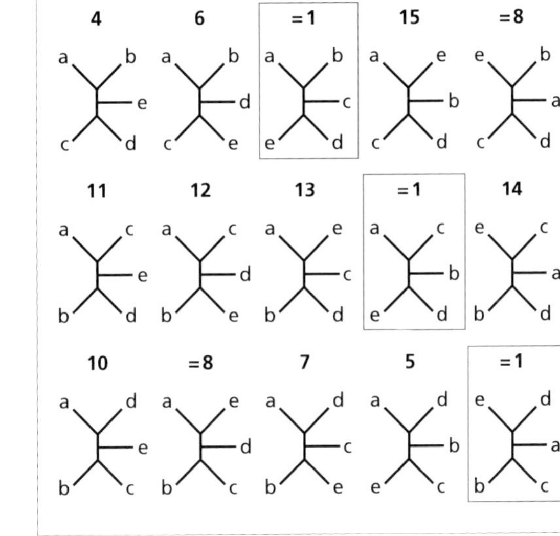

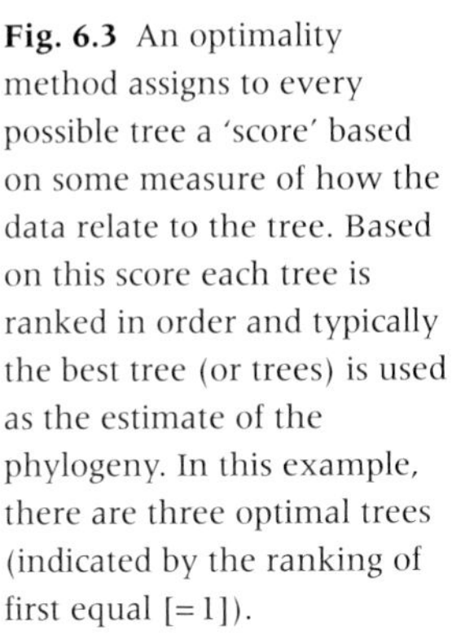

Fig. 6.3 An optimality method assigns to every possible tree a 'score' based on some measure of how the data relate to the tree. Based on this score each tree is ranked in order and typically the best tree (or trees) is used as the estimate of the phylogeny. In this example, there are three optimal trees (indicated by the ranking of first equal [=1]).

very expensive. An optimality method poses two problems that must be solved: firstly, for a given data set and a given tree, what is the value of the optimality criterion for that tree? For example, what is the minimum number of evolutionary events required to explain the observed data? Secondly, which of all the possible trees has the maximum value of this criterion? For example, which tree requires the fewest evolutionary events? The first problem is generally fairly straightforward; however, the second problem is rather more difficult as it belongs to a class of problems called NP-complete (see Box 6.1).

Box 6.1 NP-completeness and the problem of finding the best evolutionary tree

Despite the best efforts of mathematicians and computer scientists there is a set of problems called NP-complete (NP = non-deterministic polynomial) for which no efficient algorithms for their solution are known to exist. Members of this class of problems are all variants on the same problem, in that if one problem could be solved efficiently then all could be. The problem of finding the optimal evolutionary tree for a variety of criteria, including minimum evolution and maximum parsimony, is known to be NP-complete. In practical terms this means that for any reasonable number of sequences (e.g. more than 20) it is often impossible to guarantee that the optimal tree has in fact been found. Consequently, in many cases we must rely on heuristics (a euphemism for 'quick and dirty'). Trees found by such methods may turn out to be far from optimal, and the conclusions drawn from such trees may be somewhat suspect. A salutary example is the controversy over the geographic origins of human mitochondrial DNA in which different workers obtained quite different trees from different heuristic searches.

While for small numbers of sequences (e.g. no more than 20) it is often possible to find the optimal tree (or trees), in many cases this is not feasible, in which case we have to rely on heuristic methods. These are strategies designed to explore some subset of all the possible trees, in the hope that that subset will contain the optimal tree. A typical heuristic strategy is to start with a tree (which may be obtained using a simple clustering algorithm, or even chosen at random) and rearrange it, keeping any rearrangement that produces a better tree. Such algorithms are often called 'hill-climbing'. Imagine that you are in a valley and want to climb the highest hill, but you cannot see (some malicious text-book author has forced you to wear a blindfold). A reasonable strategy for climbing the highest hill would be to take a step, which may take you either a little bit downhill, a little bit uphill, or may keep you on a flat surface; if the step was downhill you are clearly not going up a peak, hence you retrace your step and try again; if the step is uphill you may be on the right track, and may wish to take another step in the same direction. If no step in any direction takes you up, but at least one takes you neither up nor down, then you are on a plateau. This plateau may be the top of the highest hill in the region, in which case you have succeeded; however, it may be merely a local high-point—being blindfolded you cannot tell.

Translating this analogy into the problem of tree searching, the 'hills' represent sets of locally optimal trees, and your steps are the tree rearrangements. Figure 6.4 shows two 'hills' or 'islands' of trees, and two possible starting trees, 1 and i. By rearranging the starting tree, and keeping only those trees that are better under our optimality criterion, we arrive at two different islands (a and b), one of which (b) is better than the other. The search that landed on island a would be unable to reach island b as to get there it would have to accept a suboptimal tree (tree 5) before reaching island b. If a set of possible trees contains more than one island then heuristic methods may land on a suboptimal island, and the optimal island goes undiscovered.

6.1.3 Subtree methods—assembling larger trees from smaller ones

Most algorithms for finding optimal phylogenetic trees use various techniques for rearranging trees to explore the phylogenetic 'landscape' comprising possible solutions (Fig. 6.4). If we can readily compute a score for each tree this approach is often reasonable. However, for some optimality criteria, such as maximum likelihood discussed below, computing a score for even a single tree is time consuming. Because the effectiveness of a heuristic search depends in part on how many trees we are willing to examine, this can hamper our search.

An alternative approach that has been used for both distance and discrete character methods is to divide the set of sequences into smaller sets and find the optimal tree for these subsets, which are then combined to form a larger tree. The smallest useful subset of an unrooted tree comprises four sequences (a **quartet**). For n sequences there are $\binom{n}{4} = n(n-1)(n-2)(n-3)/24$ possible

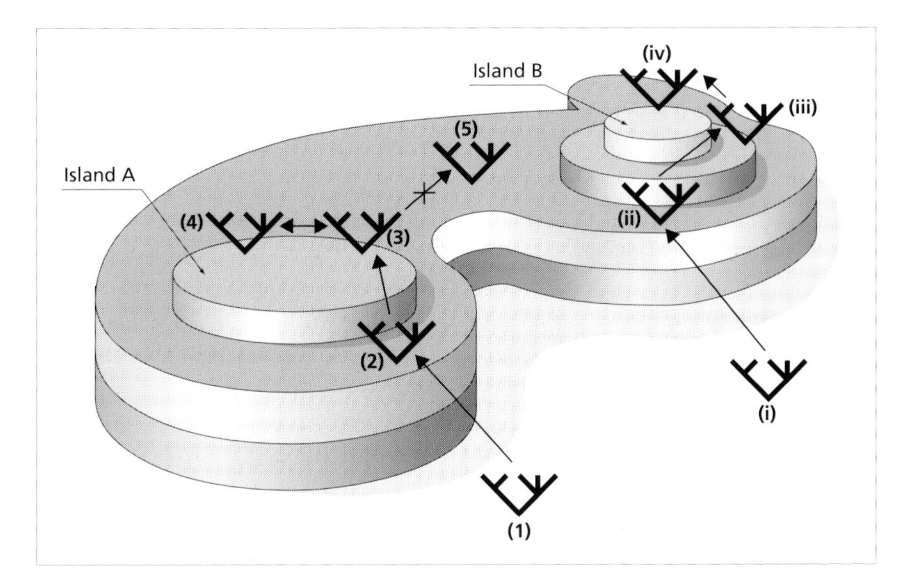

Fig. 6.4 Landscapes and the problem of islands of trees (locally optimal sets of trees). A hill climbing algorithm that started from tree 1 would succeed in finding trees 3 and 4 (which comprise island a), but would fail to discover that tree iv (island b) was even better, because to get to that tree it would have to cross a plateau of trees that were worse than trees 3 and 4. However, a search starting from tree i would succeed in finding the best tree. If the set of possible trees contains more than one island then heuristic methods may land on a suboptimal island, and the optimal island will not be discovered. After Maddison (1991).

quartets. Each quartet has three possible unrooted trees (see Table 2.1, p. 18) (Fig. 6.5). Each tree partitions the set of sequences {A, B, C, D} into two pairs: for example Q_1 corresponds to {A, B} and {C, D}. These two sets are each other's **neighbours**. Quartet-based tree building methods follow these two steps:

1 For each quartet identify which of the three possible trees is optimal for those four sequences.

2 Take all the four-sequence trees from step 1 and assemble them into a tree.

Given perfect data the second step is trivial as there will be only one tree that all the subtrees will agree on. However, given homoplasy it may be that not all subtrees can be combined into one tree, in which case we want the tree that

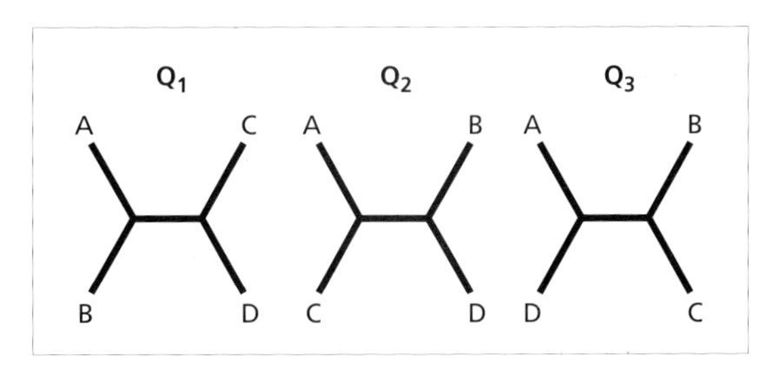

Fig. 6.5 The three possible trees for four taxa (a quartet).

accommodates the greatest number of the quartets. Finding this tree is, alas, also an NP complete problem (see Box 6.1) so it suffers from the same computational complexity that besets other methods. However, fast heuristic algorithms for assembling quartets into larger trees are available.

6.1.4 How do we compare different tree-building methods?

Based on the distinction we have made between tree-building methods that use distances versus those that use discrete characters, and methods that use a clustering algorithm versus those with an explicit optimality criterion, we can classify some commonly used methods (Fig. 6.6).

Note that methods from different classes may be related. For example, UPGMA can be considered a heuristic method for finding the best least squares ultrametric tree, and similarly neighbour joining is a heuristic method for estimating the minimum evolution tree.

Given the range of tree-building methods available, how can we decide which ones are better than others? David Penny and colleagues have suggested five desirable properties a tree-building method should have:

- **efficiency** (how fast is the method?);
- **power** (how much data does the method need to produce a reasonable result?);
- **consistency** (will it converge on the right answer given enough data?);
- **robustness** (will minor violations of the method's assumptions result in poor estimates of phylogeny?);
- **falsifiability** (will the method tell us when its assumptions are violated, i.e. that we should not be using the method at all).

Efficiency is effectively the time in which a computer program can find a tree using a given method. As we have seen above most, if not all, optimality

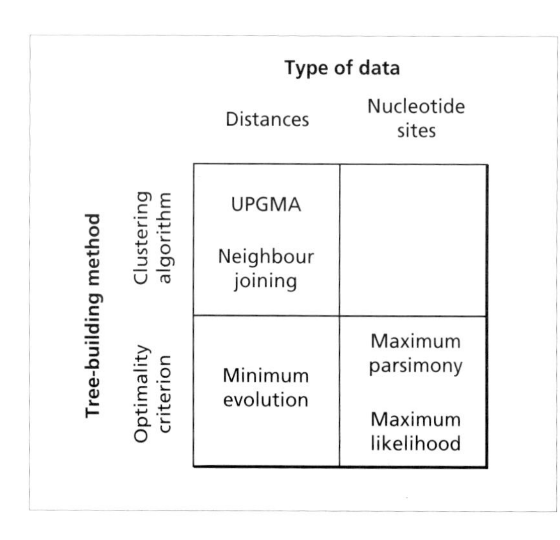

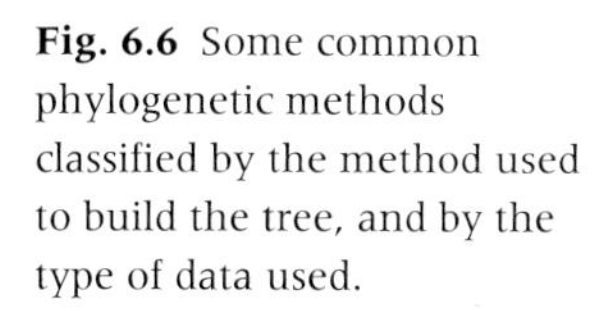
Fig. 6.6 Some common phylogenetic methods classified by the method used to build the tree, and by the type of data used.

methods are NP-complete (see Box 6.1) and hence efficient tree searching algorithms that guarantee to find the best tree are unlikely to be found. As a result, we have to rely on heuristics for all but the smallest problems. Some optimality criteria can be evaluated more quickly than others; for example, the most parsimonious set of nucleotide substitutions can be calculated orders of magnitude more quickly than the likelihood of the same tree giving rise to the same data. One practical consequence of this is that in the same period of time, heuristic searches using parsimony can explore a much larger set of trees than a search using likelihood. Subtree methods (section 6.1.3) may be used when the optimality criterion is time consuming to compute.

The *power* of a method is a measure of how much data we need to collect before we can be reasonably sure of arriving at the correct result. A method might be theoretically very appealing, but if it requires huge numbers of sites to be sequenced then it may not be of much practical use. Another consideration is whether the method will converge on the true tree as we add more data. This desirable property is *consistency*; an inconsistent method would fail even if we kept feeding it more data.

All tree-building methods make (implicit or explicit) assumptions about the evolutionary process. Violation of these assumptions may result in a method returning a poor estimate of phylogeny, for example a method that assumed a molecular clock when there was none may be very misleading about evolutionary relationships. The sensitivity of a method to violations of its underlying model is a measure of its *robustness*. Ideally, we would like to know whether these violations are sufficient to rule out a particular model, that is, the method is *falsifiable*. A falsifiable method that assumed a molecular clock when there was none would allow us to test the clock assumption; if that assumption was found wanting then we would abandon that method and use another that did not make the clock assumption.

The ideal tree-building method would meet all five criteria, but such a method does not exist, nor is it likely to. All current methods emphasise one or more of these criteria at the expense of the remainder. For example, UPGMA is extremely fast (efficient) but is not robust to its implicit assumption of a molecular clock, whereas maximum likelihood is consistent (with respect to its chosen model of evolution) but computationally very intensive. In this chapter we will evaluate each method based on these five criteria.

6.2 Distance methods

Distance methods are based on the idea that if we knew the actual evolutionary distance between all members of a set of sequences, then we could easily reconstruct the evolutionary history of those sequences. This follows from the relationships between distances and trees outlined in Chapter 2: evolutionary distance is a tree metric and hence defines a tree. In practice, however, distances are rarely, if ever, exactly tree metrics, and hence one class of 'goodness of fit' methods seeks the metric tree that best accounts for the 'observed' distances (i.e. the pairwise distances calculated between the sequences). The second class of method seeks the tree whose sum of branch lengths is the minimum ('minimum evolution').

6.2.1 Goodness of fit measures

The goodness of fit F between observed distance d_{ij} and tree distances p_{ij} (see Chapter 2) for each pair of sequences i and j is given by

$$F_{\alpha} = \sum_{1 \le i < j \le n} \left| d_{ij} - p_{ij} \right|^{\alpha} \tag{6.1}$$

where α can take various values. If $\alpha = 1$ then the criterion is Farris's f statistic, if $\alpha = 2$ then F is the least-squares-fit criterion. As an example of the latter criterion, consider the distance matrix between hominoids shown in Table 6.1 and the corresponding additive tree (Fig. 6.7).

Table 6.1 Kimura 2-parameter distances between hominoid sequences (above diagonal) and tree distances obtained by least squares (below the diagonal) for the tree shown in Fig. 6.7. Tree distances larger than the observed distances are shown in bold, tree distances smaller than the observed are shown in italics.

	Human	Chimp	Gorilla	Orang-utan	Gibbon
Human	–	0.09190	0.1083	0.1790	0.2057
Chimp	0.0919	–	0.1134	0.1940	0.2168
Gorilla	*0.1068*	**0.1151**	–	0.1882	0.2170
Orang-utan	**0.1816**	*0.1898*	**0.1893**	–	0.2172
Gibbon	**0.2078**	*0.2160*	*0.2155*	0.2172	–

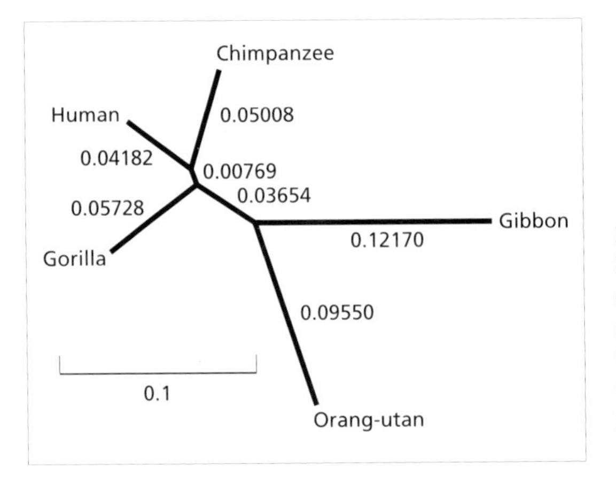

Fig. 6.7 Additive tree for hominoid mtDNA sequences showing branch lengths computed using least squares. The pairwise tree distances for this tree are given in the lower left triangle of Table 6.1.

The observed and tree distances are in close agreement, but note that some tree distances are larger than the observed, and some are less than the observed. While tree distances can be larger than observed distances due to homoplasy (i.e. multiple substitutions), the reverse is counter intuitive as tree distances less than those we observe between sequences imply that less evolutionary change took place than we actually observe. This contradiction (discussed further on p. 186) has led some workers to abandon the use of distance methods, or to use other methods of computing branch lengths from distances (see below).

In the example just given we were fitting an additive tree with $(2n - 3)$ branches to $\binom{n}{2} = n(n - 1)/2$ pairwise distances. However, measures of fit can also be applied to ultrametric trees, in which case there are $(n - 1)$ independent branch lengths to be estimated. That there are fewer parameters to be estimated for an ultrametric tree, which has one more branch than an unrooted additive tree, may seem paradoxical, but this is due to the constraint that all terminal taxa are equidistant from the root. This constraint is equivalent to postulating a 'molecular clock' (a concept which is discussed in detail in Chapter 7). For example, in the ultrametric tree shown in Fig. 6.8 the branches e_1 and e_2 must be of the same length as they both represent the same interval of time (i.e. the time since sequences A and B diverged).

If the distances between sequences are ultrametric then both the ultrametric and additive trees would fit those distances equally well (indeed, the additive tree would be identical to the ultrametric tree). The greater the departure from the molecular clock, the more the data will depart from being ultrametric, and the greater the difference in fit between additive and ultrametric trees, the additive tree always being a better fit as it lacks the constraints imposed by the ultrametric tree. This property is employed by some tests of

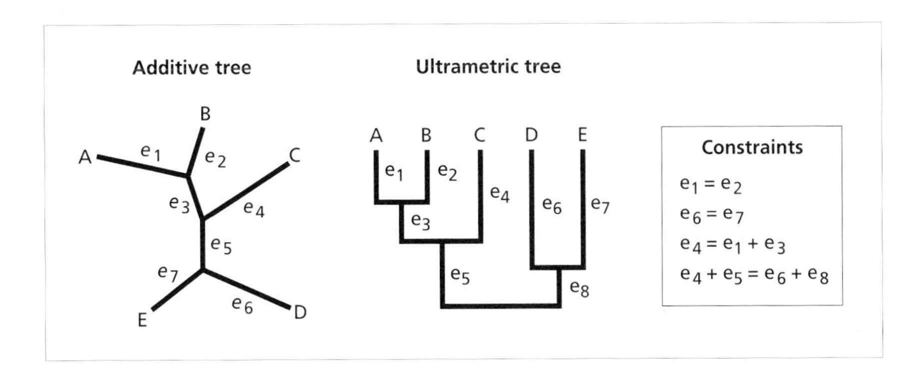

Fig. 6.8 Additive and ultrametric trees for the same sequences. Both trees specify the same cladistic relationships among the taxa, but whereas the additive tree has $(2n - 3 = 7)$ independent branches the ultrametric tree has only $(n - 1) = 4$ because some branches are constrained to be equal to others, or to combinations of others. This constraint is equivalent to a molecular clock.

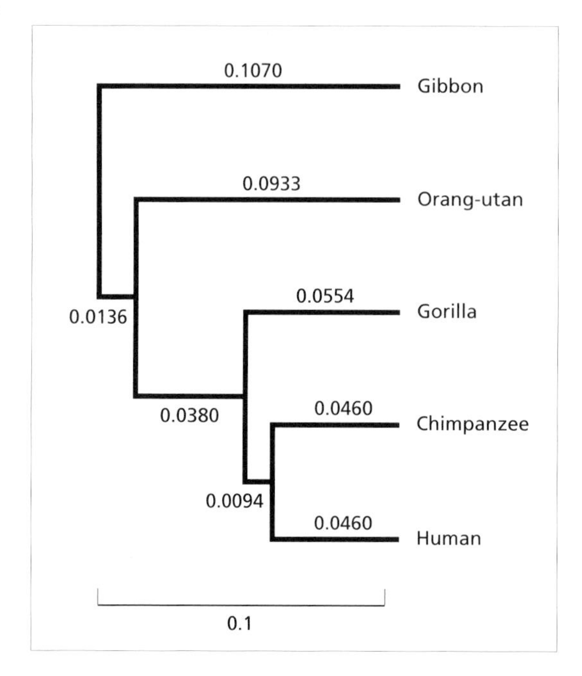

Fig. 6.9 Ultrametric tree for hominoid mtDNA with least squares branch lengths computed from the Kimura 2-parameter distances shown in the upper right triangle of Table 6.1. Compare with the additive tree in Fig. 6.7.

the molecular clock, such as the likelihood ratio test described later in this chapter (see p. 198).

Figure 6.9 shows an ultrametric tree for the same distances used to compute the additive tree in Fig. 6.7. The two trees specify the same cladistic relationships; however, the ultrametric tree also supplies a root (between the gibbon and the great apes including humans) and the branch lengths are slightly different. For instance, the additive tree apportions more evolutionary change to chimpanzee mtDNA than to human mtDNA, whereas in the ultrametric tree these two sequences are constrained to have exactly the same amount of evolutionary change since their divergence.

6.2.2 Minimum evolution

Given an unrooted metric tree for n sequences there are $(2n - 3)$ branches, each with length e_i. The sum of these branch lengths is the **length** L of the tree:

$$L = \sum_{i=1}^{2n-3} e_i \tag{6.2}$$

The minimum evolution tree (ME) is the tree which minimises L. This method is similar in spirit to parsimony, which we shall discuss below; however, the length in this case is computed from the pairwise distances between the sequences rather than from the fit of individual nucleotide sites to a tree. To use this method we need to be able to compute the length of any tree. However, these lengths are not always biologically valid (see p. 186).

Linear programming

Linear programming is a widely used mathematical technique to find the optimal solution to a problem, given a set of constraints (most spreadsheet programs can perform linear programming). When applied to finding the length of a tree, the two constraints that we must satisfy are that all the branch lengths are non-negative (i.e. $e_i \geq 0$ for all i), and that for any pair of sequences the tree distances are never less than the observed distances (i.e. $p_{i,j} \geq d_{i,j}$ for all i and j). Table 6.2 gives the observed pairwise distances for the hominoid DNA sequences, and the corresponding tree distances obtained for the tree shown in Fig. 6.10. The tree distances are in good agreement with the observed distances. Note that where the tree distances differ from the observed differences (e.g. between human and gibbon) the tree distances are always greater than the observed distances, satisfying the condition that the amount of evolutionary change inferred cannot be less than that which we have observed.

Table 6.2 Observed pairwise distances (p) between hominoid sequences (above diagonal) and tree distances computed by linear programming (below diagonal). Tree distances that differ from the observed are marked in bold. The corresponding tree is shown in Fig. 6.10.

	Human	Chimp	Gorilla	Orang-utan	Gibbon
Human	–	79	92	144	162
Chimp	79	–	95	154	169
Gorilla	92	**102**	–	150	169
Orang-utan	144	154	150	–	169
Gibbon	**163**	**173**	169	169	–

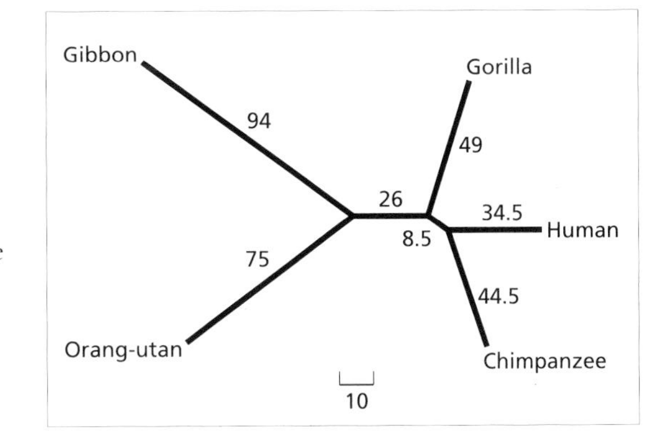

Fig. 6.10 Minimum evolution tree for the hominoid sequences with the branch lengths computed from the observed pairwise distances using linear programming. The total length of this tree is 331.5.

Least squares

More commonly, the branch lengths of the minimum evolution tree are estimated using least-squares methods. The branch lengths are estimated in the same way as for goodness of fit measures; however, rather than compare the fit of the observed distances the least squares branch lengths are added together to give the length of the tree.

6.2.3 Algorithms for finding distance trees

As with discrete characters, finding the optimal distance tree is computationally difficult. For small numbers of taxa exact methods can be used; for larger numbers we must rely on heuristics, of which there are a large number, some of which have largely been used only with distance data. We mention three here.

Neighbourliness

Neighbourliness makes use of the additive condition (see Chapter 2) where, given any four sequences, it is possible to label them A–D such that the distances between them obey this condition:

$$d(\mathrm{A, B}) + d(\mathrm{C, D}) \leq d(\mathrm{A, C}) + d(\mathrm{B, D}) = d(\mathrm{A, D}) + d(\mathrm{B, C}) \qquad (6.3)$$

Recalling from Chapter 2 that the additive condition defines a tree, this equation corresponds to the tree Q_1 shown in Fig. 6.5. On this tree A and B, and C and D are neighbours with respect to each other, hence the name of the method. Given data for n sequences that are perfectly additive (i.e. perfectly tree-like) we could use equation (6.3) to identify the additive subtrees for the data and from them construct the phylogeny for all n sequences. For actual data additivity may not hold, in which case we could seek the tree that has the greatest number of quartets for which equation (6.3) holds.

Neighbour joining

Neighbour joining (NJ) is a widely used method for tree building which combines computational speed with uniqueness of result—most implementations give a single tree. These two attributes (i.e. getting one tree, fast) have made it seem very appealing. Neighbour joining is a clustering method rather than an optimality method, and hence suffers from the limitation that it does not optimise a criterion of fit between tree and data. However, it is a good heuristic method for estimating the minimum evolution tree. One strategy for finding the ME tree is to first compute the NJ tree, then see if any local rearrangement of the NJ tree produces a shorter tree. Note that this strategy is not guaranteed to find the ME tree, for the reasons given in section 6.1.2. However, in practice the NJ tree is often the same or very similar to the ME tree.

Unweighted pair group method with arithmetic means (UPGMA)

UPGMA is one of the few tree-building methods that constructs an ultrametric tree (see Chapter 2). In an ultrametric tree all the tips are equidistant from the root of the tree, which is equivalent to assuming a molecular clock (see Chapter 7). Indeed, UPGMA is best viewed as a heuristic for finding the least squares ultrametric tree for a distance matrix.

Box 6.2 Are distance methods 'phenetic?'

A criticism often levelled at distance methods is that they are 'phenetic', and therefore inferior to other, 'phylogenetic', methods. Our view is that phenetics is a philosophy of systematics that eschews phylogenetic analysis as a search for the unknowable (because, with very few exceptions, no one has observed a phylogeny unfold over time). Hence phenetics favours summarising observed features of organisms using 'maximally informative' or 'predictive' classifications (which may or may not be actual phylogenies; if they are, this is merely a happy accident). Typically such classifications were constructed by computing a measure of similarity between organisms, then doing a cluster analysis on those similarities. Phenetics was rejected by most systematists by the late 1970s, largely because: (1) estimating evolutionary trees was thought to be a more worthwhile goal; (2) techniques (such as cladistics) existed which claimed to be able to estimate such trees; and (3) phenetics was unable to provide unambiguous reasons for choosing between a plethora of measures of similarity and clustering algorithms. However, if the goal is to reconstruct phylogeny then there are clear reasons for favouring one measure of similarity over another, and for favouring particular methods for converting distances into trees. This does not mean that all criticisms of distances raised by opponents of phenetics do not apply in this phylogenetic context, rather, we suggest that the criticism that a method is 'phenetic' is not, by itself, meaningful.

6.2.4 Objections to distance methods

In considering distance methods we should distinguish between methods for constructing the trees and methods for obtaining the distances. If the estimates of evolutionary distance are poor then the performance of a distance method may be adversely affected, which may not be a true reflection of the merits of the tree building method itself. Furthermore, many existing computer programs for finding distance trees lack the sophistication of equivalent programs for discrete data, making it difficult to determine if there is more than one optimal tree for example, or how many trees are nearly as good as the optimal tree. Again, this is a limitation of existing software, not distance methods as such. Hence in this section we consider objections to the use of distance data *per se*.

The major objections to distance methods are:

- summarising a set of sequences by a pairwise distance matrix loses information;
- branch lengths estimated by some distance methods may not be evolutionarily interpretable.

Distances lose information

If the original data are in the form of distances, such as those obtained from DNA hybridisation studies, then we have no other option but to use distances. However, if we have the sequences we have the option of analysing them directly or converting them into distances (Chapter 5). Doing the latter may lose information. For example, once converted to distances, we cannot trace the evolution of individual sites, or categories of sites on a tree, we have only an overall estimate of the relationship between tree and data. This can be clearly seen in Fig. 6.1 where the parsimony analysis of the original sequence data allows us to locate where in the tree each site changes, whereas the distance tree merely tells us how much change occurs along each branch.

Another way of expressing the loss of information inherent in converting sequences to distances is to compare the number of possible distance matrices with the number of possible data sets—the latter greatly exceed the former.

Uninterpretable branch lengths

It is possible to reconstruct branch lengths from distances that are mathematically valid but biologically difficult to interpret. For example, on p. 183 we used linear programming to fit observed distances between hominoid sequences to a tree. The length of that tree was 331.5 substitutions. This value raises two problems, the first being how to interpret 0.5 substitutions—a nucleotide can either be substituted or not; we cannot have half a substitution. One response to this difficulty is that branch lengths can represent two quite different quantities: the *expected* amount of evolutionary change and the *actual* or realised amount of change. For example, if we have a branch in a tree that corresponds to an interval of 1 Myr, and the rate of nucleotide substitution is 2.5% per Myr, then given a sequence of 100 bases in length we expect on average 2.5 substitutions to have occurred. Obviously, we cannot have half a substitution, so in reality some whole number of substitutions would have occurred, say 2 or 3 (or some other whole number). Under this interpretation, a branch length of 2.5 is entirely reasonable.

The second problem is that the figure of 331.5 substitutions is biologically not possible for this data set. As we shall see below when we discuss maximum parsimony, the minimum number of substitutions that must have occurred in the evolution of these particular sequences is 353. The tree length obtained by linear programming is internally consistent—but biologically impossible. In much the same way, when estimating branch lengths using least squares we can obtain tree distances that are less than the distances observed between the sequences (e.g. Table 6.1). Again, we cannot have less evolutionary change than we actually observe in the data.

6.3 Discrete methods

In contrast to distance methods, discrete methods operate directly on the sequences, or on functions derived from the sequences, rather than on pairwise distances. Hence they endeavour to avoid the loss of information that occurs when sequences are converted into distances. The two major discrete methods are **maximum parsimony** (MP) and **maximum likelihood** (ML). Maximum parsimony chooses the tree (or trees) that require the fewest evolutionary changes. Maximum likelihood chooses the tree (or trees) that of all trees is the one that is most likely to have produced the observed data. We also consider two methods that treat the data not as individual sites but as 'splits' (see Chapter 2): **spectral analysis** and **split decomposition**.

6.4 Maximum parsimony

The data for maximum parsimony comprise individual nucleotide sites. For each site the goal is to reconstruct the evolution of that site on a tree subject to the constraint of invoking the fewest possible evolutionary changes. Consider the following four sequences:

1 ATATT
2 ATCGT
3 GCAGT
4 GCCGT

Figure 6.11 shows the unrooted tree ((1,2),(3,4)) and two possible reconstructions of the evolution of the first site on that tree. In each reconstruction we have postulated which nucleotide the two internal nodes (ancestral sequences)

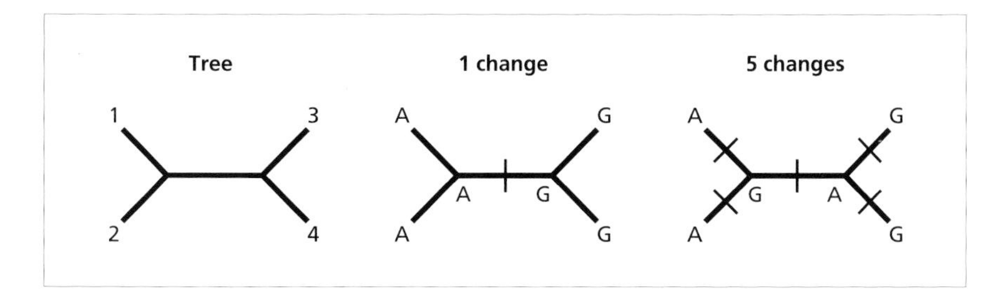

Fig. 6.11 An unrooted tree for four sequences 1–4 and two possible reconstructions of the evolution of the same site on that tree. One reconstruction requires one change, the other requires five. The former is more parsimonious.

possess. On branches where the nucleotide at each end differs we postulate a substitution. Note that our reconstruction does not specify where along that branch the change took place, merely that it did. Under the principle of parsimony the reconstruction that invokes a single substitution is preferred over the less parsimonious reconstruction that requires five substitutions. We can now find the most parsimonious reconstructions for each of the remaining sites (Fig. 6.12). Note that for the third site there are two equally parsimonious reconstructions, both requiring two steps.

The total number of evolutionary changes on a tree (often referred to as the tree's **length**) is simply the sum of the number of changes at each site. Hence, if we have k sites, each with a length of l, then the length L of the tree is given by

$$L = \sum_{i=1}^{k} l_i \tag{6.4}$$

Hence the tree in Fig. 6.11 has a length of 1 + 1 + 2 + 1 + 0 = 5 steps. The other two trees have lengths of six and seven steps (Table 6.3), so the tree ((1,2),(3,4)) is the most parsimonious tree.

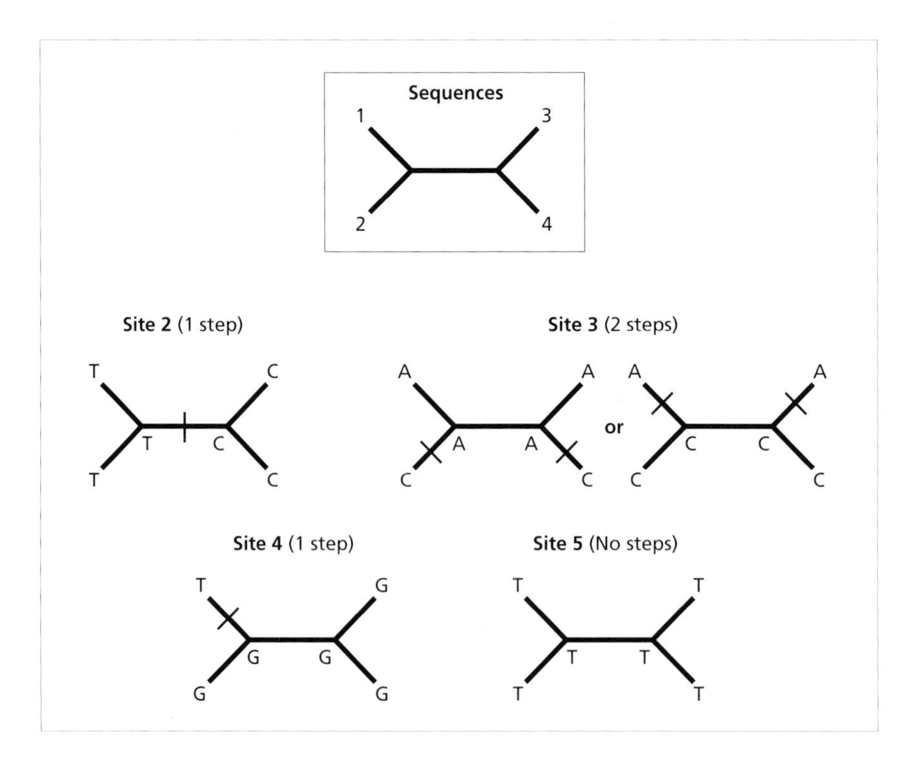

Fig. 6.12 Most parsimonious reconstructions for sites 2–5. There are two equally parsimonious reconstructions for site 3.

Table 6.3 Changes required for each site to fit the three possible trees for four sequences.

Tree	Sites 1	2	3	4	5	Total
((1,2),(3,4))	1	1	2	1	0	5
((1,3),(2,4))	2	2	1	1	0	6
((1,4),(2,3))	2	2	2	1	0	7

Notice that the number of steps at sites 4 and 5 are the same for all three trees, hence these sites do not discriminate between the three alternative trees. Such sites are termed **phylogenetically uninformative**; sites that are invariant (each sequence has the same nucleotide at the same position) or sites where only one sequence has a different nucleotide are examples of such sites.

6.4.1 Generalised parsimony

When counting changes in the above example, each substitution was accorded the same 'cost', so that the transition A → G counts as one step, as does the transversion A → C. Another way of representing this is by means of a step matrix (Fig. 6.13).

In the first model, which we have implicitly used above, each substitution has equal cost, so each cell in the corresponding step matrix has the entry '1'. Note that the cost of not changing is zero; if the nucleotide at each end of a branch is the same then we infer that no change has occurred.

The second model assigns a higher cost to transversions, in this case twice the cost of a transition. This is equivalent to saying the transversions are rarer

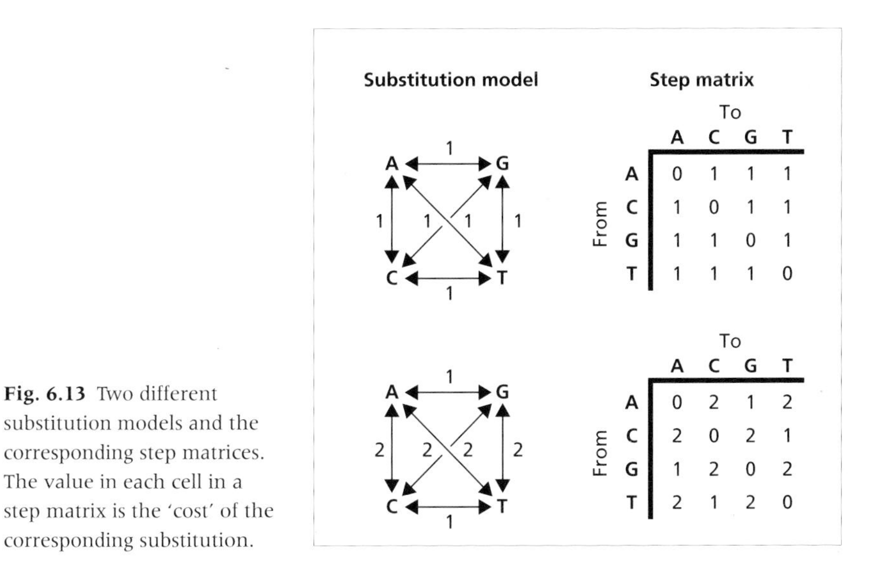

Fig. 6.13 Two different substitution models and the corresponding step matrices. The value in each cell in a step matrix is the 'cost' of the corresponding substitution.

than transitions, and therefore may be more reliable indicators of phylogeny. If we were to use this step matrix on the data discussed above, the number of steps for sites 1, 2 and 4 would not change as they all involve transitions. However, site 3 requires an A–C transversion. Trees 1 and 3 require this transversion to occur twice, hence for those sites the total cost for site 3 is 4. Tree 2 requires a single transversion with cost 2. Under this model tree 2 is now just as parsimonious as tree 1.

Generalised parsimony is very flexible in that we can specify a wide range of substitution models (see Chapter 5) using step matrices. However, it is not immediately obvious what cost to assign the different kinds of substitutions. We shall return to this point below.

6.4.2 Weighted parsimony

Not all sites may be equally phylogenetically useful. Sites that evolve very rapidly are likely to become quickly saturated so that the trace of history is overprinted and lost. Indeed, such sites may even be positively misleading. In contrast, sites that are conservative and evolve very slowly are less likely to suffer the effects of saturation. The relative value of different sites can be reflected by the **weight** w given to each site. Hence the length of a tree becomes

$$L = \sum_{i=1}^{k} w_i l_i \tag{6.5}$$

The greater the phylogenetic value of the character, the greater the weight we might wish to assign it.

As with step matrices, weighting adds flexibility to parsimony analyses, but raises the question of exactly what weights to assign different sites.

6.4.3 Justification for parsimony

Among parsimony's advantages are that it is relatively straightforward to understand, it apparently makes few assumptions about the evolutionary process, it has been extensively studied mathematically, and some very powerful software implementations are available. However, the justification for choosing the most parsimonious tree as the best estimate of phylogeny is the subject of considerable controversy. Essentially two main arguments have been presented. The first is that parsimony is a methodological convention that compels us to maximise the amount of evolutionary similarity that we can explain as homologous similarity, that is, we want to maximise the similarity that we can attribute to common ancestry. Any character which does not fit a given tree requires us to postulate that the similarity between two sequences shown by that character arose independently in the two sequences—the similarity is due to homoplasy not homology. Hypotheses of homoplasy (such as convergence or parallel evolution) may be judged *ad hoc* in that they are attempts to explain why data do not fit a particular hypothesis. The most parsimonious tree minimises the number of *ad hoc* hypotheses required, and for that reason is preferred.

The second view is that parsimony is based on an implicit assumption about evolution, namely that evolutionary change is rare. Rarity of change implies that the tree that minimises change is likely to be the best estimate of the actual phylogeny. Under this view, parsimony may be viewed as an approximation to maximum likelihood methods (discussed below), and indeed it was in this context that parsimony methods were first proposed by Edwards and Cavalli-Sforza.

The debate between advocates of these two positions has been explored in great detail by Elliot Sober (1988), to which the reader is referred for more details. Of the two positions, the latter has the advantage that it is possible to explore the circumstances under which parsimony will fail to reconstruct the correct phylogeny, and to develop a framework in which parsimony can be compared with other methods.

6.4.4 Objections to parsimony

The principal objection to parsimony is that under some models of evolution it is not consistent (see section 6.1.4), that is, even if we add more data it is possible to obtain the wrong tree. The classic scenario where this might happen has been termed 'long branches attract'.

In the tree shown in Fig. 6.14, there are two unrelated sequences that are each separated from their ancestor by a long edge. In order for parsimony to recover the correct tree ((A, B),(C, D)) there must be more sites supporting

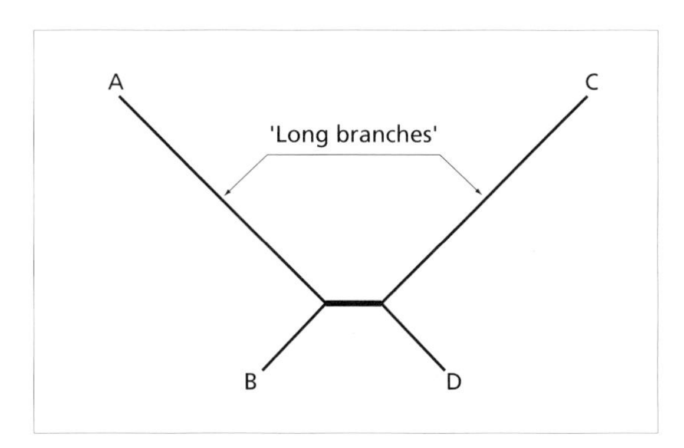

Fig. 6.14 The problem of 'long branches attracting'. The edges leading to sequences A and C are long relative to the other branches in the tree, reflecting the relatively greater number of substitutions that have occurred along those two edges. If the difference in numbers of substitutions is sufficiently great, there may be more apparent support for the split {{A, C},{B, D}} than for the split {{A, B},{C, D}}.

the split {{A, B},{C, D}} than either of the two other possible internal splits: {{A, C},{B, D}} and {{A, D},{B, C}}. If the internal edge is short relative to the long terminal edges then by chance alone A and C may acquire the same nucleotide independently. These convergences may outweigh the sites changing along the internal edge, and hence by the parsimony criterion the tree ((A, C), (B, D)) would be favoured. Under some circumstances, no matter how much data is added parsimony will obtain the wrong tree, hence it is inconsistent.

The problem of long branches attracting is most likely to occur when rates of evolution show considerable variation among sequences, or where the sequences being analysed are quite divergent. One strategy to reduce the effects of long edges is to add sequences that join onto those edges thus breaking them up (see Box 6.3).

Box 6.3 Big trees and long branches

The problem of 'long branch attraction' may be severe in the case of trees for four sequences. Paradoxically, it may be less of a problem for large phylogenies. The diagram below shows a tree for 228 angiosperm (flowering plant) 18S ribosomal DNA sequences.

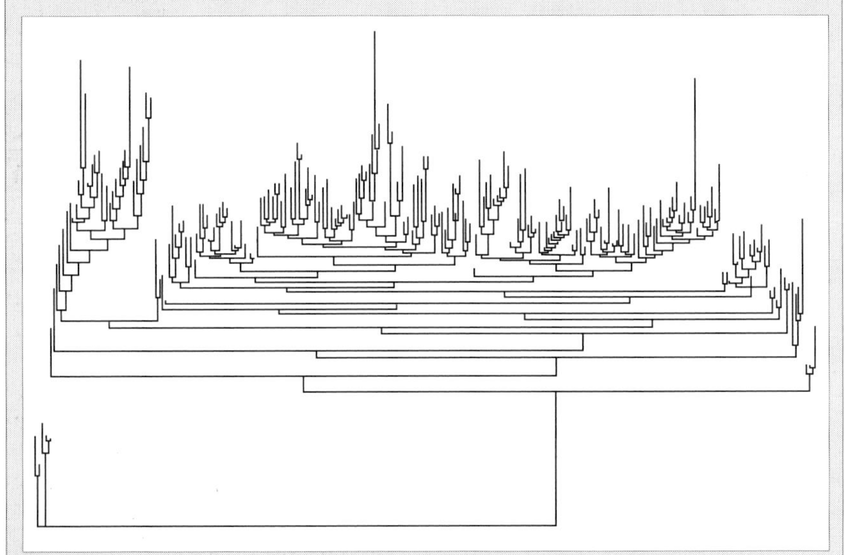

This tree contains several prominent long edges. To test our ability to recover this tree Hillis (1996) generated artificial data sets by simulating the evolution of DNA sequences of various lengths on this tree, then analysed those data sets using a range of tree-building methods (this approach of simulating data on a tree is called 'parametric bootstrapping' and is discussed in section 6.8.3). Parsimony and neighbour joining both did surprisingly well, requiring only 5000 nucleotides to recover the original tree. The crucial problem posed by

continued

Box 6.3 *continued*

'long branches' is not so much the *length* of the branches, but that the *same* substitutions have occurred along the two branches, fooling tree-building methods into joining them together. The probability of such covarying homoplasies is less if the branches are widely separated. Intuitively, if the long branches are close together then their ancestral states probably closely resembled each other, whereas branches far apart in a tree likely had very different ancestral states. It is more likely that similar substitutions will occur in two lineages that were similar to start with than in two lineages that were very different, Hence, in small trees (such as the four-taxon trees much used in simulations—see section 6.7.3), the rapid rate of evolution in the two lineages that were initially quite similar is more likely to yield the same substitutions in the two edges than in large trees.

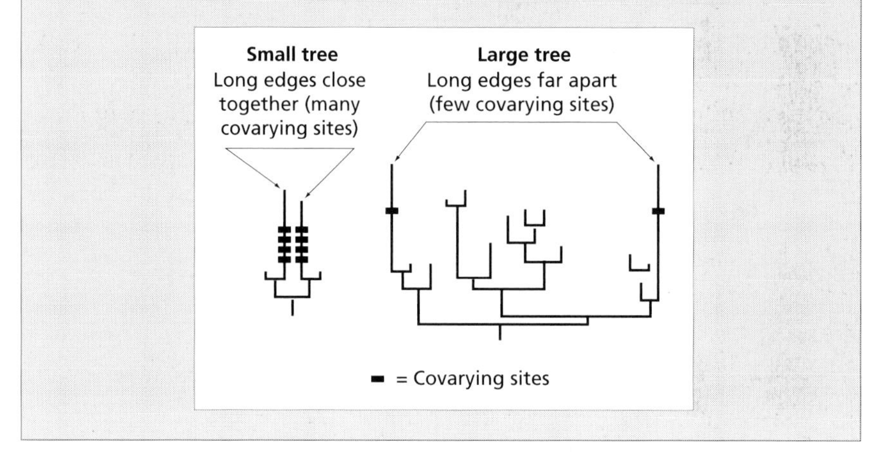

6.5 Maximum likelihood

Given competing explanations for a particular outcome, which explanation should we choose? The principle of **likelihood**, which we encountered in Chapter 5, suggests that the explanation that makes the observed outcome the most likely (i.e. the most probable) occurrence is one to be preferred. Put more formally, if given some data D, and a hypothesis H, the likelihood of that data is given by

$$L_D = \Pr(D|H) \tag{6.6}$$

which is the probability of obtaining D given H. In the context of molecular phylogenetics D is the set of sequences being compared, and H is a phylogenetic tree, hence we want to find the likelihood of obtaining the observed sequences given a particular tree. The tree that makes our data the most probable evolutionary outcome is the **maximum likelihood estimate** of the phylogeny.

Even though equation (6.6) is couched in terms of probability, it is important to distinguish between likelihoods and probabilities. Probabilities sum to 1, whereas likelihoods do not. Given a tree and a model of sequence evolution, we could in principle work out the probability of obtaining all possible data sets. The sum of these probabilities would equal 1. However, here we are interested in the probability of obtaining just one of those possible data sets (the one we actually observed). The likelihood is not the probability that the tree is the true tree, rather it is the probability that the tree has given rise to the data we collected.

Maximum likelihood is an appealing method of inference as it can incorporate explicit models of sequence evolution, and also permits statistical tests of evolutionary hypotheses.

6.5.1 Models, data and hypotheses

Maximum likelihood requires three elements, a model of sequence evolution, a tree and the observed data. Note that the tree specifies both the topology (branching order) and the branch lengths. In common with other optimality methods, maximum likelihood methods of tree building must solve two problems: (1) for a given tree topology, what set of branch lengths makes the observed data most likely (i.e. what is the maximum likelihood value for that tree); and (2) which tree of all the possible trees has the greatest likelihood.

Suppose we have two sequences, 1 and 2, separated by an average of d substitutions per site. This distance $d = \mu t$, where μ is the mutation rate and t is time.

$$1\bullet \overset{d}{- - - - -} \bullet 2$$

Given a model of substitution (see Chapter 5) for each site we can compute the probability $P_{ij}(d)$ that two sequences separated by d would have nucleotides i and j, respectively (put another way, if sequence 1 has nucleotide A, $P_{AG}(d)$ is the probability that sequence 2 would have nucleotide G). This probability is the likelihood of that site having that pair of nucleotides in the two sequences. The log likelihood of obtaining the observed sequences is the sum of the log likelihoods of each individual site:

$$\ln L = \sum_{i=1}^{k} \ln L_i \tag{6.7}$$

where k is the number of sites. Different values of d will have different likelihoods; the value that maximises equation (6.7) is the maximum likelihood estimate of d.

To compute the likelihood of a given tree we need to find the maximum likelihood estimate for the tree's branch lengths. This is equivalent to the problem just discussed, but with one important difference: in all branches at

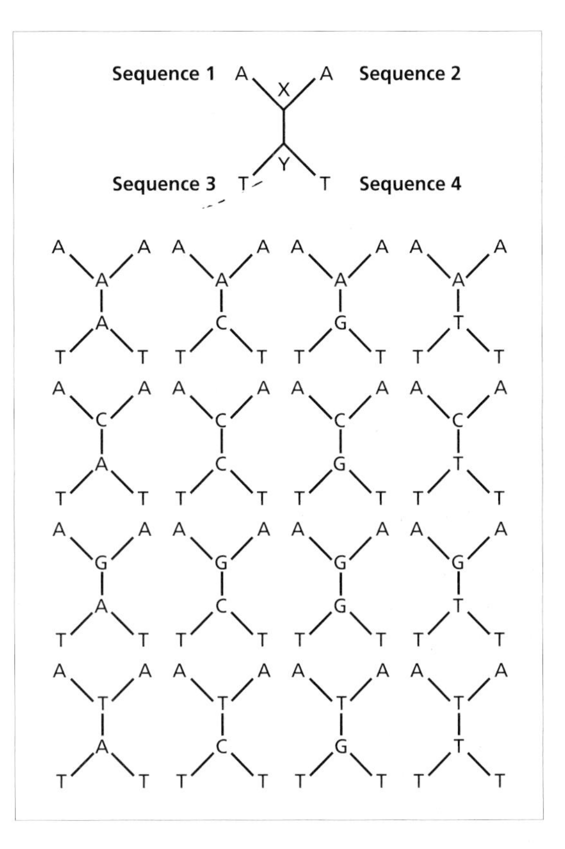

Fig. 6.15 The 16 different combinations of ancestral states for a single site on a four-taxon tree.

least one, if not both of the nodes connected by the branch are internal (ancestral) nodes with unknown sequences. Hence to compute the likelihood of a given site we need to compute the probabilities of the observed states at that site given all possible combinations of ancestral states. Figure 6.15 shows the 16 possible combinations of ancestral sites for a tree for four sequences. Obtaining the maximum likelihood estimate of branch lengths for a given tree is computationally time consuming, and in practice this has limited the application of the method to fairly small data sets.

In describing likelihoods we have glossed over the model used to compute those likelihoods. This model may include parameters for the transition/transversion ratio (TS/TV), base composition, and variation in rate among sites (Chapter 5). Which values do we use for these parameters? Unlike parsimony, which also enables us to include variation in rates of substitution, likelihood methods provide us with a criterion for choosing the best values for the parameters, namely those values that maximise the likelihood of the data. For example, in any but the simplest of models (such as the Jukes–Cantor model; see Chapter 5) the rate of transitions and transversions is different. The likelihood of a tree will vary with the transition/transversion ratio. If transitions in the data are much more common than transversions the observed

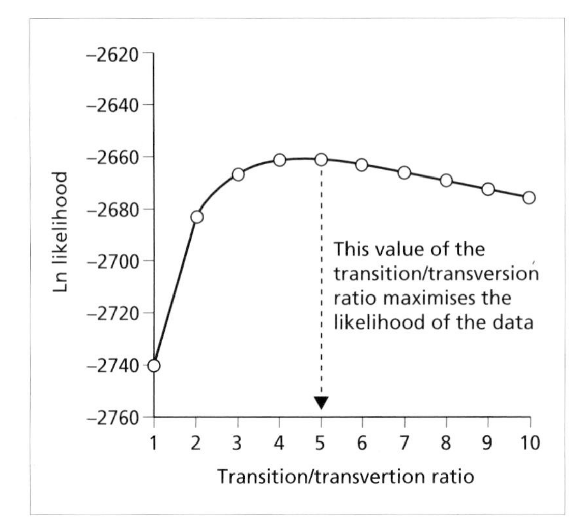

Fig. 6.16 To find the maximum likelihood estimate for the transition/transversion ratio for the hominoid data set, the likelihood of obtaining the observed data is calculated for a range of values for the ratio, holding everything else constant (including the tree). The ratio yielding the highest likelihood is the best estimate. In this case it is approximately 5.

data are more likely to be the result of a high rather than a low TS/TV ratio. The value of TS/TV that maximises the likelihood of the data is the maximum likelihood estimate of the TS/TV ratio. Figure 6.16 shows the likelihood of the hominoid data set given various values of TS/TV, showing that the best estimate of TS/TV is about 5. The same approach can be used to obtain maximum likelihood estimates of other parameters, such as rate variation among sites (see Chapter 5).

6.5.2 Likelihood ratio tests

We can test alternative hypotheses concerning the same data using a **likelihood ratio test**. The **likelihood ratio statistic** (Δ) is the ratio of the likelihood of the alternative hypothesis (H_1) to the null hypothesis (H_0). Because likelihoods are often very small it is more convenient to use log-likelihoods, so the likelihood ratio test statistic becomes

$$\Delta = \log L_1 - \log L_0 \tag{6.8}$$

where L_1 is the maximum likelihood of the alternative hypothesis H_1, and L_0 is the maximum likelihood of the null hypothesis H_0. In some cases the significance of Δ can be evaluated by consulting standard χ^2 tables. Such cases occur when H_0 is a special case of H_1, obtained by constraining one or more parameters in H_1. For such **nested** hypotheses 2Δ is approximately distributed as χ^2 with the degrees of freedom (d.f.) being the difference in the number of parameters in the two hypotheses. An example (discussed below) is the molecular clock hypothesis, which is equivalent to comparing the likelihoods of an ultrametric and an additive tree (the former is a special case of the latter). If the hypotheses are not nested then use of the χ^2 distribution is not warranted and the distribution of Δ must be evaluated by other means, such as the parametric bootstrap (see section 6.8.3).

Likelihood ratio tests can be used to test a range of hypotheses, such as whether a particular model of molecular evolution is valid, whether a molecular clock adequately describes the data, and whether one phylogenetic hypothesis is significantly better than another.

Is the model valid?

Given the dependence of maximum likelihood on an explicit model of sequence evolution, it is clearly desirable to be able to say when we would reject the chosen model. One way to evaluate the model is to measure how well it fits the observed data by comparing the likelihood a tree and model confers on the data, L_{tree}, with the theoretical best value for the likelihood, L_{max}, using a likelihood ratio test. Figure 6.17 shows this result of using this test to evaluate the adequacy of the HKY85 model (see Chapter 5) when applied to the hominoid mtDNA data set. The observed value of $\log L_{max} - \log L_{tree}$ is outside the expected distribution, indicating that the HKY85 model can be rejected for these data.

On the basis of this test the HKY85 model is not adequately describing the actual evolution of these sequences. Recall from Chapter 5 that the HKY85 model allows the ratio of transitions and transversions to vary (the test in Fig. 6.17 used the maximum likelihood value of 5 as determined in Fig. 6.16), but does not allow for variation in rates of substitution among sites. If we incorporate site variation using the gamma (Γ) distribution (see Chapter 5), then the resulting HKY85 + Γ model is a significantly better fit to the data (Yang *et al.*, 1994).

Note that, although the HKY85 + Γ model does better than HKY85 alone, using either model results in the same maximum likelihood tree. In this sense,

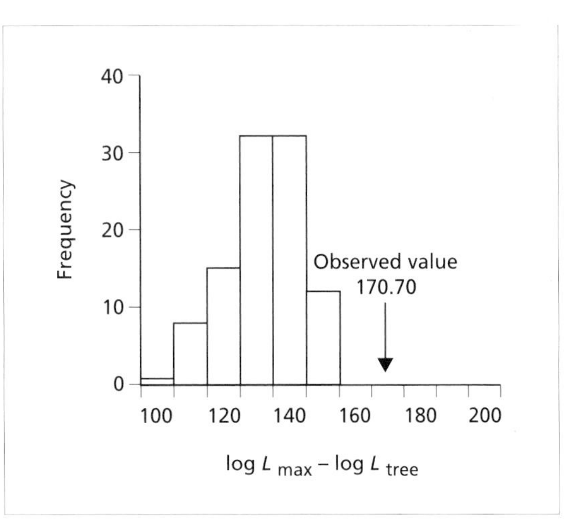

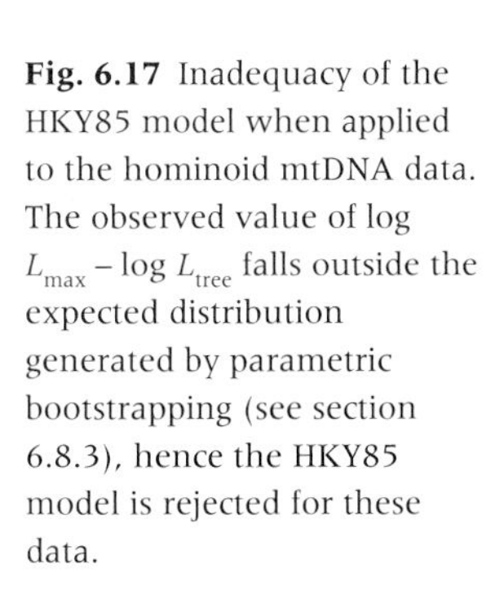

Fig. 6.17 Inadequacy of the HKY85 model when applied to the hominoid mtDNA data. The observed value of $\log L_{max} - \log L_{tree}$ falls outside the expected distribution generated by parametric bootstrapping (see section 6.8.3), hence the HKY85 model is rejected for these data.

the maximum likelihood estimation of the tree topology for the hominoids is robust to violation of assumptions of the model. However, estimates of branch lengths can be greatly affected by the choice of model, hence for questions involving rates of evolutionary change having a good model is very important.

Rate variation

Maximum likelihood methods can be used to test hypotheses about rate variation, in particular, hypotheses of 'molecular clocks'. If sequences are evolving different rates then an ultrametric tree would be a poor representation of the relationships among those sequences—an additive tree would be more appropriate. However, given a constant rate of evolution an ultrametric tree would not be significantly worse than an additive tree (see Fig. 6.8). We can test the hypothesis of a molecular clock using the likelihood ratio test

$$2\Delta = \log L_{\text{no clock}} - \log L_{\text{clock}} \tag{6.9}$$

which is distributed as χ^2 with $(n-2)$ degrees of freedom where n is the number of sequences. The degrees of freedom in this case correspond to the difference in the number of branch lengths we have to estimate in an additive and an ultrametric tree (see Fig. 6.8). Applying this test to the hominoid mitochondrial data set, Fig. 6.18 shows the maximum likelihood trees with and without the molecular clock constraint. The two trees have the same topology. For these two trees $2\Delta = 2(-2659.18 - (-2660.61)) = 2.86$, which is not significant ($\chi^2_{(3,\,0.05)} = 7.81$). Note also that in the tree inferred without the clock assumption the five hominoid sequences roughly line up, which visually confirms what the likelihood ratio test tells us—the data is consistent with a molecular clock. Two other tests of the molecular clock are presented in Chapter 7, Box 7.2.

Comparing phylogenetic hypotheses

As well as testing specific hypotheses about models of molecular evolution we

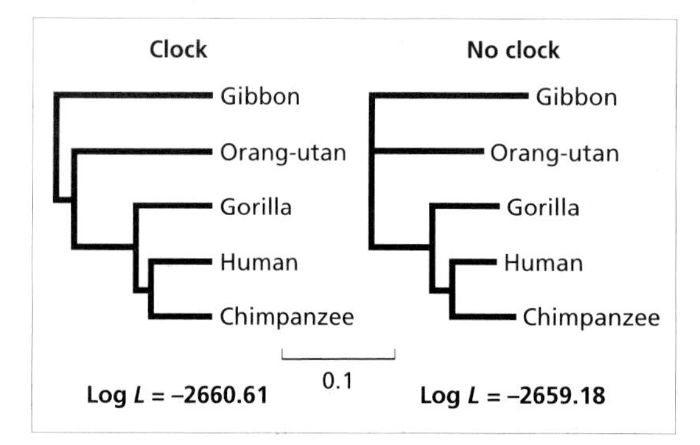

Fig. 6.18 Maximum likelihood trees for the hominoid mtDNA data set inferred with and without a molecular clock assumption. Using the likelihood ratio test the molecular clock hypothesis is not rejected, hence the data are consistent with a single rate of evolution for these sequences. The scale bar is number of substitutions per site.

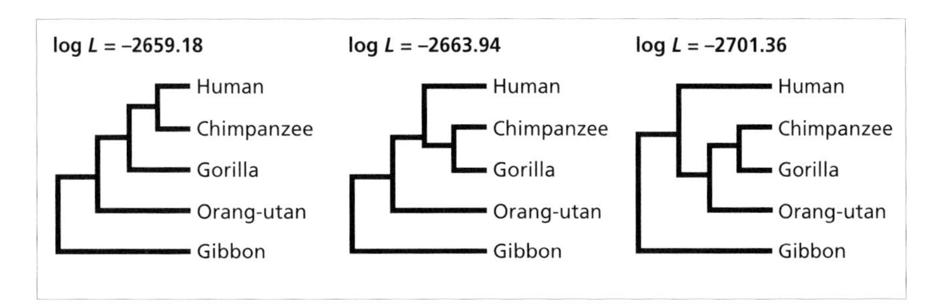

Fig. 6.19 Three different hypotheses of relationship among the hominoids and the likelihoods that each tree has given rise to the observed data.

can use likelihood ratio tests to evaluate the merits of competing hypotheses about evolutionary relationships. Given two different trees for the same data, we can ask whether one tree is significantly more likely to have given rise to the observed data than the other tree.

Kishino and Hasegawa (1989) proposed a test to compare the likelihoods of different trees that resembles the likelihood ratio test. Given two different trees, such as the first two in Fig. 6.19, the difference in log likelihoods is computed by taking each site in turn and finding the difference between the log likelihoods for that site on the two trees. Some sites may favour one tree, some sites the other tree. If the two trees are not significantly different then the sum of these likelihood differences:

$$\Delta = \sum_{i=1}^{k} \left(\log L_{(k,\mathrm{tree1})} - \log L_{(k,\mathrm{tree2})}\right) = \log L_{\mathrm{tree1}} - \log L_{\mathrm{tree2}} \quad (6.10)$$

will not be significantly different from zero. Kishino and Hasegawa's test assumes that the distribution of likelihood differences is approximately normal, so if $\Delta \pm$ its standard error (S.E.) includes zero then the two trees are not significantly different. For the first two trees in Fig. 6.19 the difference in log likelihoods is $\Delta = -4.76 \pm 7.97$, hence by this test these two trees are not significantly different. Although the data cannot distinguish between the (human, chimp) and (chimp, gorilla) hypotheses, it can reject the hypothesis that the 'great apes' (chimp, gorilla and orang) are monophyletic: $\Delta = -42.18 \pm 14.62$.

The analogous test can be applied to parsimony methods as well. In this instance the test statistic is the sum of the differences in the number of steps each site requires on the two trees. Again, the null hypothesis is that this sum is zero.

6.5.3 Parsimony and likelihood

Parsimony was originally proposed by Edwards and Cavalli-Sforza as an approximation to maximum likelihood. Faced with serious computational

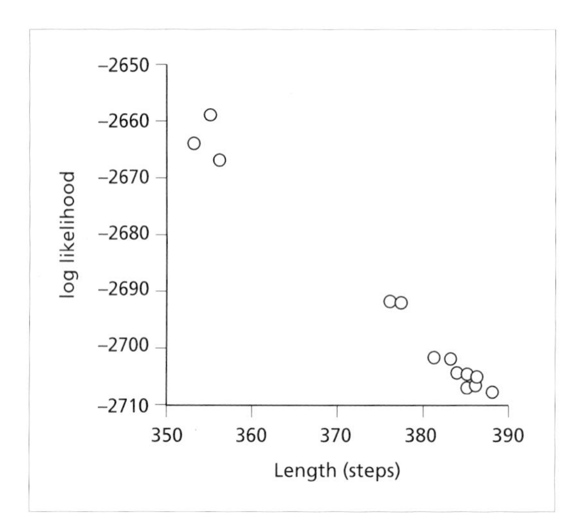

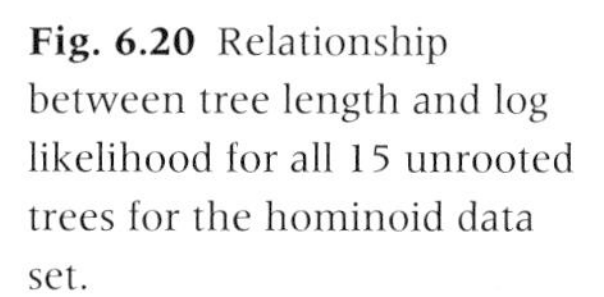
Fig. 6.20 Relationship between tree length and log likelihood for all 15 unrooted trees for the hominoid data set.

problems in computing likelihoods, they suggested that under some circumstances the shortest tree may also be the maximum likelihood tree. The most parsimonious tree is indeed the maximum likelihood tree when the expected amount of evolutionary change is very small, the tree that postulates the least amount of evolutionary change is also the maximum likelihood tree.

In practice, maximum parsimony and maximum likelihood trees can be very similar. Figure 6.20 shows the relationship between tree length and likelihood for the hominoid data set. Both criteria rank the trees in almost the same order, and both recognise a gulf between the possible resolutions of the tree (((human, chimp, gorilla), orang-utan), gibbon) and the 12 other trees. In this example, parsimony favours the (chimp, gorilla) clade, whereas likelihood prefers (human, chimp). In cases where rates of substitution are high, or are unevenly distributed over the branches of the tree, we might expect parsimony and likelihood to disagree about the relative merits of the same trees.

6.5.4 Objections to likelihood

Likelihood requires an explicit model of evolution, which may be seen as both a strength and a weakness. It is a strength because it makes us aware of the assumptions being made—in other methods these assumptions are often only implicit and hence may be overlooked. However, dependence on a model raises the question of just which model to use, and what values of the parameters, such as transition/transversion ratio, should be employed. This is a familiar phylogenetic dilemma: if we knew the probabilities of different evolutionary events we could use that insight to reconstruct evolutionary trees. Where do these probabilities come from? Well, if we knew the evolutionary tree we could obtain the probabilities from that tree. The information we require to infer the tree presupposes that we have the tree in the first place. One approach is to choose the combination of model and parameters that maximises the likelihood. However, this requires that we search for the best model as well as the best tree, greatly increasing the computational difficulty of obtaining an answer. A further technical difficulty concerns computing the likelihood itself. This is computationally time consuming, and recently it has been shown that more than one maximal likelihood value may exist for a given tree, making it difficult to guarantee that the likelihood value for that tree is actually maximal. These problems may make likelihood more useful as a tool for testing models of molecular evolution than as an all-purpose method of phylogenetic inference.

6.6 Splits and spectra

In Chapter 2 we introduced the concept of a split. Now we consider two methods of data analysis that use splits: spectral analysis and split decomposition. An advantage these methods share is that they emphasise the importance of exploring the data, and they can help uncover patterns that might otherwise be missed.

Box 6.4 Split numbers

A split can be represented as a binary character where a species is either '0' or '1', depending on which set the species belongs in. For example, the split {{gorilla, orang, gibbon}, {human, chimp}} can be written as 00011. If we treat this string as a binary number then each split can be assigned a unique number (Hendy & Penny, 1993). Below are the 16 splits for the five hominoid species and their corresponding numbers:

Human	0	1	0	1	0	1	0	1	0	1	0	1	0	1	0	1
Chimp	0	0	1	1	0	0	1	1	0	0	1	1	0	0	1	1
Gorilla	0	0	0	0	1	1	1	1	0	0	0	0	1	1	1	1
Orang	0	0	0	0	0	0	0	0	1	1	1	1	1	1	1	1
Gibbon	0	0	0	0	0	0	0	0	0	0	0	0	0	0	0	0
Number	0	1	2	3	4	5	6	7	8	9	10	11	12	13	14	15

By convention, the split numbers are in the range 0 to $2^{(n-1)}$, so the split {{gibbon},{human, chimp, gorilla, orang}} is written as 01111 = 15 rather than 10000 = 16. The advantage of this numbering scheme is that we can refer to any split by a single number rather than listing all the species in that split.

6.6.1 Spectral analysis

For n sequences there are $2^{(n-1)}$ splits, hence for the five hominoid mtDNA sequences there are 16 possible splits. Five of these splits separate a single sequence from the other four (e.g. humans from the apes), and hence do not tell us about the relationships among the sequences. Only splits that divide the sequences into two sets that both have more than one sequence can tell us about phylogenetic relationships. However, any given unrooted tree for n sequences contains only $(2n-3)$ such splits. For the hominoids this means that only two of the 11 non-trivial splits can be in the tree for these sequences. Ideally, our data would contain evidence for just mutually compatible splits, but this is rarely, if ever, the case.

Spectral analysis provides a means for visualising the support for all the splits in a data set. In its simplest form, spectral analysis consists of plotting the frequencies of each split in the data set. This is straightforward if there is a maximum of two character states per character, because then each character can be directly translated into a split. To illustrate, the first informative sites in the hominoid data set each have only two states:

Human	G	T	C	A	T	C	A	T	C	C
Chimp	A	T	T	A	C	C	A	T	T	C
Gorilla	G	T	T	G	T	T	A	T	T	A
Orang-utan	A	C	C	A	C	T	C	C	C	A
Gibbon	A	C	C	G	C	C	C	C	C	A

so we can substitute 0 and 1 for the two states:

Human	1	1	0	1	1	0	1	1	0	1
Chimp	0	1	1	1	0	0	1	1	1	1
Gorilla	1	1	1	0	1	1	1	1	1	0
Orang-utan	0	0	0	1	0	1	0	0	0	0
Gibbon	0	0	0	0	0	0	0	0	0	0
Split	5	7	6	11	5	12	7	7	6	3

Each split is numbered according to the scheme described in Box 6.4. Doing this we discover that these 10 sites correspond to seven different splits. Applying this procedure to the remaining sites, we obtain the spectrum for this data set (Fig. 6.21).

Note that every split has at least some support, hence we need some grounds for choosing which splits will be used to construct a tree (not all splits can coexist in the same tree). The five trivial splits labelled {human}, {chimp}, {gorilla}, {orang} and {gibbon} will of course be in any tree, but which of the non-trivial splits should we choose? One possible solution for this data set is to choose the two mutually compatible splits that have the most support. From Fig. 6.21 we can see that the split labelled {orang, gibbon} has the most support of any non-trivial split, and that the next best supported split is {human, chimp}, which is also compatible with {orang, gibbon}. These two splits give the tree ((human, chimp), gorilla, (orang, gibbon)). These splits are highlighted in Fig. 6.22. Note that the data contains almost equal support for the {human, chimp} and {gorilla, chimp} splits. When we analysed this data using maximum likelihood as our optimality criterion the best tree contained the {human, chimp} split, but was not significantly better than a tree containing the {chimp, gorilla}

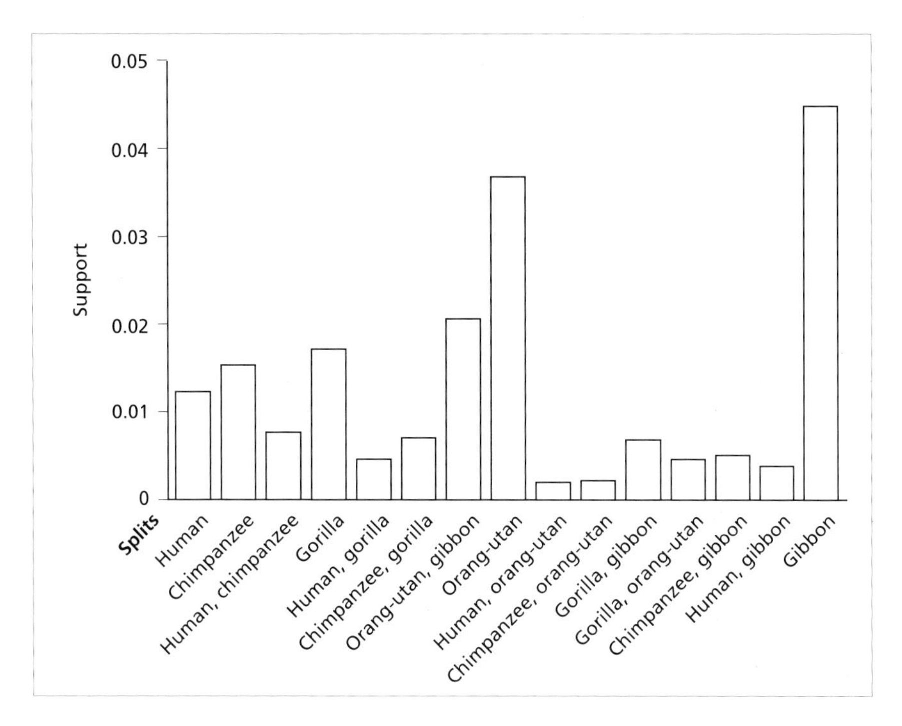

Fig. 6.21 The spectrum for the hominoid data set. Note that every split has at least some support.

split (p. 199). The inability to chose between these two is understandable given the almost equal support for these two conflicting splits in the data.

Another way of inferring a tree from a spectrum is the 'closest tree' method (see Box 6.5). For the hominoid data set the closest tree (Fig. 6.23) contains the two best supported splits.

Box 6.5 Spectral analysis and the Hadamard transform

Consider this unrooted tree labelled with the probability of observing a change along each branch.

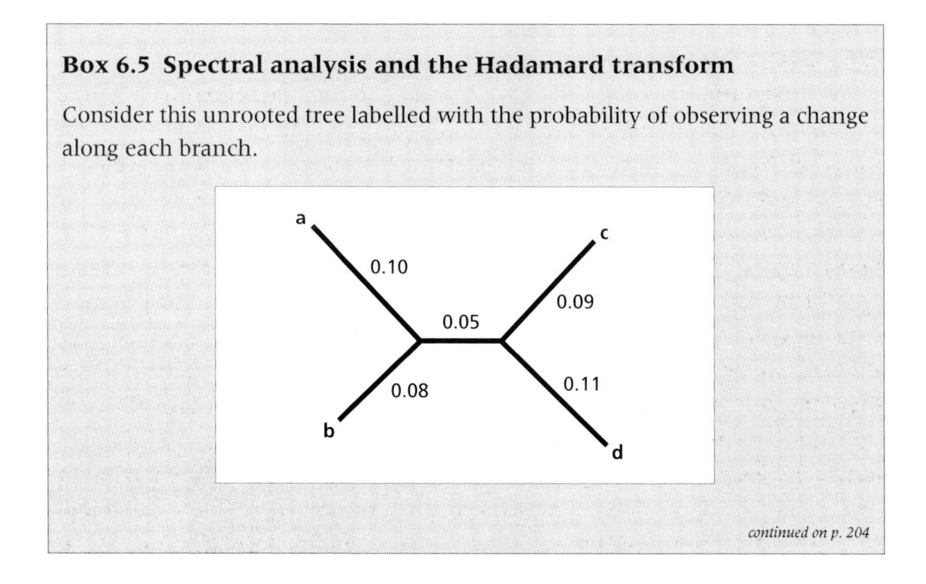

continued on p. 204

Box 6.5 *continued*

If this was the true tree, what would the sequences look like? Hendy & Penny (1993) have developed a method that uses a mathematical procedure called the Hadamard transform to generate the expected spectrum for a given tree and edge lengths. The expected spectrum for this tree is shown below, where the vertical axis is the probability of observing each split:

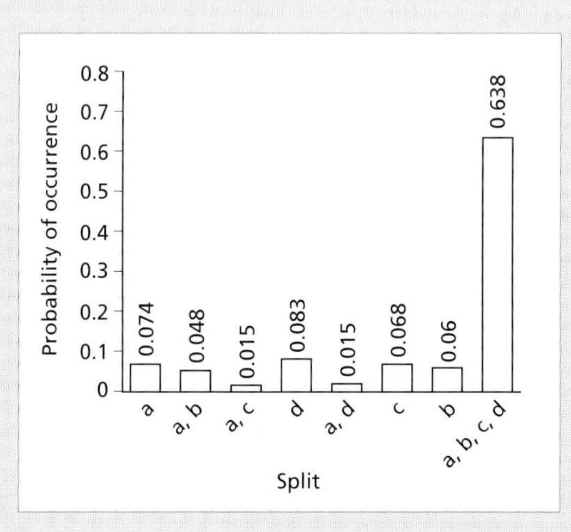

Note that while the split {{a, b}, {c, d}} has the most support of any nontrivial split, the two competing splits {{a, c}, {b, d}} and {{a, d}, {b, c}} also have some support. An attractive property of the Hadamard transformation is that it is reversible: given the spectrum above we can exactly recover the original edge lengths of the tree we used to generate the spectrum. Hence, one method for constructing a tree from a spectrum is to find the tree that would have generated a spectrum most like the actual spectrum we observed. This is the 'closest tree' method. See Hendy (1991) and Hendy and Penny (1993) for details.

Although spectral analysis is a potentially powerful technique, it has some limitations which stem from the rapid increase in the number of possible splits with increasing numbers of sequences; for 20 sequences there are over half a million splits in the spectrum. Furthermore, for discrete data the method is effectively restricted to two-state characters. Nucleotide sites may have up to four states, which have to be converted in some way into two-state characters, such as purines and pyrimidines. However, spectra can also be obtained from distance matrices which circumvents this difficulty to some extent.

6.6.2 Split decomposition

The spectrum for the hominoid data set (Fig. 6.21) shows that all possible splits have at least some support. However, this fact is lost as soon as we

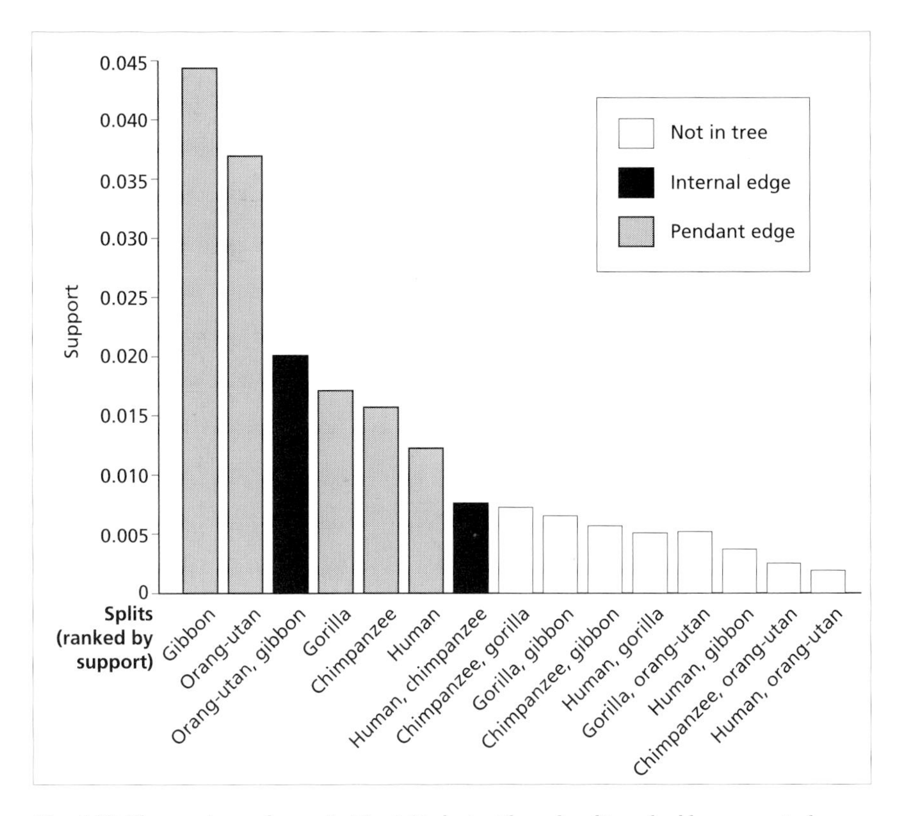

Fig. 6.22 The spectrum shown in Fig. 6.21, but with each split ranked by amount of support. The two largest non-trivial splits, {gibbon, orang-utan} and {human, chimp}, are compatible with each other and form the best tree (Fig. 6.23). Note that the bulk of the evolutionary change in this data is concentrated along the terminal branches (the 'pendant edges') and hence does not tell us about groupings among the five sequences. The {human, chimp} split is in the best tree but is only fractionally better supported than the {chimp, gorilla} split.

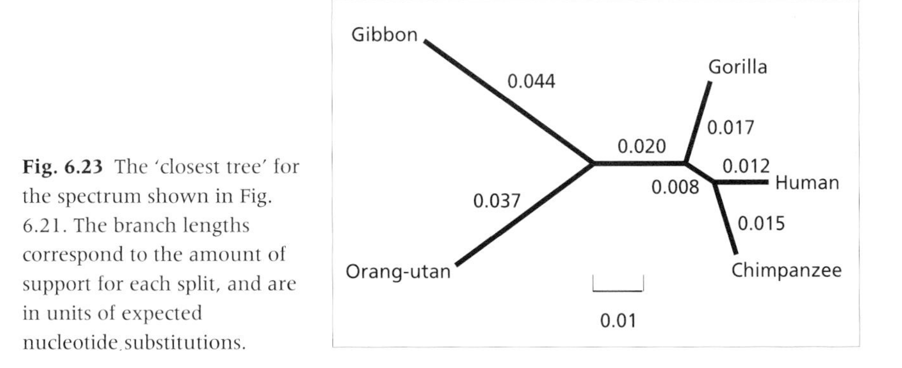

Fig. 6.23 The 'closest tree' for the spectrum shown in Fig. 6.21. The branch lengths correspond to the amount of support for each split, and are in units of expected nucleotide substitutions.

represent the spectrum by a single tree (e.g. Fig. 6.23) because the tree, by definition, contains only compatible splits. Looking at the spectrum we can see that the split {chimp, gorilla} has almost as much support as the {human, chimp} split. Split decomposition is a method that seeks to represent more of the information contained in the spectrum.

The method can be best explained by an example. Consider the following nine sites for the five hominoids.

	1	2	3	4	5	6	7	8	9
Human	T	C	C	T	T	A	A	A	A
Chimp	T	T	C	T	A	T	A	A	A
Gorilla	T	T	A	C	A	A	T	A	A
Orang-utan	C	C	A	C	A	A	A	T	A
Gibbon	C	C	A	C	A	A	A	A	T

At each site there are just two nucleotides, hence each site corresponds to a split. For example, site 1 corresponds to the split {human, chimp} {gorilla, orang, gibbon}. Taking each split in turn, we can attempt to construct a tree by combining the splits. The first site partitions the five hominoids into those with T (human, chimp, and gorilla) and those with C (orang and gibbon). In Fig. 6.24(a) this is represented by dividing the set of five hominoids into two sets, connected by a line labelled by the site that supports this split. The second site corresponds to the split {chimp, gorilla} {human, orang, gibbon}. These two splits are compatible, so we now have the tree ((human, (chimp, gorilla)), (orang, gibbon)) (Fig. 6.24b). However, site 3 poses a problem. The split {human, chimp} {gorilla, orang, gibbon} cannot be combined with the split supported by site 2. If we were to continue to build a tree we would have to decide which of these splits to accept. However, we can attempt to represent both splits using a network. In this example, we can depict both splits 2 and 3 by introducing a parallelogram to indicate that there are two alternative splits (Fig. 6.24c). Site 4 supports the same split as site 3, hence adding that to the network does not change the network topology (Fig. 6.24d). The remaining sites (5–9) are all trivial splits that each partition a single sequence from all the other sequences (Fig. 6.24e).

Figure 6.24(f) summarises the network for these nine sites. The length of each link in the graph is proportional to the number of sites supporting each split. By collapsing the parallelogram linking human, chimp and gorilla, we could obtain either the tree ((human, chimp), gorilla) or (human, (chimp, gorilla)). Hence, in a sense the network represents a set of possible trees in a single diagram. Applying this method to the complete hominoid data set, we obtain the diagrams shown in Fig. 6.25.

Although split decomposition diagrams can be difficult to interpret at first, they do offer a way of representing more of the information in a data set, and promise to be a useful tool for exploring phylogenetic spectra.

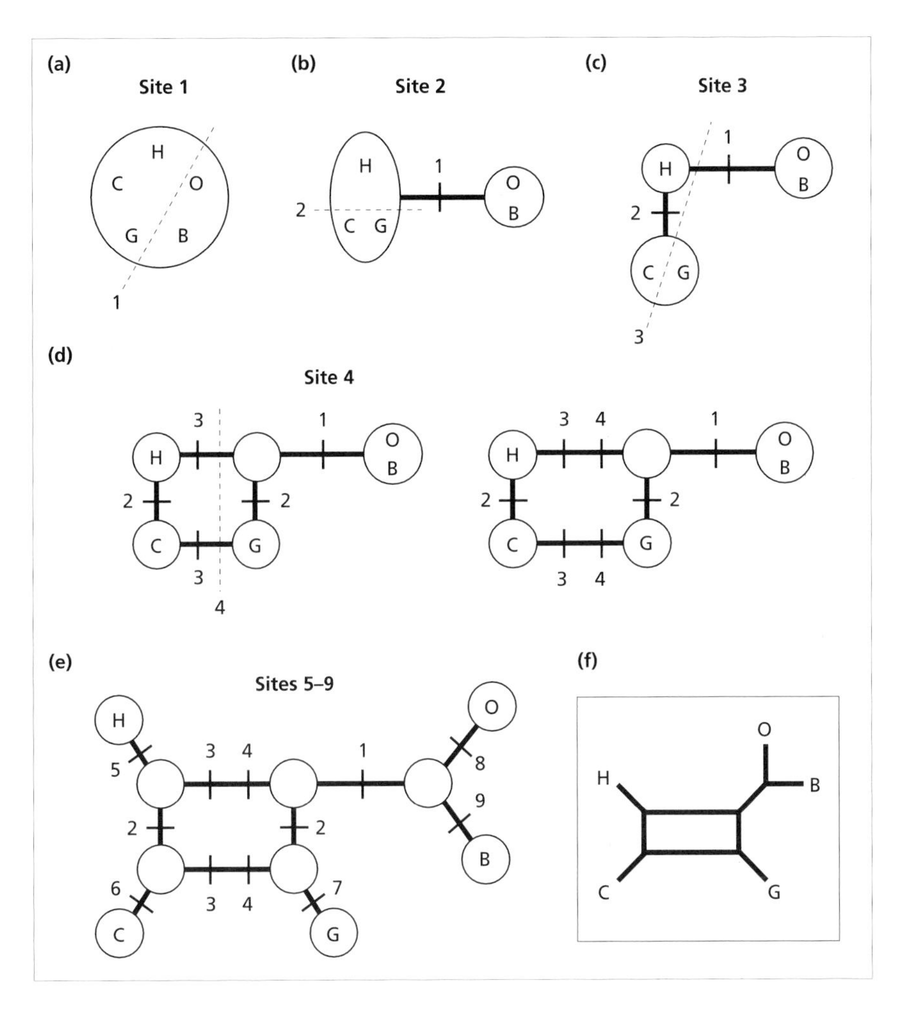

Fig. 6.24 Example of constructing a network from incompatible splits.

6.6.3 Is a network better than a tree?

In the discussion of split decomposition above, we have assumed that the sequences are related by a tree, even though there may be conflicts among different sites. Hence the split decomposition network depicts the conflicting signals rather than an actual picture of how the sequences are related. However, the assumption that a tree is an appropriate representation of relationships among the sequences may be incorrect. If competing splits have significant support, then perhaps the relationships among the sequences are better represented by a network. This may be particularly true where the sequences have undergone recombination.

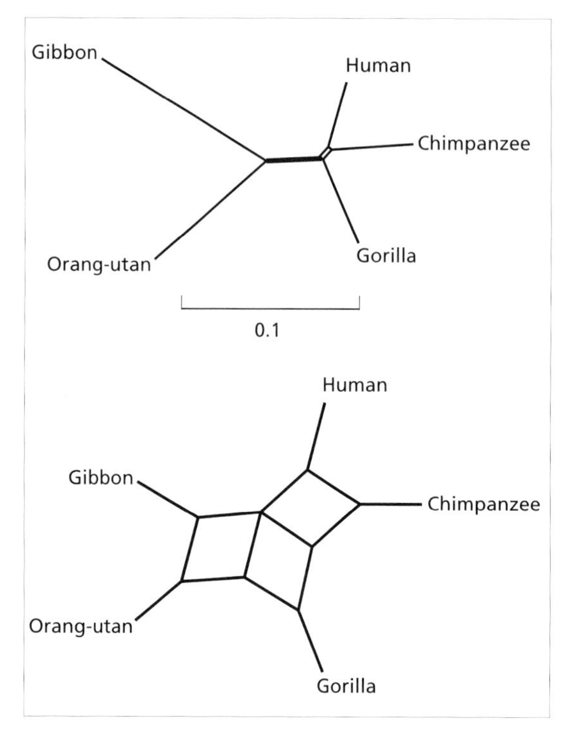

Fig. 6.25 Two views of the splits for the hominoid mtDNA data set. In the top diagram the lengths of each line are proportional to the support for the corresponding split; in the bottom diagram each split is drawn at the same size. Note the conflicting splits involving the human, chimp and gorilla sequences.

Box 6.6 The difference between precision and accuracy

It is important to distinguish between **precision** and **accuracy**. When thinking of trees, accuracy is proximity to the true tree; precision is how many alternative trees are excluded. A method that finds only one tree is very precise, but if that tree is wildly different from the true tree, then the method is inaccurate. The difference may be clearer if you consider two thermometers, A and B, that are both used to measure the temperature of water at boiling point (100°C). The temperature according to thermometer A is 101°C, whereas according to B it is 98.76°C; thermometer A is more accurate but less precise than B.

A related consideration is that some optimality criteria are real numbers (e.g. $\ln L = -2965.3098$) and others are integers (e.g. 395 steps). A method that assigns an integer score to each tree (such as parsimony) is more likely to assign the same length to different trees simply because there are a finite number of integer tree lengths for a given data set, but an infinite number of real numbers. Hence, tree criteria that use real numbers may give a false sense of precision simply because it is almost impossible to obtain more than one tree with exactly the same value of the criterion.

6.7 Have we got the true tree?

Given the range of possible methods for inferring phylogeny, we naturally want to know if they work, that is, do they recover the actual evolutionary relationships among nucleotide sequences? Several approaches have been developed to answer this question: analysis, simulation, known phylogenies, and congruence.

6.7.1 Analysis

In some cases, the phylogenetic method is simple enough that we can establish mathematically the exact conditions under which that method would fail. A good example of this is UPGMA, which requires a molecular clock. This condition can be expressed more formally in terms of branch lengths for a three-taxon tree (Fig. 6.26). This amounts to requiring that the two most closely related taxa are more similar to each other than they are to any other taxa. If this is not the case, then using UPGMA will generate an erroneous tree (Fig. 6.27).

Parsimony methods have been much debated and much studied. Early on it was shown by Felsenstein that under a very simple model of evolution with

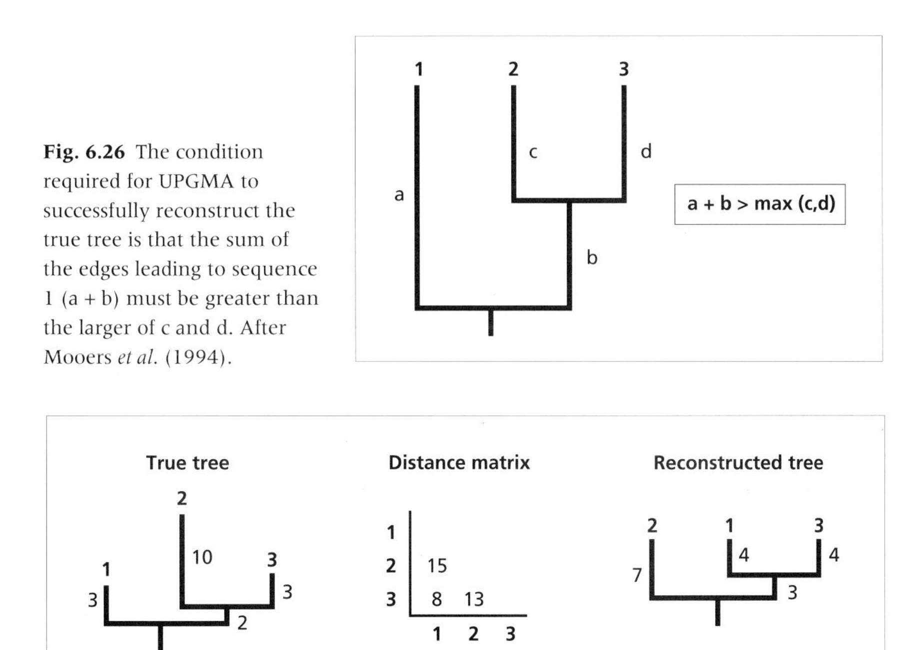

Fig. 6.26 The condition required for UPGMA to successfully reconstruct the true tree is that the sum of the edges leading to sequence 1 (a + b) must be greater than the larger of c and d. After Mooers *et al.* (1994).

Fig. 6.27 An example where UPGMA will reconstruct the wrong tree. The edge lengths on the true tree violate the condition shown in Fig. 6.26, as $a + b = 3 + 2 < \max(c, d) = 10$. Sequence 2 has evolved more rapidly than the other two sequences, so that sequence 1 and 3 are more similar to each other than either is to 2.

just two character states, parsimony could be inconsistent (see section 6.4.4). Originally it was thought that this inconsistency was due to the unequal rates of evolution in Felsenstein's example, hence parsimony might be consistent if rates of evolution were more equal. However, counter-examples have been found for five or more sequences where parsimony is inconsistent even if rates of evolution are constant. These examples serve to emphasise that it is not unequal rates of evolution *per se*, but the distribution of edge lengths that can cause problems for parsimony (see also Box 6.3).

While analytical methods are elegant and can yield important insights into phylogenetic methods, for more than a few sequences and for more complicated models of evolution, analytical methods become increasingly intractable. This has prompted the use of other strategies, such as experimental phylogenies and simulation.

6.7.2 Known phylogenies

The most compelling evidence for the success of our tree-building methods would be that they could reconstruct known phylogenies, that is, phylogenies which we knew to be true from other evidence. Unfortunately, known phylogenies are very rare. Typically the only 'known' phylogenies are those of laboratory animals and crop plants, and even these are often suspect. David Hillis and colleagues have addressed this problem by creating 'experimental' phylogenies in the laboratory. By subdividing cultures of bacteriophage T7 grown in the presence of a mutagen they constructed a known phylogeny (Fig. 6.28). At the end of the experiment, they obtained restriction site maps and nucleotide sequences for the eight T7 cultures. This data was then input into various tree-building programs, the output of which was compared with the actual T7 tree.

For the restriction site data, UPGMA, neighbour-joining, Fitch & Margoliash, Cavalli & Sforza and parsimony all recovered the topology of the actual tree. Because they had the actual ancestor cultures, Hillis *et al.* were also able to compare those ancestors with the reconstructions obtained by parsimony. Fully 97.3% of the ancestral states were correctly inferred.

Although encouraging, this particular study is perhaps less informative than

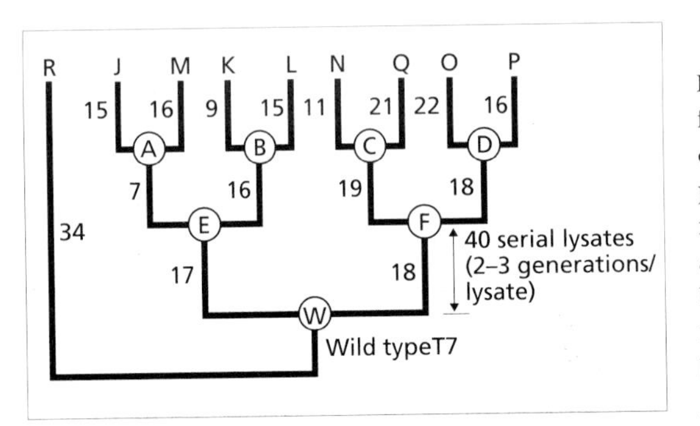

Fig. 6.28 Artificial phylogeny for bacteriophage T7 constructed by culturing the phage in the presence of a mutagen. Letters in circles are ancestors, numbers on branches are numbers of restriction site differences between phages at each node. After Hillis *et al.* (1992).

might have been hoped. The tree is well balanced and all nodes are accompanied by numerous changes making the tree a relatively easy one to reconstruct; indeed, every method tried was successful. More will be gained by fabricating more difficult trees and discovering under what circumstances different methods fail.

6.7.3 Simulation

Very rarely do we have known evolutionary trees for actual organisms or sequences, whether obtained from nature or 'grown' in a laboratory. An alternative source of known trees is computer simulation. In this case, we supply a computer program with a tree and 'evolve' DNA sequences along the branches of the tree according to some model. We then provide the resulting sequences as data for a range of tree-building methods, and determine which tree-building methods succeed in recovering the original tree. An advantage of this approach is that we can explore the effects of a wide range of parameters on the performance of tree-reconstruction methods. A disadvantage is that the models used to generate the artificial sequences may be unrealistic, making it difficult to generalise the results of the simulations to actual data sets (although see 'parametric bootstrapping' in section 6.8.3). In particular, we need to avoid biasing the model towards a particular method. For example, if we used the model implicit in one tree-building method to generate the sequences, then we would expect that method should perform very well, perhaps better than any other method. However, this result does not allow us to claim that this method is therefore the best tree-building method—all we can say is that under the chosen model the method works well.

A group of researchers led by David Hillis has used simulation to explore the relative merits of a range of tree-building methods for the simplest phylogenetic problem—finding the unrooted tree for four sequences. They varied the overall rate of nucleotide substitution, and the rate of substitution in different parts of the tree. These two parameters can be represented by a diagram showing the simulation space (Fig. 6.29).

By generating many thousands of data sets for different combinations of branch lengths and recording the success of various methods in recovering the actual trees upon which the data was generated, a picture emerges of the overall performance of the different methods.

Figure 6.30 shows the relative performance of UPGMA and parsimony under the same conditions. As we would expect, UPGMA does best when the rate of evolution is relatively constant, hence the high degree of success along the diagonal line in the chart. However, the further a tree departs from this diagonal line the less successful UPGMA is in recovering it. Parsimony does not require a molecular clock, and this is reflected in the wider range of branch lengths that parsimony can accommodate. However, there is a region in the top left of the parameter space, corresponding to a short internal edge and two long terminal edges. In this region, often called the 'Felsenstein zone', is where

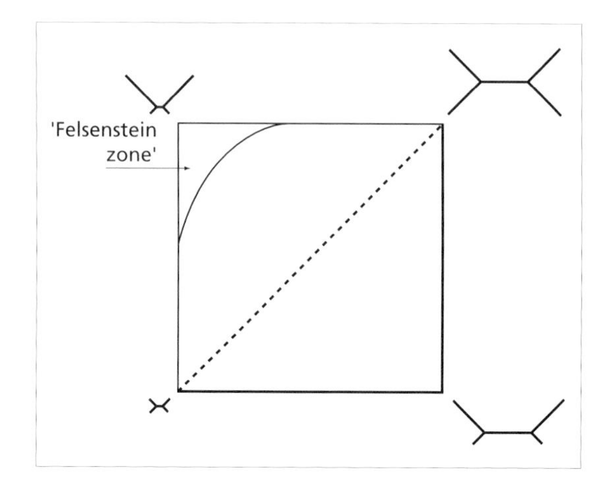

Fig. 6.29 The tree space explored by the computer simulations analysed in Fig. 6.30. At each corner of the square the corresponding tree is drawn. Along the dotted line the rate of substitution is the same on each edge in the tree. The trees at top left and bottom right show two extremes in rate variation, differing in how much change occurs along the central edge. The region at the top left-hand corner has become known as the 'Felsenstein zone', and corresponds to parameters Felsenstein (1978) used to show that parsimony could be inconsistent. After Huelsenbeck and Hillis (1993: Fig. 4).

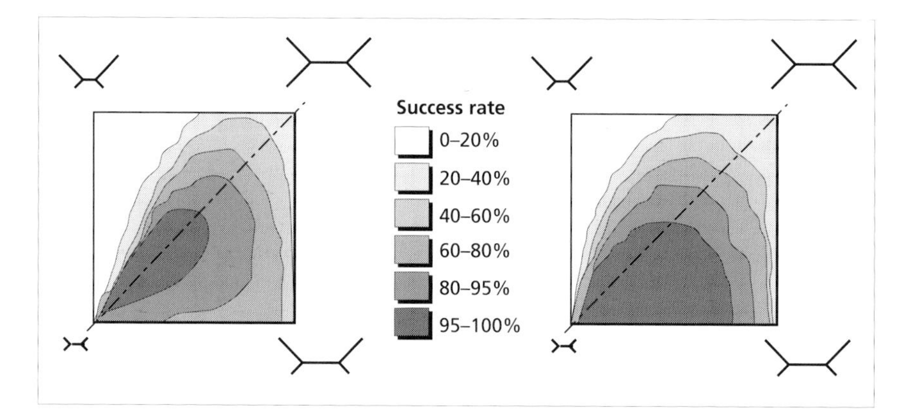

Fig. 6.30 Performance of UPGMA (left) and parsimony (right) in recovering the true tree under the range of parameters shown in Fig. 6.29. The success rate is the percentage of times that the correct tree was recovered in that region of the parameter space. Note the white region in the top left of the two diagrams (the 'Felsenstein zone') where neither method performs well. After Huelsenbeck and Hillis (1993).

'long branches attract' (see section 6.4.4), misleading parsimony into inferring the wrong tree.

Another variable affecting the performance of phylogenetic methods is the amount of data available. By creating artificial data sets of different sizes, it is

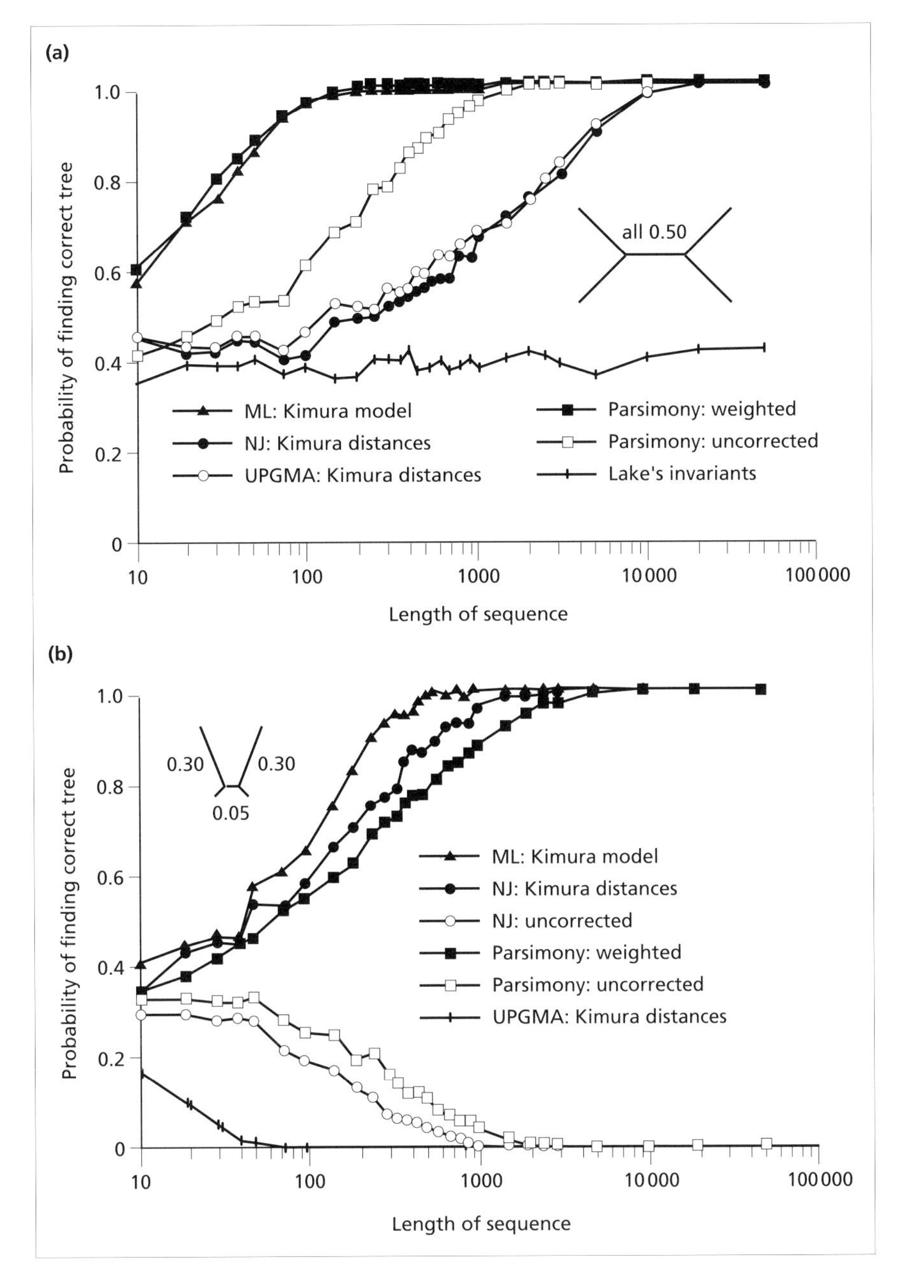

Fig. 6.31 The accuracy of several different phylogenetic methods in reconstructing two four-taxon trees with (a) all edges equal in length and (b) with a short internal edge and two long terminal edges. In each graph the proportion of analyses that recovered the correct tree is plotted against the length of the simulated sequences. From Huelsenbeck *et al.* (1996: Figs 2 and 3).

possible to see how if adding more data improves a method's rate of success. Figure 6.31 shows the results of two simulation experiments to evaluate the accuracy of several different phylogenetic methods in reconstructing four-taxon trees with all edges equal in length and with a short internal edge and two long terminal edges. In the case of equal branches all the methods are consistent, but differ in their efficiency (Lake's method, which is not discussed here, requires about 10 million base pairs to converge!). The tree with unequal rates of evolution requires more data to accurately reconstruct, and methods such as unweighted parsimony, neighbour joining on uncorrected distances, and UPGMA all converge on the wrong answer and are hence inconsistent. Hence the accuracy of different tree-building methods can vary depending on the tree being reconstructed.

The four-taxon tree studied by Hillis and colleagues is the simplest possible, and may be unrepresentative of the problems faced when we try and infer larger phylogenies (see Box 6.3). The difficulty of simulation experiments on larger trees is that as the trees get larger the numbers of combinations of tree shape and edge lengths also increases, prohibiting the kind of thorough investigation possible for the four-taxon case.

6.7.4 Congruence

Congruence is the agreement between estimates of phylogeny based on different characters. If the data sets are independent then the probability of obtaining the same or even similar trees for the same organisms using the different data sets is vanishingly small for any reasonable number of species. This is a simple consequence of the large numbers of possible phylogenies (see Chapter 2). Hence, if different data sets give us similar trees this gives us confidence that both reflect the same underlying cause, namely they reflect the same evolutionary history.

The power of the congruence test stems from the assumption that there is a single phylogeny and that each data set being analysed should reflect that history, coupled with the improbability of obtaining similar phylogenies due to chance alone (Box 6.7). Congruence has been used in two related, but distinct ways: validating a method of phylogenetic inference, and validating a particular source of data.

Validating a method of inference

If two data sets for the same set of taxa contain phylogenetic information then we might expect the trees inferred from those two data sets to be similar, if not the same. This follows from the assumption that each data set has the same evolutionary history. Hence, a method that consistently recovered similar trees from different data sets would be preferred to a method that produced different trees from different data sets. An early illustration of this approach using DNA

Box 6.7 How similar are two trees?

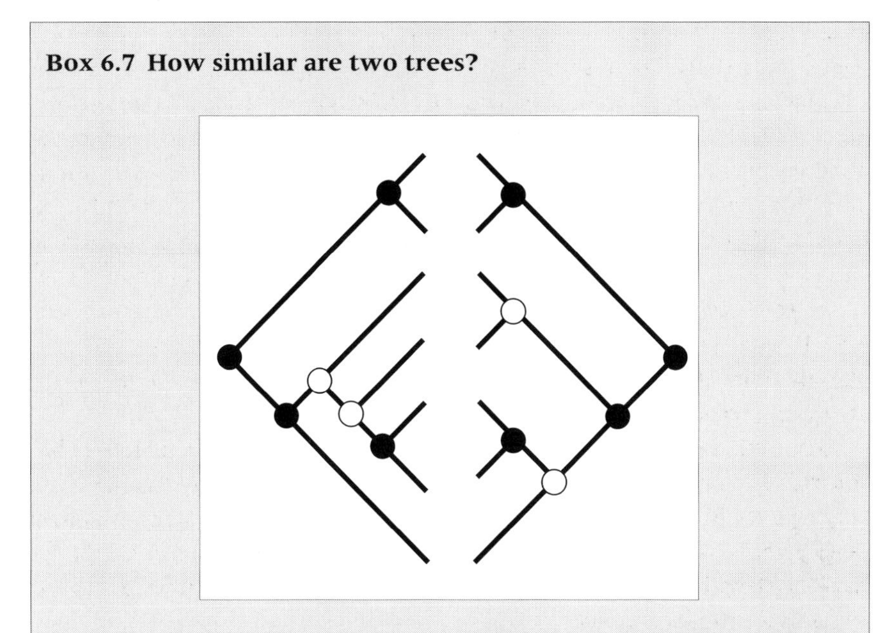

In this example the two trees each share four identical splits (solid circles) and each have two unique splits (open circles), making a total of four unique splits. Hence the partition distance between the two trees is four. As each tree has five internal nodes, two maximally dissimilar trees would have a partition distance of 10. The chart below shows the distribution of this measure of tree dissimilarity if pairs of trees are drawn at random from the set of all possible trees for seven sequences. The bulk of trees are as different as it is possible to be, and fewer than 0.6% of trees share as much in common as the two trees shown above.

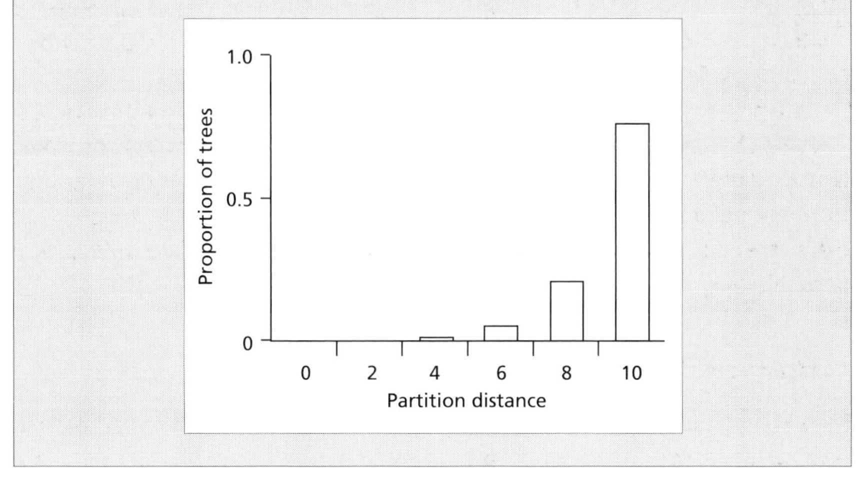

sequences is the study by Penny and colleagues. Using parsimony they found minimal and near-minimal length trees for five different proteins (cytochrome c, fibrinopeptide A and B, haemoglobin α and β) and asked whether the trees

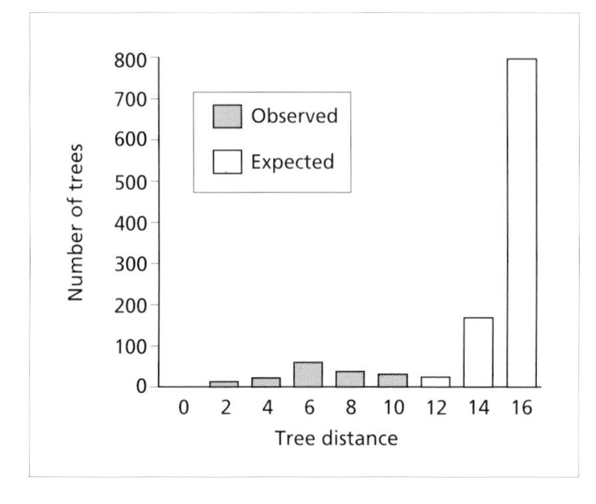

Fig. 6.32 Observed distribution of pairwise distances between 39 parsimony trees for 11 mammals inferred from five genes and the expected distribution if the trees were sampled at random. Note that the trees from the five genes are much more similar than could be expected due to chance alone.

were more similar to each other than could be expected due to chance alone using the partition distance (Box 6.7). The trees parsimony analysis obtained from the different genes were much more similar than could be expected due to chance alone (Fig. 6.32).

Validating a new source of data

As an example of the second use of congruence, we could ask whether a newly sequenced gene, or a new measure of genetic variation, contained phylogenetic information by comparing trees constructed using those data with trees from other data sources. Indeed, this test can be applied to any potential source of phylogenetic information. Figure 6.33 shows trees for modern humans based on genetic data and languages. The trees are different, but more similar than could be expected due to chance alone, suggesting that some common cause underlies this similarity. The obvious possibility is that both the genes and the language of modern human populations reflects the historical relationships among those populations.

6.8 Putting confidence limits on phylogenies

In the previous section we discussed accuracy, that is, how close we are to the truth. Now we turn to precision. In molecular phylogenetics our goal is to recover 'the one true tree' for a set of sequences. A measure of the phylogenetic precision is how many trees we can reject as candidates for the true tree. If we reject all but one, then we have maximum precision. Of course, this sole surviving tree may not be the true tree for our method might have misled us (remember that precision is not the same as accuracy; see Box 6.6, p. 208). The greater the number of trees we cannot reject, the

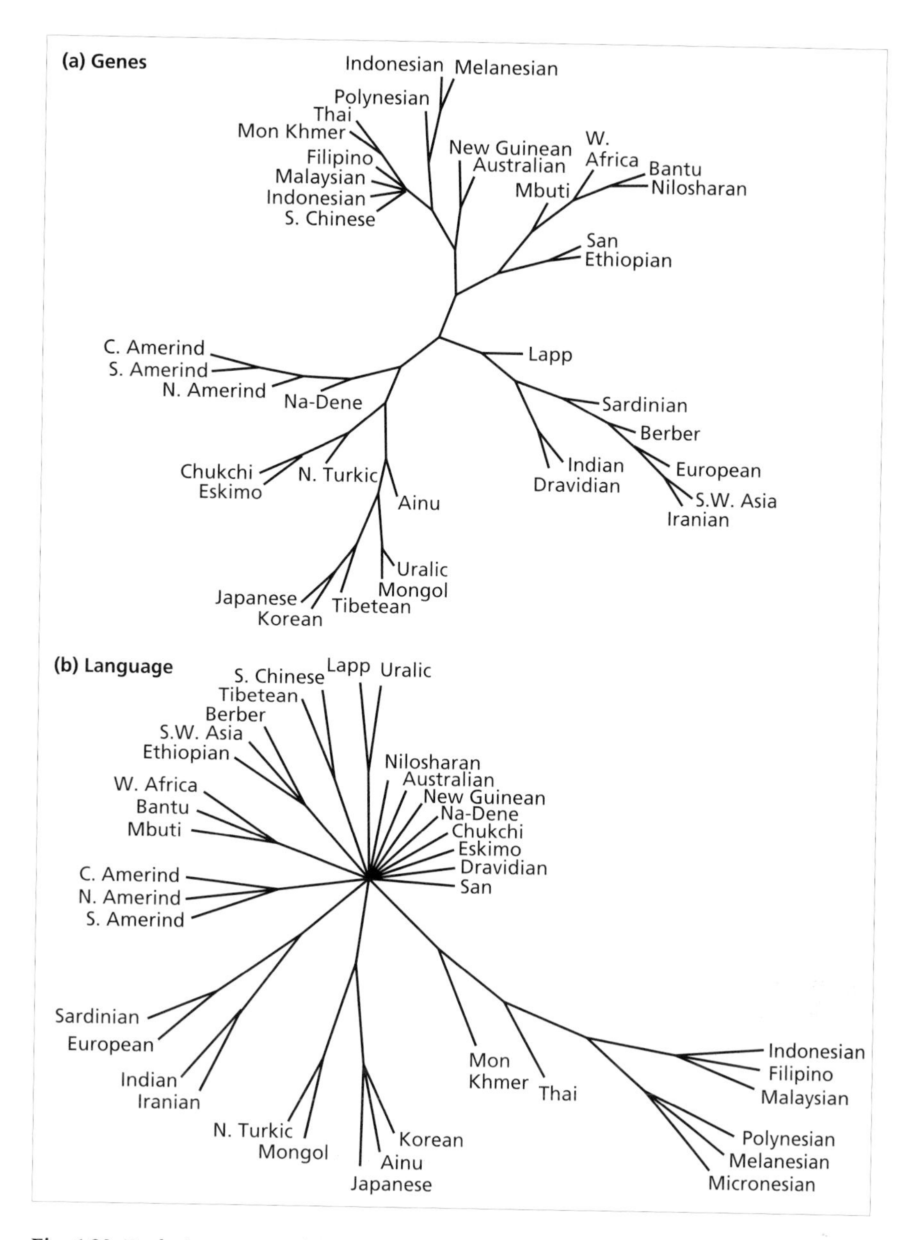

Fig. 6.33 Evolutionary trees for modern humans derived from genes and from language characteristics. The two trees are different but more similar than two randomly chosen trees. This suggests that both genes and language reflect the history of modern humans. After Penny *et al.* (1993).

lower the precision. Most scientific measurements are accompanied by some estimate of precision, such as 13.5 ± 0.2 mm. Phylogenetics should be no different, hence estimates of phylogeny should be accompanied by some indication of confidence limits.

Box 6.8 Congruence and total evidence

Suppose that the trees shown in Box. 6.7 are for the same organisms but obtained from different genes. The trees show points of similarity and points of difference. If our goal is to obtain a tree for the organisms themselves we have a dilemma—what is the best estimate of the phylogeny of these organisms? Two different solutions are to combine the two data sets and analyse them as one data set ('total evidence') or to analyse the data separately and combine the resulting trees for each data set using a consensus tree (see Chapter 2).

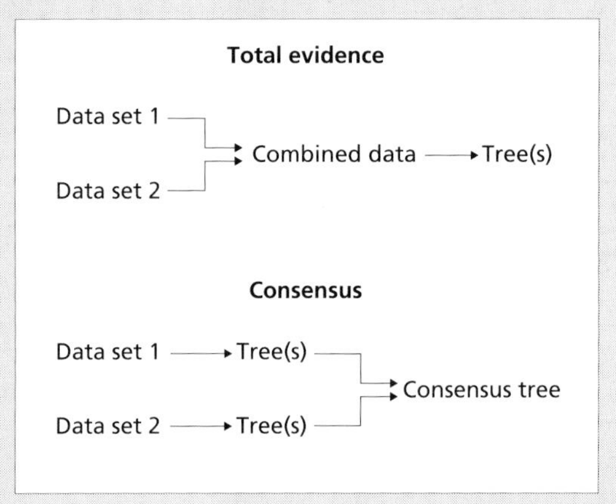

There are arguments for and against both views. Combining data sets makes use of all the data, whereas combining trees can result in less than optimal solutions. However, analysing data sets separately can reveal whether the different data sets support similar trees. If they do not then it may be that the genes have quite different histories and the data sets should not be combined to estimate organismal phylogeny (Bull, 1993) (see Chapter 8 for the gene tree/species tree problem).

6.8.1 Sampling error

One reason for a poor estimate of a phylogeny may not be the method used, but the data itself. If a data set contains homoplasy then different nucleotide sites support different trees, hence which tree (or trees) a given data set supports will depend on which characters have been sampled. As a consequence, estimates of phylogeny based on samples will be accompanied by sampling error.

As an example, consider the 896 base pairs of mitochondrial DNA used by Brown and colleagues to study hominoid phylogeny. Of these 896 sites, 90 are phylogenetically informative using parsimony (Fig. 6.34). The most parsimonious tree for all 90 sites is ((human, (chimp, gorilla)), orang, gibbon). However, consider what would happen if we sampled only the first 31 base pairs, recovering just five informative sites. The first such site supports (human,

```
     Human GTCATCATCCTTCTTTTTTTAGCAATTTCCTCACCTTCTCCGTCACGCTC 50
Chimpanzee A.T.C...T.C.T....CCCC...T.C...CTG......T.A.T.T.TCT 50
   Gorilla ..TG-T..TACCTCCC...C.A...CCC.T.TGTT.CAC.TA..G..TC. 50
Orang-utan AC..CTCC.ACC...CC.CCTAAG.C.CA.A...TCAACT..C...A.CT 50
    Gibbon AC.GC.CC.A.C.CC.CCC.CAAGTCC.ATC..T.CAA..TACTGTA..T 50

     Human TCGCCGCTCTCACTCCCCTTATTTTCTTGTCCGGTGACCG 90
Chimpanzee C.....T..C..T.TT...C........ACT.A....... 90
   Gorilla C..T.AT..CA...TT.......C.T.C.C.TA....TTA 90
Orang-utan CTATTA.CT.AGTC..TACCGCC.AGCCA.TTCACACTAA 90
    Gibbon .TA.TA.CT.AG.C..TACAGCCCAGCCAAA..ACACTAA 90
```

Fig. 6.34 The 90 phylogenetically informative sites from the Brown *et al.* (1982) mitochondrial DNA data for hominoids.

gorilla), the second site supports (human, chimp, gorilla), the third site contradicts the first and supports (chimp, gorilla), site four groups gorilla and orang, and site five again supports human + gorilla. The most parsimonious tree for this subset is (((human, gorilla), chimp), orang, gibbon), which is not the most parsimonious tree for the complete data set. This example is rather contrived, but it illustrates the point that results may depend on the sample of sequences. Indeed, the 896 sites sequenced by Brown *et al.* represent in themselves only a small fraction of the approximately 16 000 base pairs that comprise the mammalian mitochondrial genome, and more extensive mtDNA data sets support the tree (((human, chimp), gorilla), orang, gibbon).

The effects of sampling error can be seen by comparing trees for different mitochondrial genes. The mitochondrial genome is inherited without recombination, and so every gene has the same phylogenetic history. Árnason and colleagues recently compared trees for different mitochondrial genes from mammals for which the complete mtDNA sequence was known. While most genes agreed on the same tree, others supported different trees (Fig. 6.35).

Just as we often have limited samples of DNA, we may also have limited samples of taxa. In the example just given, mammalian relationships were being inferred based on just six species, a tiny fraction of the total extant Mammalia. Relationships among sequences can change if additional sequences are added. Indeed, for clades with a good fossil record, morphological data may prove superior to molecular data because the evidence from the extinct taxa is crucial to recovering the actual phylogeny.

6.8.2 Estimating sampling error: the bootstrap

One way to measure sampling error is to take multiple samples from the population being studied and compare the estimates obtained from the different samples. The spread of those estimates gives us an indication of the extent of sampling error, that is, how much our conclusions would vary depending on

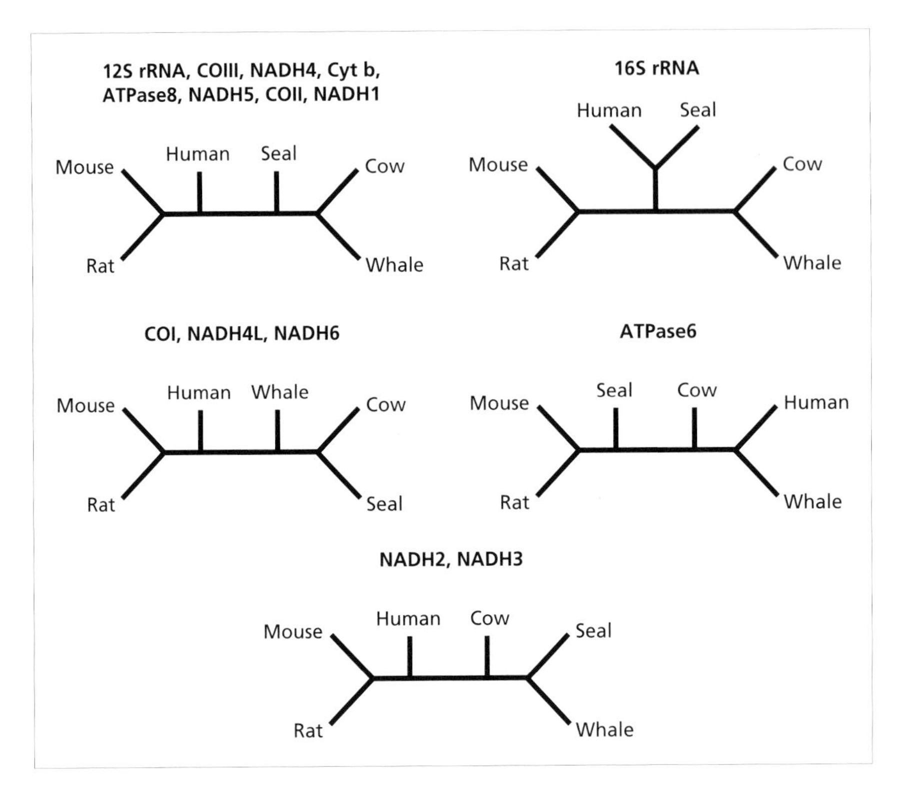

Fig. 6.35 Phylogenies for the same six mammals based on 15 different mitochondrial genes. After Årnason and Johnsson (1992).

what sample we took. Given that repeated sampling is often expensive, it would be nice to be able to calculate sampling error without having to take multiple samples. For many simple distributions (for example, the normal distribution) there are simple equations for calculating confidence intervals around an estimate (for example, the standard error of a mean). Trees are rather complicated structures, and it is extremely difficult to develop equations for confidence intervals around a phylogeny. Hence the reliance on other means of measuring confidence intervals, such as the computer intensive methods of jack-knifing and bootstrapping. Of these two the bootstrap is most often used in phylogenetics.

In a sense the bootstrap mimics the first method of estimating sampling error, but instead of sampling from the population it resamples from our sample. Each resampling is a **pseudoreplicate**. From each pseudoreplicate we derive an estimate of the parameter we are trying to measure, such as the mean height of a population. The variation among estimates derived from each pseudoreplicate is a measure of the sampling error associated with the parameter. For example, we could use the standard deviation of the means of the pseudoreplicates as an estimate of the standard error of our original estimate of the mean (Fig. 6.36).

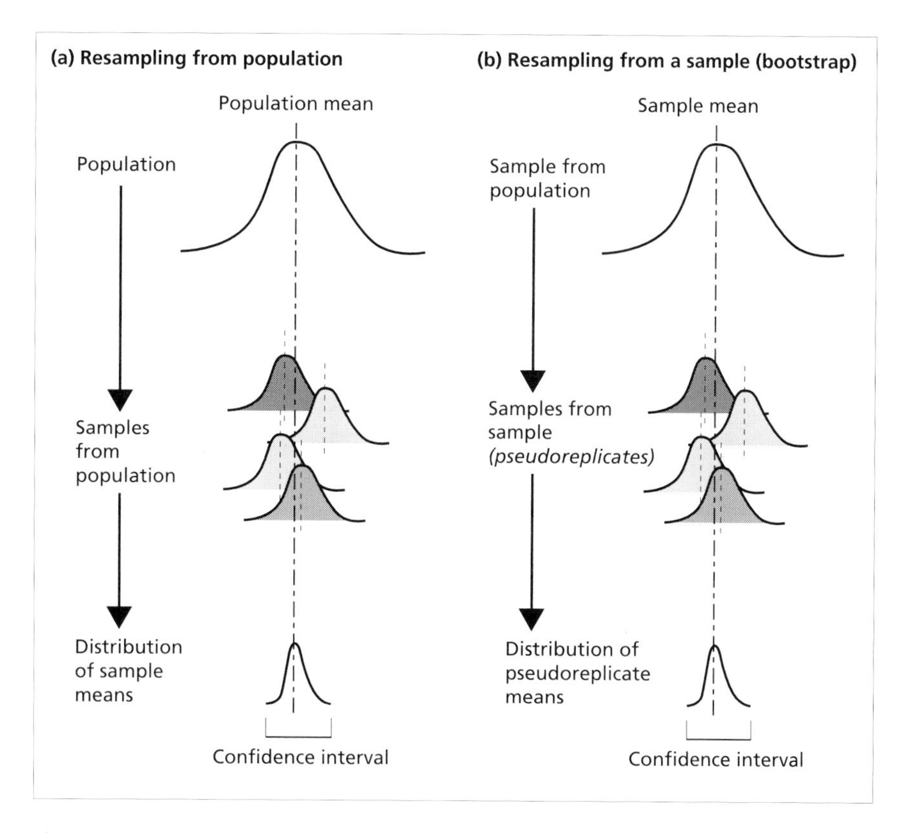

Fig. 6.36 Two methods for estimating the sampling error of the estimate of population mean. One method (left) is to take repeated samples from the population and obtain the distribution of the mean of each sample. An alternative method is to apply the same method but to a single sample (right). Instead of resampling from the original population, the single sample is itself repeatedly sampled to generate pseudoreplicates, and the distribution of the mean of each pseudoreplicate is used to estimate the sampling error of the original estimate of the population mean.

Bootstrapping can be applied to phylogenies by generating pseudoreplicates from the sequence data. For example, given the 896 nucleotide hominoid data set, we could generate a single pseudoreplicate by sampling at random and with replacement from the original data set until we had a new data set comprising 896 nucleotide sites. Because we are sampling with replacement (i.e. any site sampled is 'returned' to the data set before the next sample is taken) some sites may occur more than once in the pseudoreplicate, while others may not be represented at all. Hence the pseudoreplicate will resemble the original data set in that it contains only sites found in that data set, but it will differ in the frequencies of different sites. From this pseudoreplicate we would then build a tree using any of the methods described in this chapter. We then repeat this two-step process a large number of times (anywhere from 100- to 1000-fold), resulting in a set of bootstrap trees. This set of trees contains information on the sampling error associated with our sample.

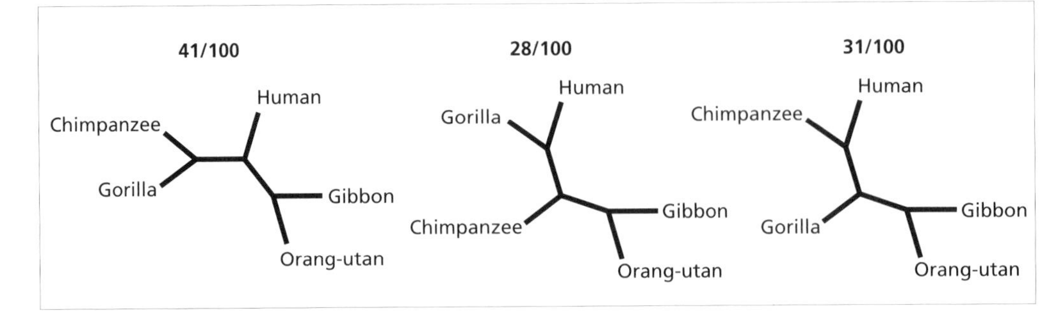

Fig. 6.37 The three trees for the hominoid sequence data that are obtained from 100 bootstrap pseudoreplicates and their relative frequencies. All three trees have the split {{orang, gibbon} {human, chimp, gorilla}} but they disagree about relationships between humans and the African apes.

Bootstrapping the hominoid data 100-fold produces 100 trees, each of which is one of the phylogenies shown in Fig. 6.37. The split {orang, gibbon} {human, chimp, gorilla} occurs in all the trees and hence has a bootstrap value of 100%. However, there is clearly conflict about relationships among the humans and African apes, with all three possible hypotheses receiving almost equal support. This result suggests that, on the basis of this data set, we lack sufficient information to discriminate between these three trees, and that different samples may favour different resolutions of human–chimp–gorilla relationships purely due to accidents of sampling.

For small numbers of sequences, such as the five hominoid mtDNA sequences, it is feasible to show the frequencies of the bootstrap trees. However, for larger numbers of sequences this becomes impractical. Instead, the most common splits found among the bootstrap trees can be assembled into a bootstrap consensus tree (see Chapter 2). These are often drawn with each node labelled with its frequency of occurrence among the bootstrap trees. When originally applying the bootstrap to phylogenies, Felsenstein suggested that only nodes with bootstrap values above 95% should be accepted as well supported. It is important to stress that bootstrap values estimate precision, not accuracy (see Box 6.6). A node may have a high bootstrap value but be completely wrong. In particular, if a tree-building method infers the wrong tree for a given data set, bootstrapping that data set may yield a robust, but wrong answer.

6.8.3 Parametric bootstrapping

'Parametric bootstrapping' resembles a cross between the bootstrap and simulation. Like simulations, it involves generating artificial data using a computer. However, whereas the simulations discussed in section 6.7.3 were conducted using made up trees and simple evolutionary models, in parametric

Box 6.9 Caveats concerning the bootstrap

One important caveat concerning the bootstrap is that this technique makes the assumption that nucleotide sites are independent and identically distributed (i.d.d.). This means that each site is independent of every other site, and that there is a single distribution of rate of evolutionary change across all the sites. In other words, if we were sampling from the mitochondrial genome this assumption means we could treat the genome as a single, homogeneous, entity. Cummins *et al.* (1995) made a simple test of this hypothesis by comparing the tree for 10 complete mitochondrial genomes with trees computed from two types of samples from those genomes: contiguous blocks of sequences, and randomly chosen sites scattered throughout the genome. The first method mimics how the mtDNA genome is sampled in practice; typically a single gene, or part of that gene is sequenced. If the i.i.d. assumption is valid then both kinds of samples should yield equally good approximations to the tree for the whole genome — in fact the randomly scattered sites performed better. Hence, the i.i.d. assumption is not valid for mtDNA. It is not known how robust the phylogenetic bootstrap is to violation of this assumption.

A further consideration is that the results of bootstrapping are often summarised using a majority-rule consensus tree (consensus trees are discussed in Chapter 2) showing the frequency of each split that occurs in at least half of bootstrap trees (splits that are compatible with these splits but occur in less than half the trees may be subsequently added to this consensus tree). If one or more sequences have uncertain relationships they may appear in very different positions in the bootstrap trees (i.e. they 'float' over the tree), resulting in a general lowering of bootstrap values for those parts of the tree over which the sequences float. Hence, parts of the tree which are actually quite robust may have spuriously low bootstrap values. This problem can be addressed using other consensus methods (Page, 1996a; Wilkinson, 1996).

bootstrapping the evolution of the DNA sequences is simulated using parameters (including the tree) estimated from real data. For example, analyses of 18S rRNA sequences suggests that birds and mammals are sister taxa (each other's closest living relative), whereas morphological evidence from both living and fossil taxa groups birds with crocodiles. One possible reason for this conflict between molecules and morphology is that the 18S rRNA data has suffered from long branch attraction (section 6.4.4). Certainly the tree resulting from parsimony analysis of these data shows long edges leading to both the birds and mammals (tree 1, Fig. 6.38), but this is not of itself evidence for long branch attraction. What we want to know is even if tree 3 was the real tree for the 18S rRNA data (as suggested by morphology) could our analyses mistakenly conclude that the bird–mammal tree (tree 1) was best for these data? If this was the case then this would be evidence for long branch attraction.

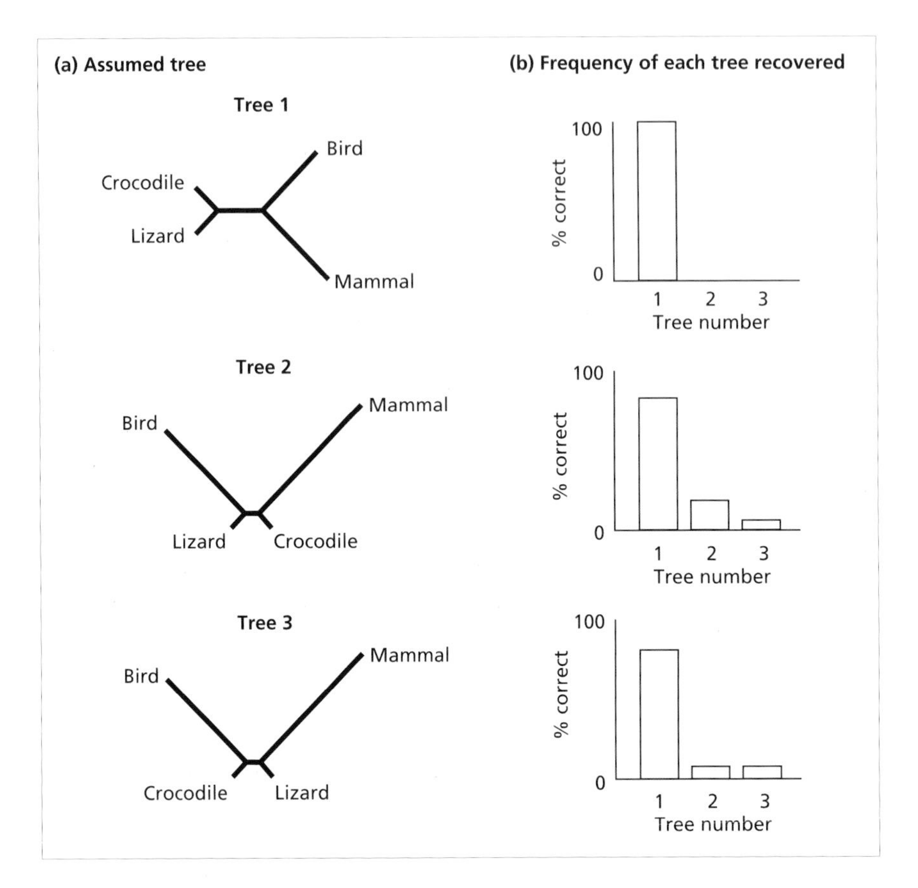

Fig. 6.38 Parametric bootstrapping. Three alternative trees for 18S rRNA sequences from a bird, mammal, crocodile and lizard are shown on the left. For each tree 1000 artificial data sets of the same length as the original 18S rRNA data were generated using parameters derived from that tree. On the right is shown the proportion of times each tree was the most parsimonious tree for the data sets derived from each tree. Note that no matter which tree was used to generate the data, tree 1 is most often recovered as the most parsimonious tree. After Huelsenbeck *et al.* (1996).

This question can be investigated by simulating the evolution of 18S rRNA on the three possible trees for the four taxa and seeing how often a tree-building method recovered the correct tree. Huelsenbeck *et al.* (1996) performed this analysis by first estimating the branch lengths, shape parameter of the gamma distribution (see Chapter 5), and transition/transversion ratio for the real 18S rRNA data set on each of the possible trees for the four taxa. They then used these parameters to generate 1000 artificial data sets of the same size as the original 18S rRNA data set. The results (Fig. 6.38) show that no matter which tree the sequences were evolved on, tree 1 was recovered at least 85% of the time. Even if tree 3 is the true tree (as suggested by morphology), the 18S rRNA data would support tree 1. This suggests that the 18S rRNA data may indeed have been affected by long branch attraction.

6.8.4 What can go wrong?

There are several sources of error that may confound our attempts to reconstruct the true evolutionary relationships among a set of sequences. Many of these have been covered above, but we will summarise them here.

Sampling error

Almost all phylogenetic inference is based on samples of characters, hence the conclusions obtained will in part reflect the vagaries of sampling. This is especially true given the ubiquitous presence of homoplasy, hence the importance of trying to gauge the confidence limits around a tree. Although our focus has been on sequence relationship as opposed to organismal phylogeny, the latter is one of the major uses of molecular phylogenies. In this context a further sampling problem arises, namely the requirement that species relationships are based only on orthologous sequences. If this requirement is not met, organismal and gene trees can become confounded. This topic is considered further in Chapter 8.

Incorrect model of sequence evolution

Methods of phylogenetic inference make assumptions, either implicit or explicit, about the evolutionary process. If these assumptions are violated then we may be misled. Basic assumptions made include variation of rate of substitution among sites, among branches in the tree, and among different nucleotides. To give but one example, the frequencies of the four nucleotides can vary considerably among organisms. If, for example, a gene is AT-rich then most substitutions are likely to involve these two nucleotides simply due to chance. If another, unrelated organism, is similarly AT-rich then it is quite likely that they will share similarities due to independent parallel substitutions, and consequently may be incorrectly grouped together due to this similarity in base composition.

Tree structure

In some cases evolutionary history itself may conspire to thwart our efforts to recover it. Rapid successive cladogenesis, widely differing rates of divergence and extinction can all compromise our ability to accurately reconstruct evolutionary trees. Both simulations and analytical studies have shown that there are situations when current tree-building methods perform poorly. Despite our best efforts, the trace of history may have been eroded by subsequent evolution.

6.9 Summary

Different methods may be more appropriate in different circumstances. This conclusion may seem disturbingly inconclusive, but it reflects the nature of the problem. Maximum likelihood is a very attractive method but it is computationally very expensive, which limits its usefulness as an optimality criterion for trees. Its strength lies in its explicit statistical basis, making it a powerful tool for analysing models of evolution. For all but relatively small data sets, maximum likelihood's main use is likely to be investigating molecular evolution on given phylogenies, rather than as a tool for finding those phylogenies. While advocates of maximum likelihood often look askance at methods such as parsimony, in reality the latter often performs very well, and it generally allows a broader exploration of alternative trees than does maximum likelihood. Rapid methods of tree assembly, such as neighbour joining, come to the fore in the analysis of very large data sets of many sequences where other methods might prove too sluggish. Newly emerging methods such as spectral analysis and split decomposition emphasise visualising and exploring the data rather than simply finding the 'best' tree. Trees are hypotheses of evolutionary relationship, and are inferences made on limited data using often greatly simplified models of evolution that nevertheless pose immensely challenging computational problems. Finding the 'one true tree' is important, but so is knowledge of the range of signals in the data and the appropriateness of the model used.

6.10 Further reading

The sources given here are by no means exhaustive, but instead offer entry points into a large (and growing) literature on inferring evolutionary trees. Useful reviews include Swofford *et al.* (1996) and Penny *et al.* (1992). For a treatment of phenetics see Ridley (1986); Hull (1988) provides a stimulating overview of the history of the debates about phenetics and cladistics. The classic critique of distance methods is Farris (1981); Penny (1982) provides additional objections. Sober (1988) provides an excellent discussion of the foundations of parsimony and its relationship to maximum likelihood. Felsenstein (1981) has long been a proponent of maximum likelihood methods. For detailed evaluation of the merits of different likelihood models see Goldman (1993) and Yang (1994). For a technical treatment of the question of whether the maximum likelihood value for a given tree is unique see Steel (1994). Huelsenbeck and Rannala (1997) give a recent review of likelihood ratio tests. A good introduction to spectral analysis is given by Lento *et al.* (1995). Split decomposition is described by Bandelt and Dress (1992) and Dopazo *et al.* (1993). Hillis *et al.* (1992) describes their elegant work on experimental phylogenies using bacteriophage T7. Leitner *et al.* (1996) give an example of using a known phylogeny for HIV to evaluate different phylogenetic methods. For a critique of experimental work see Sober (1993). Hillis's group have also made extensive simulation studies (Huelsenbeck and Hillis, 1993) and have recently applied parametric bootstrapping to phylogenetic studies (Huelsenbeck *et al.*, 1996). Sanderson (1995) reviews the debate about the merits of bootstrapping in phylogenetics.

Chapter 7
Models of Molecular Evolution

7.1 Models of the evolutionary process

Haemoglobin is one of the most studied of all proteins. Its role is to bind oxygen from the lungs (or gills) and deliver it to the tissues where it can be released, and to bring carbon dioxide, a by-product of oxidation, back to the lungs where it is then removed from the body. As we have already seen, haemoglobin in most vertebrates is made up of four polypeptide chains, two encoded by genes from the α-globin family and two from the β-globin family. Chapter 3 described some of the ways in which the organisation of this multigene family has changed through time. But how have the amino acid sequences themselves evolved? One way to approach this question is simply to count up the number of amino acid changes between different species using some of the methods described in Chapter 5. We have done this for a selection of vertebrate α-chains—comparing each with human—and plotted the results in Fig. 7.1. What we see is striking: as we compare more and more distantly related species, the number of amino acid changes (shown as Dayhoff distances) steadily increases.

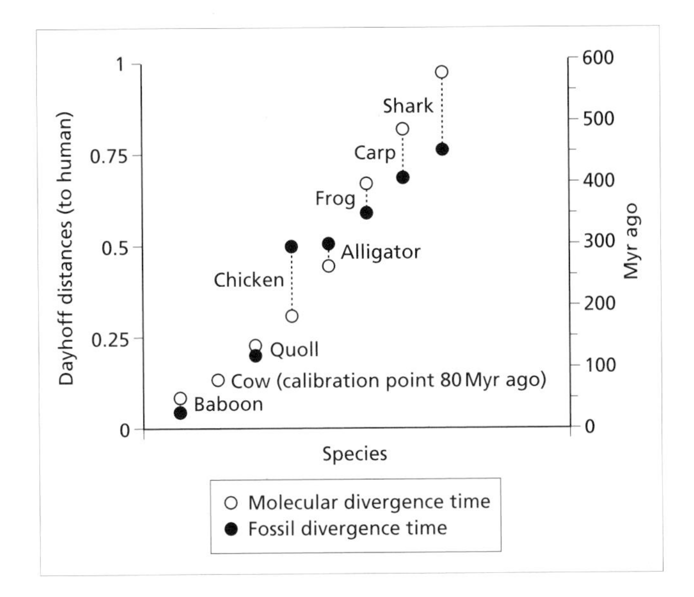

Fig. 7.1 The α-globin molecular clock. The distances between the amino acid sequences of humans and a number of other vertebrates are used to estimate species divergence times, using the split between humans (primates) and cows (artiodactyls) at approximately 80 Myr ago as a calibration point. These molecular estimates (open circles) are then compared with those taken from the fossil record (closed circles). The difference between the two estimates is indicated by the dotted line. The Quoll is a marsupial cat.

In fact, the increase is so steady that it appears that mutations are accumulating at a constant rate.

Now let us assume that we know the divergence time of one of the points on the graph, say that between humans and cows at approximately 80 million years ago. With this information in hand we can calculate the divergence times of humans and the other species. For example, humans and cows differ at 17 of the 149 amino acid sites in the alignment (Dayhoff distance of 0.131), whereas human and alligator have 47 differences (Dayhoff distance of 0.445), 3.4 times as many. This suggests that humans and alligators diverged 3.4 times longer ago than humans and cows, which works out at 270 million years ago. Remarkably, the fossil record suggests that alligators and humans diverged at about 300 million years ago. So the α-globin sequences have kept a fairly accurate record of when these two species split. If we do this for every point on the graph we find that the divergence times stored in α-globin are generally close to those suggested by the fossil record. There are some exceptions—the divergence time of chicken is greatly underestimated and that of shark greatly overestimated—but overall the sequences of this protein are surprisingly good time-keepers. In other words, α-haemoglobin behaves like a **molecular clock**.

But why should such a clock exist and how accurate is it? The answers to these questions will tell us a great deal about the underlying **processes** of molecular evolution—*how* nucleotide and amino acid sequences evolve. This is the subject of this chapter.

Since the 1960s there have been two conflicting models of how molecular evolution takes place: one that it is dominated by the genetic drift of neutral mutations, the other that the natural selection of advantageous mutations is the more important force. Because of the historical importance and impact of this debate, we will use it as the framework to understand some other aspects of sequence evolution. Knowing which of these models best explains molecular evolution is also important because it will ultimately lead to the development of more realistic models of DNA substitution, so improving the accuracy of phylogenetic analyses. Conversely, phylogenetic trees are being increasingly used as a way of testing which of these two models best fits the data.

7.1.1 The classical and balance schools of population genetics

The foundations of the neutralist–selectionist debate were laid in the 1950s in the debate between the classical and balance schools of population genetics. The dispute here was over how much genetic variation existed within populations. The **classical school** believed that natural selection was predominantly a negative or purifying force, removing deleterious alleles, so that there was little genetic variability within populations. Most individuals were therefore thought to be homozygous for a common wild-type allele at the majority of loci. For example, the Nobel laureate H.J. Muller famously predicted that in an average human only 1 locus in every 1000 would be heterozygous. The **balance school**, on the other hand, claimed that levels of genetic variation in populations were so high that most loci were polymorphic and that individuals were heterozygous at a large number of loci. These high levels of variation were thought to be maintained by different forms of balancing selection such as that favouring heterozygotes (overdominant selection—see Chapter 4). Although both schools were agreed that natural selection was the driving force of evolution, there was no evidence for the central issue which divided them: how much genetic variation existed within and between species?

The question of how much genetic variation existed between species was eventually answered during the 1950s and 1960s when the new molecular technique of protein sequencing uncovered substantial diversity in the polypeptide chains of various proteins. The globins were an early example and, like many other proteins, revealed a huge variation in sequence between different species (see the amino acid alignment in Fig. 7.6). The same was also found to be true of genetic variation within species when, in 1966, a series of papers using protein electrophoresis, another new technique, showed that natural populations harboured far more genetic variation than was previously anticipated.

Within populations genetic diversity in proteins (allozymes) can be quantified as the number of loci in an individual that are polymorphic and, at each

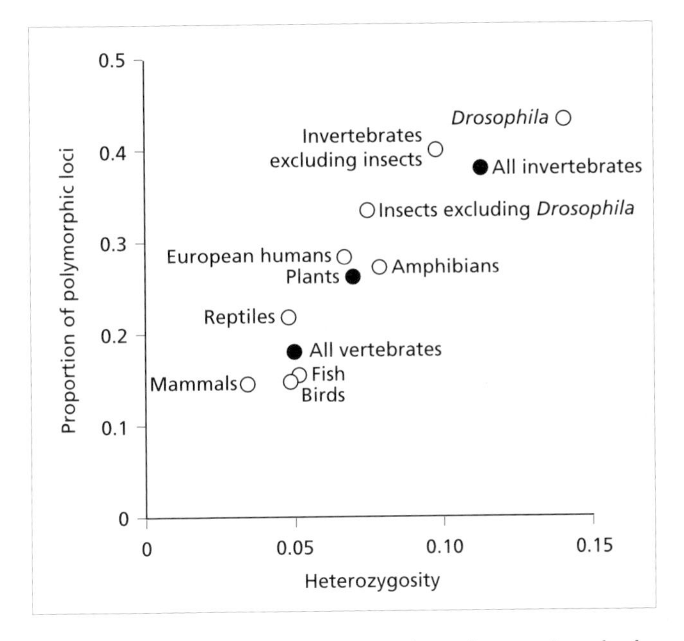

Fig. 7.2 Summary of estimated levels of heterozygosity and proportion of polymorphic loci in allozymes from different species groups. Closed circles denote average values for major taxonomic groups. It should be noted that the human values are artificially large compared with those from other mammals because highly polymorphic proteins were over represented in this case. Data from Nevo (1978). Adapted from Hartl and Clark (1989).

polymorphic locus, the average heterozygosity (described in Chapter 4). Lewontin and Hubby (1966) surveyed 18 loci in five populations of *Drosophila pseudoobscura* and found an average of 30% to be polymorphic with a mean heterozygosity of 11%. In the human populations studied by Harris (1966), the comparable values were 30% polymorphic and 9.9% average heterozygosity. Similarly high levels of variation have been subsequently found in other taxa (Fig. 7.2). These figures are also likely to be underestimates because mutations which do change protein weight or charge, and hence electrophoretic mobility, escape detection.

7.1.2 The cost of natural selection and the rise of the neutral theory

By the mid-1960s protein sequencing and electrophoresis had shown that the fuel of evolution, genetic variation, was available in abundance within and between species. The balance school was right in that populations were generally very variable at the genetic level. But these results also posed a problem. If, as it was assumed, natural selection had produced all this diversity, would not it also be true that individuals with inferior alleles would be selectively removed from the population? Could the population even go extinct with all this 'selective death'? This has been called the **cost of natural selection**.

The cost that natural selection imposes when it acts can be thought of as part of the overall **genetic load** carried by a population. Genetic load describes the extent to which the fitness of any individual is below the optimum for the population as a whole because of the deleterious alleles they carry. In the 1950s and 1960s genetic load was thought to be a serious problem and was one of the reasons why the classical school thought genetic variation was limited. It can arise in two ways. **Substitutional load**, as we have just seen, occurs when a mutation replaces (substitutes) the previously most frequent allele in a population because it has a selective advantage (positive selection). The great geneticist J.B.S. Haldane (1957) calculated that there would be an upper limit on the number of mutations that can be fixed in this way in a population at any time, estimated to be one allele substitution every 300 generations, otherwise the level of selective death (the load) would be too high and the population would go extinct.

Segregational load, on the other hand, occurs when a polymorphism is maintained through overdominant selection. A famous and highly illustrative example of segregational load is provided by human sickle-cell haemoglobin. An individual homozygous for the Hb^A haemoglobin allele produces the normal protein. However, in some parts of the world individuals with this genotype are susceptible to infection with the parasite *Plasmodium falciparum* which causes malaria. Conversely, individuals who are homozygous for the sickle-cell allele, Hb^S, produced by a single amino acid change in β-haemoglobin, will develop the disease sickle-cell anaemia. About 80% of those with this genotype will die before reproduction. Crucially, those heterozygous for the normal and sickle-cell alleles do not develop the pathogenic sickle-cell form and are much less susceptible to malaria, although the precise reasons why are unknown. Clearly there is a selective advantage for individuals to be heterozygous at this locus and this heterozygous advantage is known to maintain a high frequency of the sickle-cell allele in areas endemic for malaria (the Hb^S allele reaches a frequency of 40% in some West African populations). Unfortunately, the laws of Mendelian segregation do not allow for the production of only heterozygotes: in an $Hb^A \times Hb^S$ cross only 50% of individuals on average will be heterozygous. Individuals who are Hb^AHb^A homozygous (susceptible to malaria) and Hb^SHb^S homozygous (subject to sickle-cell anaemia) are still produced as long as heterozygotes are favoured, so that the population will suffer a genetic load. But if this was the situation at every polymorphic locus how could populations possibly survive?

There was, however, a way in which high levels of genetic variation could be explained without encountering excessive selective death: simply by dropping the assumption that natural selection was the driving force of molecular evolution and instead allowing the majority of the mutations fixed to have *no* effect on fitness. Mutations like this would obviously confer no selective cost. These mutations, which are not subject to natural selection, are called **neutral mutations** and are lost (usually) or fixed (very occasionally) by the process of genetic drift, caused by random sampling of gametes during sexual reproduction (see Chapter 4). This **neutral theory of molecular evolution** was first proposed by Motoo Kimura in 1968 and Jack King and Tom Jukes in 1969 and has become one of the most important and controversial theories in evolutionary biology.

To neutralists most mutations are either deleterious and selectively removed, or neutral, in which case there is a small probability that they are fixed. Natural selection is therefore incorporated into the neutral theory but only as a negative, purifying force, removing deleterious mutations from the genome, and plays only a very minor role in the fixation of mutants (Fig. 7.3). Although the vast majority of neutral mutations are lost by chance, a tiny minority, which will be important in evolutionary time, will eventually be fixed (see Chapter 4 for more about the probability of fixation of neutral mutations). Because neutralists believe that natural selection does not play an important role in the fixation of mutant alleles, we can think of the neutral theory as one which gives the dual processes of mutation and genetic drift predominance over that of selection at the molecular level. Selectionists, on the other hand, believe that substitutions are fixed because they confer a selective advantage and that neutral mutations are rare, although they agree that overall most mutations are likely to be deleterious and so removed from populations. These opposing schools therefore have very different views of the evolutionary process: to neutralists it is dominated by chance, to selectionists necessity is the key.

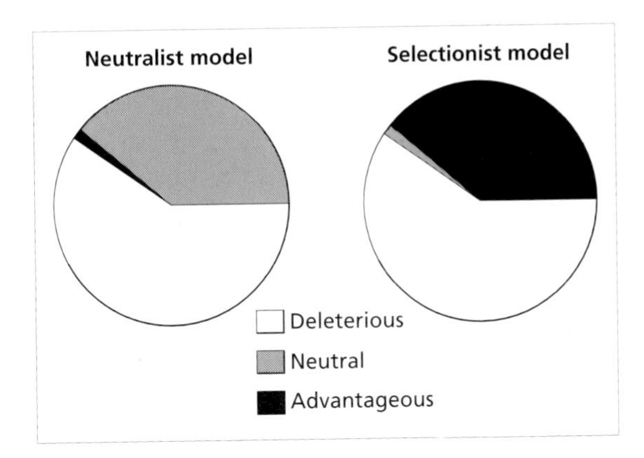

Fig. 7.3 Neutralist and selectionist models of molecular evolution. The pie represents the total number of mutations arising in a gene. In the neutralist model the majority of these mutations are deleterious and removed by negative selection but of those fixed most are neutral, with only a small number advantageous. Although the selectionist model agrees that most mutations are deleterious and lost, it differs in predicting that a majority of those fixed are selectively advantageous with neutral mutations being rare. The proportions given for each type of mutation are designed to highlight the differences between the two schools, *not* to accurately reflect the number of mutations in each category (for example, neutralists would probably give an even smaller slice to advantageous mutations).

7.1.3 The neutralist–selectionist debate

When it was first proposed the neutral theory was seen as a challenge to one of the central pillars of neo-Darwinism because it seemed to say that some characteristics of a species could be non-adaptive. However, this does not mean that the neutral theory attempts to replace the idea of evolution by natural selection. Rather, it claims that the fixation of mutants with a selective advantage, acknowledged to be the main process of morphological evolution, occurs at such low frequencies at the molecular level to make it irrelevant on a large-scale. Indeed, neutralists effectively believe that most genes and proteins are so well adapted, because of the past action of natural selection, that it is hard to improve upon them. Most new mutations will therefore be either deleterious or neutral.

Early debates also spent a lot of time on genetic loads. For example, even with the limited amount of amino acid sequence data available in the late 1960s it was obvious that rates of amino acid substitution were far higher than Haldane's upper limit of one change every 300 generations. In the original paper proposing the neutral theory, Kimura estimated an average rate of one substitution every two years in mammalian populations and argues that natural selection could not be responsible for such a rapid rate because the number of selective deaths, given Haldane's calculation, would be too large. His solution to this paradox was to propose that most substitutions were neutral. Since this time the amino acid sequences of many more proteins have been determined and most evolve at rates higher than the maximum envisaged by Haldane (Fig. 7.4).

However, a number of unrealistic assumptions were made when calculating the cost of natural selection which undermine the use of high rates of substitution as evidence in support of the neutral theory. First, it was assumed that each gene acted independently so that the load could be worked out by simply multiplying the figures for one gene by the total number of loci. However, because of the complexity of metabolic and developmental pathways genes will often act together so that the number of independent genotypes will be less than the number used to calculate selective costs. It was also assumed that individuals with an inferior allele will die before reproduction because of a direct consequence of that allele and in doing so generate extra mortality in the population—a process known as **hard selection**. But it is also possible that the death of an individual who carries this particular allele is caused by intraspecific competition, which happens in any circumstance, so that no extra mortality is caused. This is **soft selection**. If most mortality in a population is the outcome of soft rather than hard selection, the overall cost of natural selection will be greatly reduced as the number of deaths caused by gene substitution will be lower. Put together, these relaxed assumptions allow the maximum rate of substitution due to natural selection to greatly increase (by reducing the cost) and in so doing

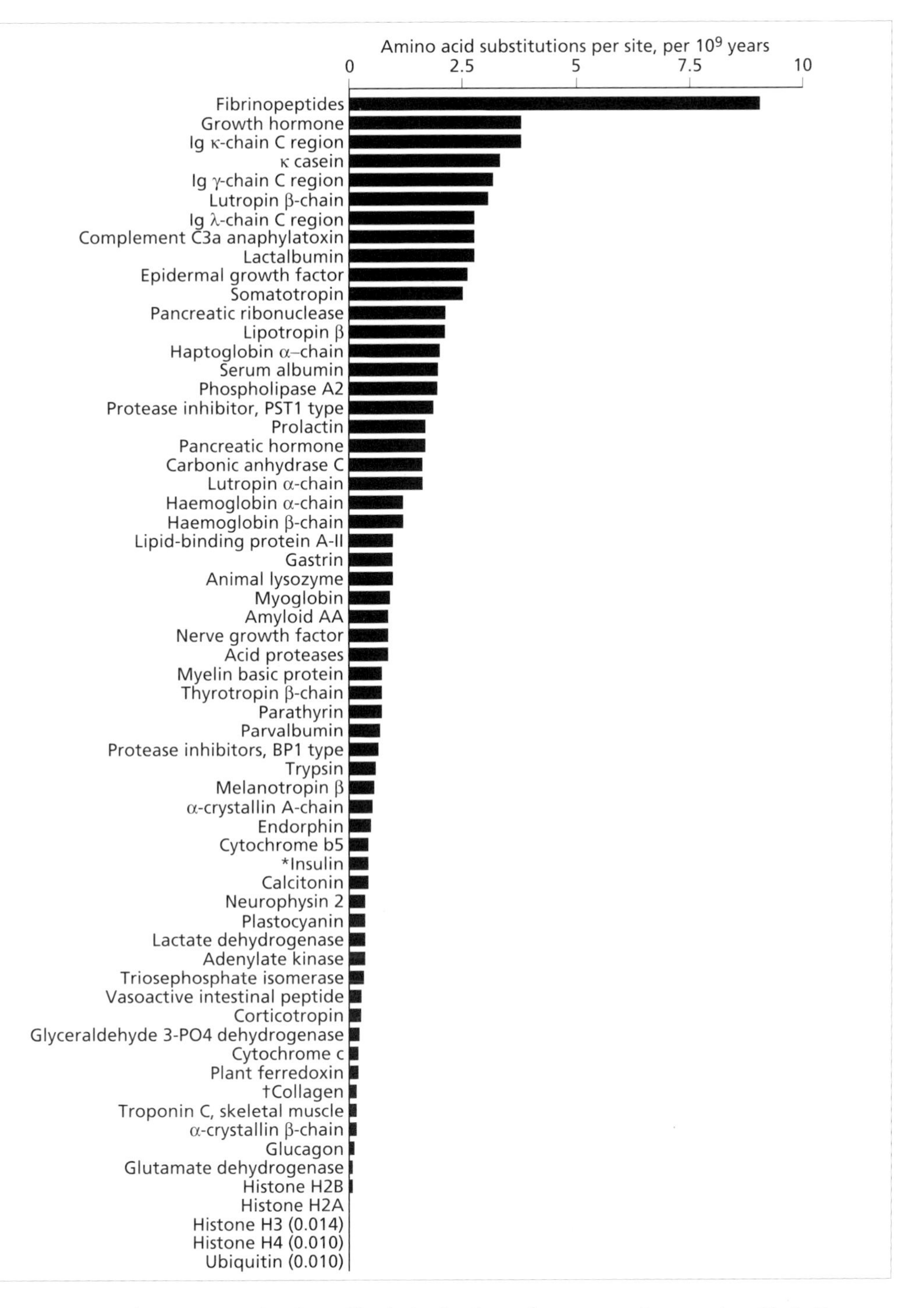

Fig. 7.4 Rates of amino acid substitution for various mammalian proteins. *Excluding guinea pig and coypu, the phylogenetic position of which has been debated (see section 7.4); † Excluding non-repetitive ends. Data from Nei (1987).

remove a potential barrier to the importance of natural selection in molecular evolution.

The same is also true of the excess genetic load supposedly produced when natural selection acts to maintain the high levels of genetic variation within species: because of the flaws in the load argument, too high a selective cost cannot be used as a way in which to discount the power of natural selection in maintaining polymorphisms. Furthermore, heterozygous advantage is not the only way natural selection can preserve genetic variation within species and other than sickle-cell haemoglobin there are few reported examples. A possible alternative is frequency-dependent selection in which the fitness of an allele is a function of its frequency in the population (see Chapter 4). Selection of this type may produce little or no genetic load.

Today the debate surrounding the neutral theory has moved on from criticisms of neo-Darwinism and the weighty theory of genetic load and is instead concerned with finding the most suitable explanation for the large-scale patterns of molecular evolution. In the sections that follow we will discuss four important predictions about the nature of molecular evolution made by the neutral theory. If we can determine whether these predictions are right or not, we will learn a great deal about how genes evolve. These predictions are that:

1 There is an inverse correlation between the rate of substitution and the degree of functional constraint acting on a gene, such that functionally constrained genes, or regions within genes, evolve at the lowest rates (and vice versa).

2 Patterns of base composition and codon usage reflect mutational rather than selective processes.

3 There is a constant rate, or molecular clock, of sequence evolution.

4 The level of within species genetic variation is the product of only population size and mutation rate and is correlated to the level of sequence divergence between species.

7.2 Functional constraint and the rate of substitution

7.2.1 Functional constraint at the amino acid level

As we can see from Fig. 7.4, not only are rates of amino acid substitution relatively high but they are often extremely variable: fibrinopeptides, for example, evolve about 900 times faster than ubiquitin. To neutralists this variation in rate *between* different proteins (or between different regions of the same protein) is explained by differences in selective constraint rather than in positive selection. According to the theory, genes (or proteins) differ in what proportion of their possible mutants will be deleterious and selectively removed, or neutral and fixed with a small probability (advantageous mutations occur so rarely that they can be ignored). The more functionally constrained the gene, the greater the probability that a mutation will be deleterious rather

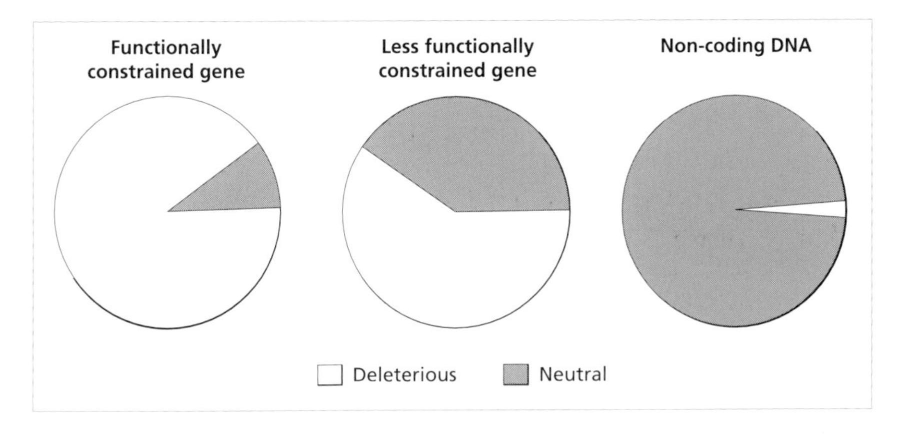

Fig. 7.5 The effects of functional constraint on the relative frequency of deleterious and neutral mutations. In a functionally constrained gene (or region within a gene) most of the mutations that occur will be deleterious and so are removed by negative selection. As only a few mutations will be neutral the rate of substitution will be low. For genes with less functional constraint the number of mutations that are neutral is higher (reaching a peak in non-coding DNA) so that the rate of substitution increases. Once again, the actual proportions used are arbitrary.

than neutral and so the lower the rate of nucleotide substitution (Fig. 7.5). For less functionally constrained genes fewer of the mutations that occur will be deleterious and more will be neutral so the rate of substitution will increase. In this way differences in functional constraint—that is, in the strength of negative selection—will also lead to differences in substitution rate. The large-scale differences in rates of amino acid substitution between proteins shown in Fig. 7.4 are compatible with differences in functional constraint. For example, proteins like histones that interact structurally with DNA so that any amino acid change may be deleterious, evolve at the lowest rates whilst those that only interact with other proteins, such as members of the immune system or hormones, evolve at the highest rates presumably because there is a great deal more flexibility in which amino acids are functionally suitable. To many, this correlation between functional constraint and rate of substitution represents the best evidence for the neutral theory.

The globins provide a good example of the relationship between functional constraint and the rate of substitution. The most functionally important positions in haemoglobin are those where haem is bound. There is a remarkable conservation of the amino acids occupying these sites over millions of years of evolutionary history. For example, the amino acids at positions 44 and 94 in the sequence alignment shown in Fig. 7.6 (F, phenylalanine and H, histidine, respectively; indicated by arrows), both of which are involved in haem-binding, are conserved in all known functional α- and β-globin sequences, as well as in myoglobin, the invertebrate globins and the leghaemoglobins found in some plants. Other sites of functional importance and which are highly

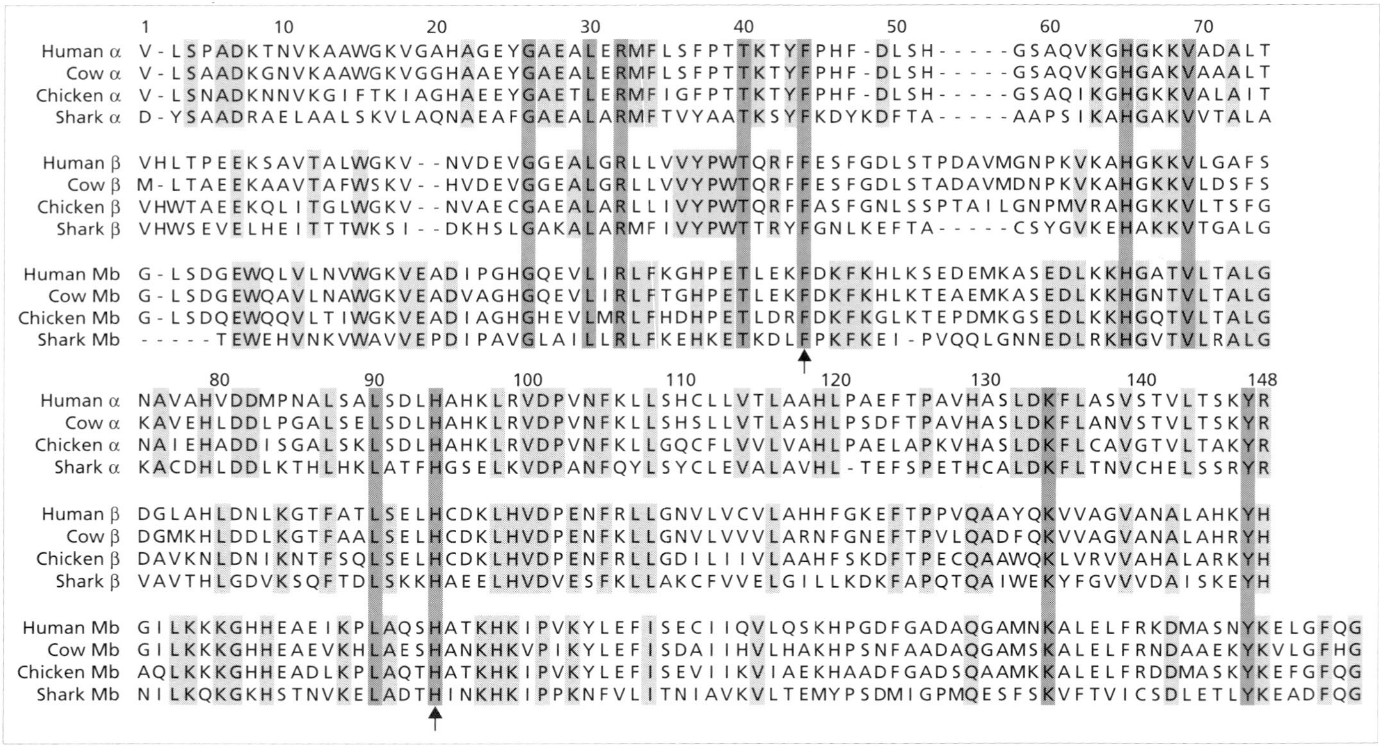

```
             1        10        20        30        40        50        60        70
Human α      V-LSPADKTNVKAAWGKVGAHAGEYGAEALERMFLSFPTTKTYFPHF-DLSH-----GSAQVKGHGKKVADALT
Cow α        V-LSAADKGNVKAAWGKVGGHAAEYGAEALERMFLSFPTTKTYFPHF-DLSH-----GSAQVKGHGAKVAAALT
Chicken α    V-LSNADKNNVKGIFTKIAGHAEEYGAETLERMFIGFPTTKTYFPHF-DLSH-----GSAQIKGHGKKVALAIT
Shark α      D-YSAADRAELAALSKVLAQNAEAFGAEALARMFTVYAATKSYFKDYKDFTA-----AAPSIKAHGAKVVTALA

Human β      VHLTPEEKSAVTALWGKV--NVDEVGGEALGRLLVVYPWTQRFFESFGDLSTPDAVMGNPKVKAHGKKVLGAFS
Cow β        M-LTAEEKAAVTAFWSKV--HVDEVGGEALGRLLVVYPWTQRFFESFGDLSTADAVMDNPKVKAHGKKVLDSFS
Chicken β    VHWTAEEKQLITGLWGKV--NVAECGAEALARLLIVYPWTQRFFASFGNLSSPTAILGNPMVRAHGKKVLTSFG
Shark β      VHWSEVELHEITTTWKSI--DKHSLGAKALARMFIVYPWTTRYFGNLKEFTA-----CSYGVKEHAKKVTGALG

Human Mb     G-LSDGEWQLVLNVWGKVEADIPGHGQEVLIRLFKGHPETLEKFDKFKHLKSEDEMKASEDLKKHGATVLTALG
Cow Mb       G-LSDGEWQAVLNAWGKVEADVAGHGQEVLIRLFTGHPETLEKFDKFKHLKTEAEMKASEDLKKHGNTVLTALG
Chicken Mb   G-LSDQEWQQVLTIWGKVEADIAGHGHEVLMRLFHDHPETLDRFDKFKGLKTEPDMKGSEDLKKHGQTVLTALG
Shark Mb     -----TEWEHVNKVWAVVEPDIPAVGLAILLRLFKEHKETKDLFPKFKEI-PVQQLGNNEDLRKHGVTVLRALG
                                                        ↑

                  80        90        100       110       120       130       140     148
Human α      NAVAHVDDMPNALSALSDLHAHKLRVDPVNFKLLSHCLLVTLAAHLPAEFTPAVHASLDKFLASVSTVLTSKYR
Cow α        KAVEHLDDLPGALSELSDLHAHKLRVDPVNFKLLSHSLLVTLASHLPSDFTPAVHASLDKFLANVSTVLTSKYR
Chicken α    NAIEHADDISGALSKLSDLHAHKLRVDPVNFKLLGQCFLVVLVAHLPAELAPKVHASLDKFLCAVGTVLTAKYR
Shark α      KACDHLDDLKTHLHKLATFHGSELKVDPANFQYLSYCLEVALAVHL-TEFSPETHCALDKFLTNVCHELSSRYR

Human β      DGLAHLDNLKGTFATLSELHCDKLHVDPENFRLLGNVLVCVLAHHFGKEFTPPVQAAYQKVVAGVANALAHKYH
Cow β        DGMKHLDDLKGTFAALSELHCDKLHVDPENFKLLGNVLVVVLARNFGNEFTPVLQADFQKVVAGVANALAHRYH
Chicken β    DAVKNLDNIKNTFSQLSELHCDKLHVDPENFRLLGDILIIVLAAHFSKDFTPECQAAWQKLVRVVAHALARKYH
Shark β      VAVTHLGDVKSQFTDLSKKHAEELHVDVESFKLLAKCFVVELGILLKDKFAPQTQAIWEKYFGVVVDAISKEYH

Human Mb     GILKKKGHHEAEIKPLAQSHATKHKIPVKYLEFISECIIQVLQSKHPGDFGADAQGAMNKALELFRKDMASNYKELGFQG
Cow Mb       GILKKKGHHEAEVKHLAESHANKHKVPIKYLEFISDAIIHVLHAKHPSNFAADAQGAMSKALELFRNDAAEKYKVLGFHG
Chicken Mb   AQLKKKGHHEADLKPLAQTHATKHKIPVKYLEFISEVIIKVIAEKHAADFGADSQAAMKKALELFRDDMASKYKEFGFQG
Shark Mb     NILKQKGKHSTNVKELADTHINKHKIPPKNFVLITNIAVKVLTEMYPSDMIGPMQESFSKVFTVICSDLETLYKEADFQG
                                ↑
```

Fig. 7.6 Alignment of amino acid sequences from adult α- and β-haemoglobin and myoglobin (Mb) proteins from four vertebrate species. Light-shaded residues are those conserved between all four sequences from each protein while heavy-shaded residues are those conserved across all proteins. Arrows indicate residues which are also conserved in invertebrate and plant globins. For ease of presentation a repetitive sequence of STSTSTS was removed from the start of the shark α-haemoglobin sequence.

conserved are those involved in binding with D-2,3-biphosphoglycerate (DPG) (a heterotropic ligand which reduces oxygen affinity so that more can be released into the tissues), those which make up salt bridges and those which provide contacts between the α- and β-chains of the molecule.

7.2.2 Functional constraint at the nucleotide level

More evidence for the influence of functional constraint on rates of substitution is found at the nucleotide level. As we saw in Chapter 3, most eukaryotes have far more DNA than is needed to make or regulate proteins, so that large parts of the genome appear to lack a function. As predicted by the neutralists even before DNA sequences became available, this non-coding DNA evolves faster than parts of the genome which encode proteins and is characterised by insertions and deletions which are generally rare in coding regions. This is evidence that most non-coding DNA is neutral. Pseudogenes—duplicated but inactivated copies of genes (see Chapter 3)—provide another example of the relationship between functional constraint and substitution rate. Although pseudogenes were originally derived from working genes they are free from functional constraints and, in accord with neutral theory, evolve at higher rates. For example, the rate of substitution for the human ψα1-globin pseudogene is higher than that of all three codon positions of the functional α-globin gene from which it was derived (Table 7.1).

While it is clear that DNA without a function evolves at the highest rate, presumably because it is completely neutral, whether this is also true of nucleotide positions within fully functional genes is more controversial. Because of the degeneracy of the genetic code, mutations in different codon positions differ in the probability that they will alter an amino acid, so that the three codon positions might also differ in functional constraint. For example, one-half of third positions are fourfold degenerate whereas no second positions are

Table 7.1 Rates of nucleotide substitution per site, per year $\times 10^{-9}$ for mammalian globin pseudogenes and their functional homologues. The rate of substitution for each codon position within the functional genes has been estimated separately. Adapted from Nei (1987). Copyright 1987, Columbia University Press. Reprinted with permission of the publisher.

		Functional genes		
Gene	Pseudogene	Position 1	Position 2	Position 3
Mouse ψα3	5.0	0.75	0.68	2.65
Human ψα1	5.1	0.75	0.68	2.65
Rabbit ψβ2	4.1	0.94	0.71	2.02
Goat $\psi\beta^x$ and ψ^z	4.4	0.94	0.71	2.02
Average	4.7	0.85	0.70	2.34

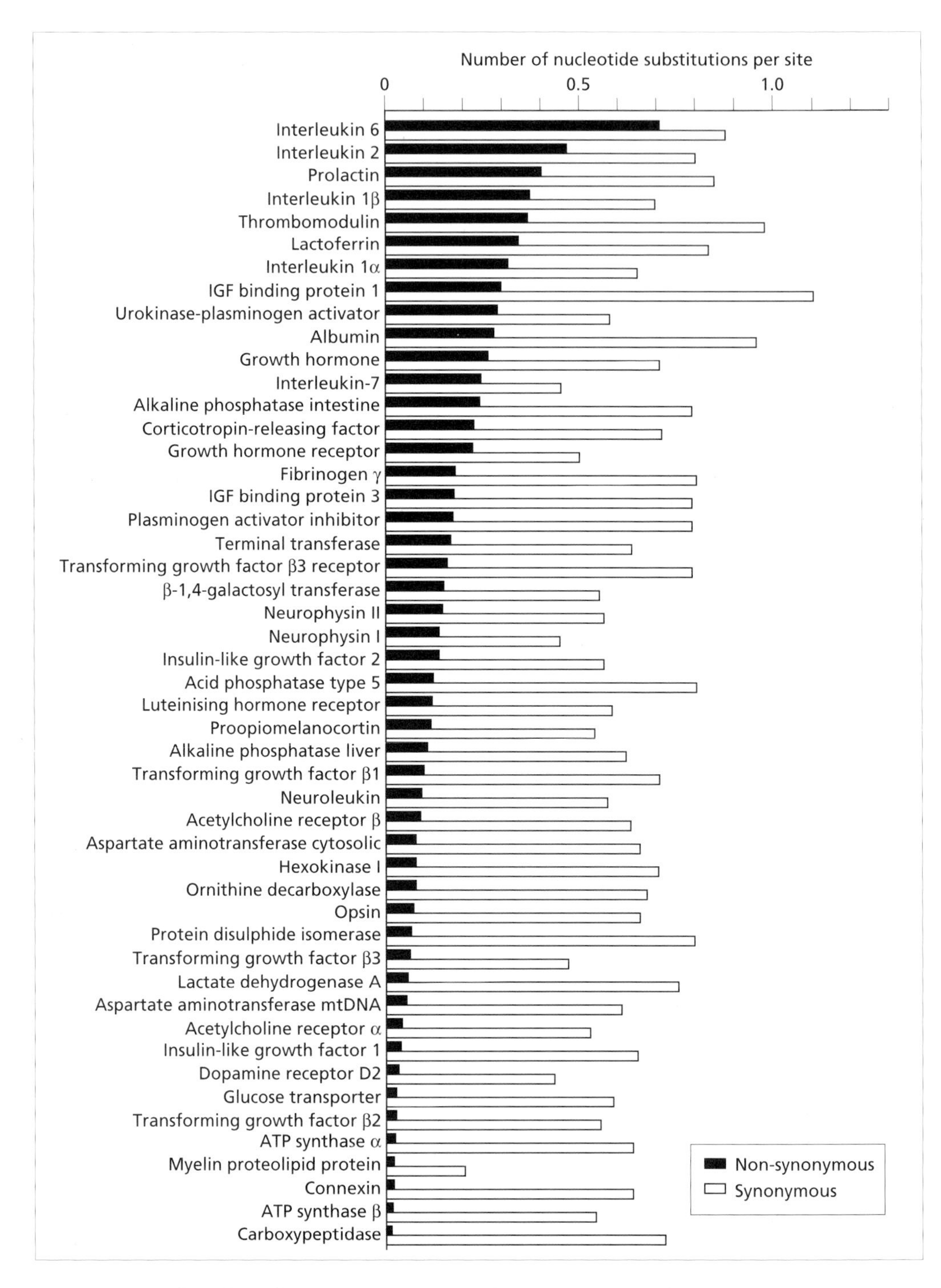

Fig. 7.7 Estimated numbers of non-synonymous (d_N) and synonymous (d_S) substitutions per site for 49 genes from primates, rodents and artiodactyls combined. The data are organised according to the number of non-synonymous changes (highest at top, lowest at bottom). Compare this to the rates of amino acid change shown in Fig. 7.4. Data from Ohta (1995).

of this type. Sequence analysis has shown that this prediction is upheld in the vast majority of genes: in the globin genes presented in Table 7.1, for example, the third codon positions are evolving approximately threefold faster than the first two positions although not as quickly as pseudogenes, suggesting they are still subject to weak constraints. More generally it is also clear that synonymous (silent) sites evolve faster than non-synonymous (amino acid changing) sites within the same gene and that the latter are much more variable in rate, again as predicted by neutral theory (Fig. 7.7; also see Box 7.1). Although this topic is addressed more fully in the next section, it looks as if synonymous sites in mammalian genomes are evolving in a generally neutral manner. On the other hand, the question of whether non-synonymous substitutions are neutral or advantageous in mammals remains open. In fact, instead of being strictly neutral, many non-synonymous substitutions may only very slightly reduce or improve fitness because they only make minor alterations to protein structure and function. Substitutions of this sort are called **nearly neutral** and may be very important in molecular evolution. A refinement of the neutral theory based on the idea that most non-synonymous or amino acid changes are nearly neutral is outlined in section 7.5.

7.2.3 Changes in functional constraint

It is important to remember that functional constraints are not fixed entities: an amino acid which is selectively important in one environment may be neutral in another (and vice versa). Changes in functional constraint over time are often cited as a way in which the neutral theory can explain variable rates of substitution because this can also be thought of as a change in the proportion of sites that are neutral and thus in the number of substitutions which can accumulate.

Changes in functional constraint can be measured by the ratio of non-synonymous to synonymous substitutions (d_N/d_S) (Box 7.1). We can turn to the human and simian immunodeficiency viruses, HIV and SIV, for an example (Fig. 7.8). It is likely that the highly pathogenic viruses found in humans (HIV-1) and macaque monkeys (SIV_{MAC}) emerged through recent cross-species transmission events from primates in which these viruses are much less pathogenic, such as SIV_{AGM} from African green monkeys and SIV_{SMM} from sooty mangabeys. HIV-1 and SIV_{MAC} elicit a strong immune response in their new hosts and there is positive selection for the envelope proteins of the virus to change to evade this response. This is reflected in the d_N/d_S ratio of sequences from the envelope protein gp120 which is higher for HIV-1 and SIV_{MAC} than for SIV_{AGM} and SIV_{SMM}. The lower d_N/d_S ratio in the non-pathogenic monkey viruses, which have probably been associated with their host species for longer, occurs because they generate a weaker immune response and so are under less selection pressure for change in the viral envelope. HIV-2, a mainly West African virus and another recent invader into human populations, has an intermediate level of pathogenicity and an intermediate d_N/d_S ratio.

Box 7.1 Testing the neutrality of mutations

Although a variety of tests of the neutrality of mutations has been proposed, we focus on three of the most common, all of which use nucleotide sequence data.

1 *Levels of genetic variation within and between species*

Because the neutral theory predicts that the rate of substitution is only dependent on the underlying mutation rate, genes that evolve quickly will exhibit more genetic variation within and between species than genes that evolve more slowly. Therefore, levels of variation within and between species should be correlated under neutrality (assuming recombination is absent). This can be tested in a number of ways.

The **McDonald–Kreitman test** examines whether the ratio of non-synonymous to synonymous substitutions differs within and between species. If both types of substitution are neutral, so that their rate of fixation is determined only by the mutation rate, their ratio will remain the same within and between species. This can be assessed with a 2×2 test of independence (a *G*-test; see Fig. 7.22). Although this test assumes that the neutral mutation rate is constant, it does not require population size to be so. The more complex **Hudson–Kreitman–Aguadé (HKA)** test assesses whether sequence variation in different gene regions (such as flanking and coding regions) is the same within and between species. Unlike that of McDonald–Kreitman, the HKA test assumes that population sizes have not changed through time.

2 *Tajima's D statistic*

One drawback with the McDonald–Kreitman and HKA tests is that they require data from at least two species. In contrast, Tajima's *D* statistic allows the neutrality of mutations to be tested using sequence (polymorphism) data from a single species.

Under neutral theory, the expected amount of genetic variation per nucleotide is based on $4N_e\mu$, usually denoted θ. θ can be estimated using either the number of segregating sites, *S*, or the average number of pairwise differences, Π, between the sequences in the sample (see Chapter 4, Box 4.1). If evolution is neutral then estimates based on *S* and Π give the *same* value of θ. However, natural selection changes the structure of genetic variation within populations because alleles persist for different times under selection compared with drift. Because *S* and Π consider different aspects of population variation—Π assesses the *frequency* of mutant alleles in the population whereas *S* examines the *number* of polymorphic sites—they will be affected in different ways by natural selection and so give *different* estimates of θ. This led Tajima (1989) to propose that the difference (*D*) between the estimates of θ given by *S* and Π could serve as a test of the neutrality of mutations ($D = 0$ under neutrality).

continued

Box 7.1 *continued*

Another way to think about how genetic drift and selection change the genetic structure of populations is with the frequency spectrum of segregating sites (the frequencies of alleles at each polymorphic site). For example, hitchhiking produces an excess of segregating sites with alleles at low frequencies (negative D value), whilst balancing selection will produce an excess of segregating sites with alleles at high frequencies (positive D value).

3 *Comparing rates of non-synonymous and synonymous substitution*
If we assume that synonymous substitutions are neutral, their rate of substitution (fixation) is equal to the rate at which they are produced by mutation. If non-synonymous sites are also evolving in a neutral manner then their rate of substitution should be the same as that at synonymous sites ($d_N = d_S$). In most genes, however, non-synonymous sites will be under functional constraint (negative selection) so that their rate of substitution will be less than that at synonymous sites ($d_N < d_S$). In contrast, under positive selection more non-synonymous than synonymous substitutions will occur ($d_N > d_S$) because the rate at which natural selection fixes advantageous changes is greater than the rate at which they are produced by mutation. $d_N > d_S$ therefore rejects neutrality.

Nevertheless, there are problems associated with testing neutrality in this manner. First, the test is highly conservative in that it is difficult to detect $d_N > d_S$ if selection only involves a small number of changes. Second, synonymous sites are assumed to be neutral which may not be the case if constraints are imposed by codon usage or secondary structure. Also, precise estimations of d_N and d_S are often difficult because they depend on the accurate correction of multiple substitutions (see Chapter 5). Finally, analyses of d_N and d_S are usually based on pairwise comparisons of sequences, ignoring the phylogeny linking them. This means that certain branches (especially those near the root of the tree) will be analysed many times, thereby biasing the results. Estimates of d_N and d_S which take into account the phylogenetic relationships of sequences are therefore a more informative way to detect positive natural selection (as in the case of *jingwei*—section 7.8).

7.3 Patterns of base composition and codon usage

7.3.1 The isochore structure of vertebrate genomes

With the development of the polymerase chain reaction (PCR) in the mid-1980s, there was a rapid increase in the amount of DNA sequence data available to analyse. This enabled evolutionary biologists to ask important questions about the large-scale patterns of genome evolution. One question of particular interest is why patterns of base composition—the frequencies of the four

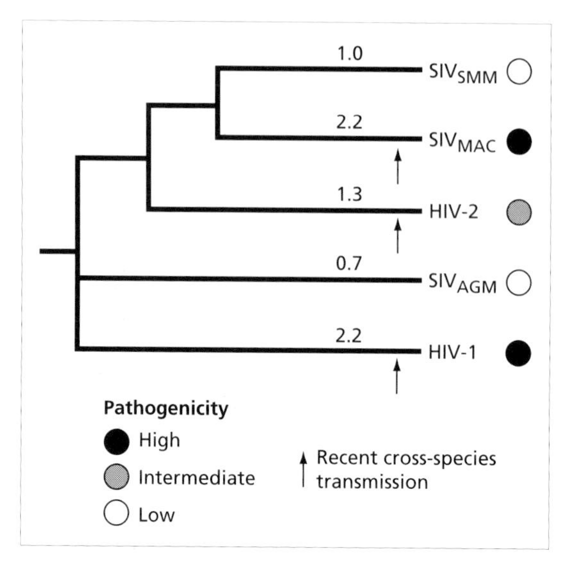

Fig. 7.8 Ratio of non-synonymous (d_N) to synonymous substitutions (d_S) for the envelope protein gp120 from different primate immunodeficiency viruses. The phylogeny presented is a simplified interpretation of the evolutionary history of these viruses (see Chapter 8). HIV-1 and HIV-2 are from humans, SIV_{AGM} from African green monkeys, SIV_{MAC} from macaques and SIV_{SMM} from sooty mangabeys. The higher the ratio, the higher the relative rate of non-synonymous substitution and a ratio > 1.0 indicates the action of positive natural selection (see Box 7.1). Data from Shpaer and Mullins (1993).

bases and of the codons used to specify amino acids—differ between genomes? The answer to this question will not only provide us with clues as to the underlying mechanisms of molecular evolution, but also improve methods of phylogenetic reconstruction because uneven base composition is one way in which errors can creep into trees (see Chapter 6).

As soon as the first DNA sequences became available from different organisms it was apparent that there was great variation in the frequencies with which the four bases were used. The usual way to express differences in base composition is the frequency of G + C nucleotides compared to A + T nucleotides. In bacteria, for example, mean G + C content varies from approximately 25% to 75% between species, but there is little intragenome variation. The genomes of vertebrates, on the other hand, internally have a much greater range of G + C values. This variation in base composition is caused by the presence of **isochores**—continuous sections of DNA sequence > 300 kb in length, each of which has a fairly uniform G + C content. The homogeneity of base composition within each isochore extends to non-coding regions, introns and exons, and even the amino acid composition of proteins reflects the underlying base composition.

The G + C content of isochores also varies between species. Those in most cold-blooded vertebrates tend to have a narrow range of G + C content,

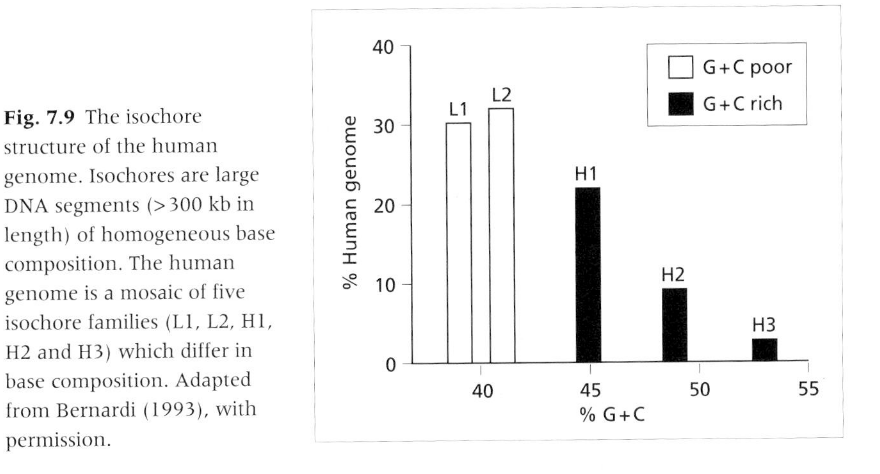

Fig. 7.9 The isochore structure of the human genome. Isochores are large DNA segments (>300 kb in length) of homogeneous base composition. The human genome is a mosaic of five isochore families (L1, L2, H1, H2 and H3) which differ in base composition. Adapted from Bernardi (1993), with permission.

centred around 40%. The base composition in the isochores of warm-blooded vertebrates (birds and mammals) is much broader, with both G + C rich and G + C poor isochores found within the same genome. In humans, for example, five different isochore families have been identified (L1, L2, H1, H2 and H3) with a G + C content ranging from 39% to 53% (Fig. 7.9). The L1 and L2 families have low G + C content and together make up 62% of the genome, whereas isochores H1, H2 and H3 have a high G + C content and constitute 22%, 9% and 3% of the genome, respectively (the final 4% of the genome is made up of satellite and ribosomal DNAs). A similar pattern is found in most other mammals although rodents seem to lack the heavily G + C-rich H3 family. Isochores have also been found in higher plants with, in the angiosperms, monocots having higher G + C contents than dicots.

Not only do isochores reflect base composition, but they also appear to be associated with other aspects of genome organisation, such as the density and type of genes found, replication time in the cell cycle, the presence of transposable elements and the frequency of chiasmata (the sites of crossing-over) (Table 7.2).

Both neutralist and selectionist theories have been put forward for the existence of isochores. A selectionist hypothesis proposed by Bernardi *et al.* (1985) is that G + C-rich isochores, predominantly found in birds and mammals, are an adaptation to the high body temperatures of these species. This occurs because G + C nucleotides have an extra hydrogen bond between them and may therefore be more thermally stable than A + T nucleotides at increased temperatures. To put it another way, the evolution of high body temperatures in birds and mammals increased the possibility of thermal damage to their DNA so that placing genes in G + C-rich isochores was selectively advantageous. This explanation may also apply to other taxa—plant species which live in arid climates also favour G + C-rich isochores. Evidence put forward in support of this hypothesis is that the development G + C-rich isochores

Table 7.2 The properties of isochores in vertebrates. Adapted from Holmquist and Filipski (1994), with permission.

G + C-rich isochores	A + T-rich isochores
Correlate with reverse Giemsa bands (R bands) on chromosomes	Correlate with Giemsa bands (G bands) on chromosomes
Early replicating	Late replicating
High density of genes, including both housekeeping and tissue-specific	Low frequency of genes, only tissue-specific
SINEs present	LINEs present
CpG islands in genes	No CpG islands
High G + C content at third codon position	High A + T content at third codon position
High frequency of retroviral sequences	Low frequency of retroviral sequences
High frequency of chiasmata	Low frequency of chiasmata

More details about the evolution of genome organisation are given in Chapter 3.

must have occurred independently, as birds and mammals do not share an immediate common ancestor, which suggests that it is linked to their high body temperatures (the similar genome structures in birds and mammals also mean that they have proteins more similar than expected, which may explain the small distance between the human and chicken α-globin sequences seen in Fig. 7.1, and may even make them group together by mistake in phylogenetic trees). However, it is also the case that some thermophilic (heat-loving) bacteria are A + T-rich, so that the association between G +C content and thermal stability is tenuous, and it is hard to visualise how selection would act upon changes in base composition which, individually, would have a very small effect on fitness.

As noted previously, it is possible to think of neutralism as an evolutionary theory based on the importance of mutation. Consequently the neutralist explanation for the existence of isochores is that they simply reflect variation in the process of mutation across the genome. For example, Wen-Hsiung Li and colleagues looked at mutation patterns in the argininosuccinate synthetase-processed pseudogenes (and their flanking regions) from anthropoid primates. These pseudogenes were derived from the same ancestral functional gene at around the same time but then inserted into different regions of the genome. Yet, despite their common ancestry, they now differ in base composition. Because pseudogenes are unlikely to be subject to natural selection, the differences in base composition must have been caused by regional variation in mutation patterns. But why should mutation patterns vary across genomes? There are three mechanisms by which this could occur, involving DNA replication, DNA repair and recombination.

According to the **replication hypothesis** there is a correlation between replication time and base composition so that genes which replicate early in the cell cycle are more G + C-rich than genes that replicate at later times.

This is thought to occur because the G and C nucleotide precursor pools (dNTPs), which contain the nucleotides to be inserted into the replicating DNA molecule, are larger than the A + T pools during the early phase of replication so that polymerase errors are more likely to result in the insertion of G + C at this time. Although the extent of the correlation between replication time and base composition is disputed, the sizes of isochores are compatible with the amount of chromosome replicated at a particular time, suggesting that patterns of replication may indeed influence patterns of genome evolution.

In contrast, the **repair hypothesis** is based on the assumption that the efficiency of DNA repair varies across the genome and that certain DNA errors are repaired more efficiently than others. This variation may be an outcome of transcriptionally active areas of the genome being repaired more efficiently than genomic regions that are transcribed less often. Further, CpG islands—regions of the genome with a high frequency of CpG dinucleotides and which are associated with the presence of genes (see Chapter 3)—are maintained by a special repair system, suggesting that the efficiency of DNA repair is dependent on genome location. However, it is uncertain whether these repair differences can extend over large enough sections of the genome to produce isochores. The repair hypothesis is described in more detail in section 7.4.

Finally, the **recombination hypothesis** claims that the isochore structure of vertebrate genomes is the outcome of differences in the pattern and frequency of recombination, with low G + C localities associated with regions of reduced recombination. There are a number of pieces of evidence for this in mammals. First, and most obvious, genes with low rates of recombination have low G + C values. The second piece of evidence cited is a more dramatic extension of this: the large non-recombining part of the Y chromosome likewise has a low G + C base composition. Finally, there is a positive correlation between G + C content and chiasmata density, a marker of recombination rate. Although the mechanics by which the correlation between base composition and recombination rate takes place are unclear, the fact that recombination does seem to play a major role in structuring the eukaryotic genome (as we saw in Chapter 3), makes this an attractive hypothesis for the existence of isochores.

Although it is currently undetermined which of these hypotheses, singly or in combination, explains why patterns and rates of mutation vary across the vertebrate genome, such a neutralist interpretation presently seems a more likely explanation for the existence of isochores than theories based on selective differences between nucleotides.

7.3.2 What determines codon usage?

Another important aspect of genome organisation is the observation that species and genes differ in patterns of **codon usage**. The degeneracy of the genetic code means that most amino acids can be specified in a number of different triplet combinations of nucleotides. The null hypothesis is that all codons for a particular amino acid are used with equal frequency. This was refuted when nucleotide sequences became available for a wide range of organisms and genes and extreme biases were found in the usage of codons. For example, the amino acid glutamic acid is specified by two codons, GAA and GAG. In human nuclear genes GAG is used more often than GAA (frequencies of approximately 40 and 28 per 1000 codons, respectively) whereas in *E. coli* the relationship is reversed. The full codon usage patterns of human nuclear genes and *E. coli* are shown in Fig. 7.10. There has been a great deal of debate about how the 'choice' of a particular codon is made, especially at those codons which differ only in the third position so that they code for the same amino acid—the **synonymous codons**.

A selectionist argument to explain non-random patterns of codon usage is that the choice of particular synonymous codons some how increases the fitness of the organism. For example, in some unicellular organisms, such as *E. coli* and yeast, highly expressed genes show more codon bias than lowly expressed genes because the former require more translational efficiency (rapid production with low error) than the latter. Natural selection therefore favours codons that match the set of tRNA molecules required to transport them to the ribosome and which have themselves coevolved to match the codon population (Fig. 7.11). In lowly expressed genes there is more flexibility in which tRNAs can be used and therefore less codon bias.

Although the differences in fitness between synonymous codons will be very small, the large population sizes of unicellular organisms mean that even these small differences can lead to particular codons being favoured (remember that natural selection works best in large populations). There is even evidence from *E. coli* that the selection for codon choice may partly determine which amino acids are used because changing amino acids might provide a more or less translationally efficient codon. However, because so few unicellular genomes have been studied in detail it is premature to say how widespread selection for codon choice is.

Although the codon–tRNA coevolution model is a selectionist explanation of codon choice, it also supports the neutralist prediction that there is a relationship between functional constraint and the rate of nucleotide substitution: high expression–high codon bias genes have a lower rate of synonymous substitution than low expression–low codon bias genes because there are fewer potentially neutral changes in the former group (Fig. 7.11). A correlation between codon bias and substitution rate is also found in *Drosophila* genes (Fig. 7.12). For example, the *Adh* and *Adhr* genes are located very close to each other (on chromosome II) and obviously arose from a gene duplication, yet they differ greatly in levels of expression: *Adh* is very highly expressed but *Adhr* only at low levels. As predicted, *Adh* has a strong codon bias and a low rate of silent substitution whereas *Adhr* is only weakly codon biased and has a high rate of silent change. Such a correlation, along with direct measures that

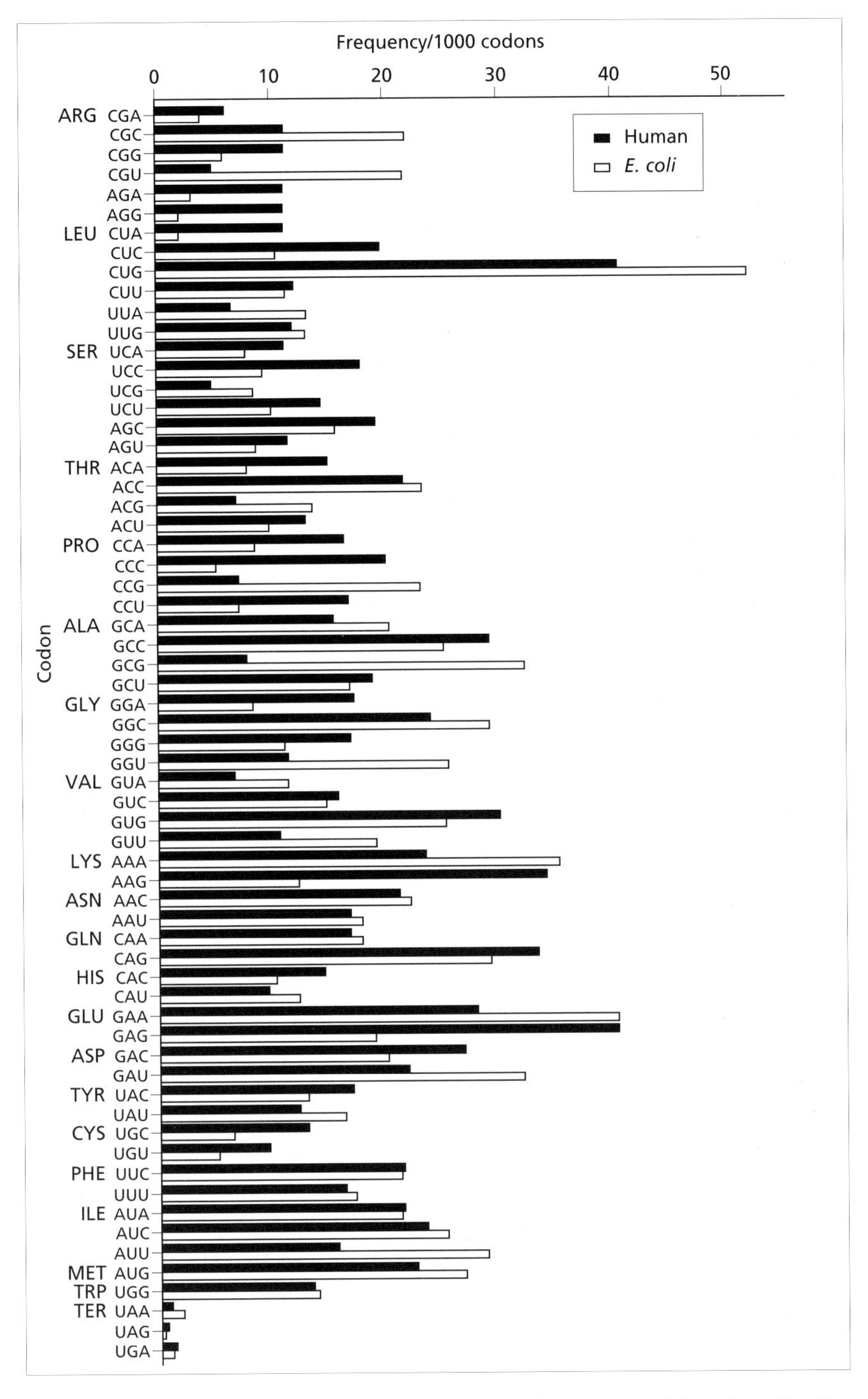

Fig. 7.10 Overall patterns of codon usage in human nuclear genes and those from *E. coli*. The human data is based on 8701 genes and the *E. coli* data from 5976 genes. It should be noted that these patterns vary greatly between genes (see text).

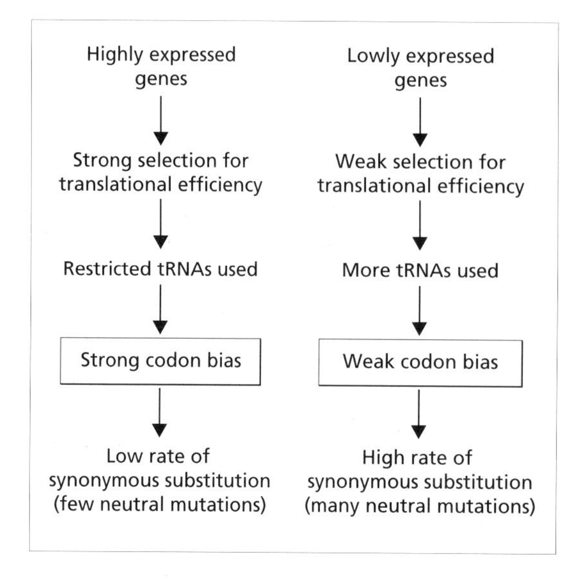

Fig. 7.11 How gene expression in some unicellular organisms determines the extent of codon bias and the rate of synonymous substitution.

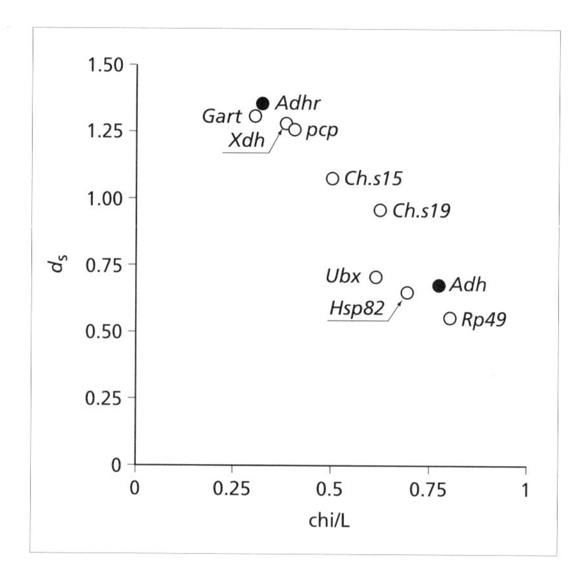

Fig. 7.12 Relationship between the number of silent substitutions per site (d_S) and synonymous codon bias (chi/L) in 10 *Drosophila* genes. Each point represents the comparison of one gene from the *D. melanogaster* species group (usually *D. melanogaster*) with one from the *D. obscura* species group (usually *D. pseudoobscura*). Chi/L represents a χ^2 for deviation from equal use of synonymous codons, divided by gene length (L), averaged for each species pair. *Adh* and *Adhr* are highlighted by closed circles. Adapted from Sharp and Li (1989).

highly expressed genes are the most biased and that codon choice matches tRNA abundance, suggests that natural selection plays an important role in determining patterns of codon bias in *Drosophila*. It is also possible that codon choice is selected in some plant species. In maize (*Zea mays*), for example, highly expressed genes tend to be more G + C-rich.

Different explanations have been put forward to explain patterns of codon usage in mammalian genomes. In particular, codon bias seems more likely to be determined by differences in (neutral) mutational pressure than natural selection. Evidence for this is that the base composition of synonymous codons does not differ significantly from that of the surrounding non-coding DNA and introns which are assumed to be neutral, so that the codons used in any particular gene correspond to the local mutational bias: genes in G + C-rich regions of the genome tend to use G + C-rich codons. This contrasts with what is seen in unicellular organisms where codon choice can be selected and where the base composition at silent sites may be unlike that seen in surrounding non-coding regions. Furthermore, patterns of codon usage in mammals do not appear to be associated with either the level or timing of gene expression or with tissue type, as might be expected if natural selection were in operation. Natural selection is also less likely to be the explanation for codon bias in mammalian genomes because the smaller population sizes of these species mean that selection pressures would have to be considerably stronger than those observed in *E. coli* to explain the differences in codon bias. As with intragenomic differences in base composition, variation in the patterns of codon usage along the mammalian genome seem best explained by varying patterns of mutation.

7.4 The molecular clock

Perhaps the most exciting, unexpected and controversial claim in molecular evolution, and one that we encountered at the start of this chapter, is that genes evolve at a constant rate, giving rise to a molecular clock. Neutralists have taken the clock as support for their theory because to them it demonstrates the constancy of the underlying neutral mutation rate, the driving force of molecular evolution. But the molecular clock is important not only because of its potential support for the neutral theory. It may also represent a general evolutionary rule, and one which enables us to estimate times of species divergence simply by comparing their gene sequences. At the same time, methods of phylogenetic reconstruction are more likely to be accurate if genes evolve at a constant rate (Chapter 6).

When the neutral theory was first proposed in the late 1960s, many of the protein sequences available appeared to show a picture of long-term rate constancy. We have already seen that this is generally true of α-globin. This clock-like behaviour is very different to evolution at the morphological level which is characterised by extreme rate variation. Examples of low rates of evolutionary change are provided by 'living fossils' such as the coelacanth *Latimeria* which has seemingly remained little changed over millions of years, while adaptive radiations, such as the diversification of the orders of eutherian (placental) mammals at the end of the Cretaceous period (approximately 80 million years ago), show evolution at higher speeds.

The concept of a molecular clock does not mean, however, that all genes and proteins evolve at exactly the *same* rate, as it was clear from the earliest studies that there was great variation between proteins (see Fig. 7.4). As we have already pointed out, variation in rate among genes and proteins is compatible with the neutral theory if the underlying causes are differences in selective constraint. It is also clear that different genetic systems evolve at different rates: for example, vertebrate mitochondrial DNA evolves much faster than vertebrate nuclear DNA. The debate concerning the existence of a molecular clock, and thus the validity of the neutral theory, therefore focuses on whether rates of substitution are constant *within* genes across evolutionary time. Furthermore, a long-term *average* rate does not exclude the possibility of short-term rate fluctuations.

7.4.1 Neutral theory and the molecular clock

At face value, the neutral theory has a simple explanation for existence of a molecular clock. The rate of nucleotide substitution (or fixation) at a nucleotide site *per year*, k, in a diploid population of size $2N$ is equal to the number of new mutations (neutral, deleterious or advantageous) arising *per year*, μ, multiplied by their probability of fixation, u. Therefore:

$$k = 2N\mu u \tag{7.1}$$

For a neutral mutation, the probability of fixation is simply the reciprocal of the population size (see Chapter 4):

$$u = 1/2N \tag{7.2}$$

so that the rate of substitution of a neutral mutation is given by:

$$k = (2N)(1/2N)\mu \tag{7.3}$$

The fact that both $2N$ and $1/2N$ are included in this equation means that the population size part can be cancelled out, so that:

$$k = \mu \tag{7.4}$$

This is one of the most important formulae in molecular evolution because it means that the rate of substitution of a neutral mutation is dependent *only* on the underlying mutation rate and independent of other factors such as population size. This also holds for mutants where the selective coefficient, s, is so small that it is less than the reciprocal of the effective population size ($s < 1/2N_e$) (note that when we are dealing with natural selection we need to consider the effective population size, N_e, rather than the census size N, because the probability of fixing an advantageous mutant depends on it arising in a subset of the population that are able to reproduce).

The situation is different for selectively advantageous mutations. Although the rate of substitution will depend on the precise type of selective process assumed, Kimura showed using a simple model that the probability of fixation of a selectively advantageous mutation with codominance is:

$$u = 2sN_e/N \tag{7.5}$$

i.e. roughly twice the selective coefficient (although see Gillespie (1995) for a discussion of some more complex selective models). If this is substituted into equation (7.1) we get:

$$k = 4N_e s\mu \tag{7.6}$$

Therefore the rate of substitution for an advantageous mutation depends on the population size and the magnitude of the selective advantage, as well as the mutation rate. For natural selection to produce a molecular clock it is then necessary for three variables, N_e, s and μ, incorporating ecological, mutational and selective events, to be the same across evolutionary time. This makes a molecular clock highly unlikely with natural selection.

The fact that the neutral theory appeared to predict a molecular clock and that this clock was observed in the protein sequence data available in the late 1960s initially led Kimura to cite rate constancy as the best evidence for the neutrality of mutations. However, during the 1970s, as more comparative protein sequence data accumulated and particularly from mammalian genes, examples of rate variation started to appear and a debate surrounding the constancy of the molecular clock began.

The most common way to test the molecular clock at this time was with the **dispersion index, *R(t)***, which measures whether there is more variation between lineages than expected under a Poisson process, the statistical distribution which describes the accumulation of substitutions under neutral theory. This test recognises that a molecular clock driven by neutral mutation is not a metronome, perfectly in time, but that there will be some random statistical fluctuations in the rate of substitution because the genetic drift of neutral mutations is itself a random sampling process. We can therefore think of neutral theory as predicting a **Poisson clock**. If the data fit a Poisson clock (and therefore support the neutral theory) the variance in substitution rate should be no greater than the mean substitution rate, resulting in an *R(t)* not significantly greater than 1.0 (see Box 7.2). However, it was soon evident that in many mammalian proteins values of *R(t)* were greater than 1.0, with an average of around 2.5 (Table 7.3). This represented a serious problem for the neutralists and led to claims that natural selection was in fact the more important force in molecular evolution.

How could the neutral theory explain this rate variation? One possibility is that rates do not vary at all but they have been miscalculated because mistakes have been made concerning the phylogenetic relationships or times of divergence of the species analysed. For example, anomalously high rates of substitution

Protein	Species (n)	Amino acids	$R(t)$
Haemoglobin α-chain	6	141	1.17
Haemoglobin β-chain	6	146	3.04
Myoglobin	6	153	1.60
Cytochrome c	4	104	3.22
Ribonuclease	4	123	2.15
α-Crystallin	6	175	2.71

Table 7.3 Estimates of the dispersion index, $R(t)$, for different mammalian proteins. Data from Kimura (1983) and Gillespie (1986b).

are found in guinea pigs and other histricomorph rodents which if true pose a problem for the neutral theory. However, it now seems likely that the guinea pig is not a rodent at all but in fact a much more divergent lineage of mammals. The large genetic distance between the guinea pig and other mammals is then due to its ancient divergence rather than an increase in substitution rate. One way to reduce the problem of phylogenetic error in rate estimation is to make use of the **relative rate test** (Box 7.2), which only requires sequence information from three species, although accurately determining rates of substitution is likely to remain a problem in molecular evolution.

Box 7.2 Testing the molecular clock

Testing the accuracy of the molecular clock has formed an important part of the study of molecular evolution. Here we describe two tests that are often applied to sequence data. A likelihood ratio test of the molecular clock is described in Chapter 6.

1 *The dispersion index, R(t)*

Historically, the most important way to test the molecular clock has been to examine whether there is more rate variation between lineages than expected under a Poisson process, the statistical distribution used to describe the accumulation of mutations under neutral theory. If the data fit a Poisson process, then the variance in the number of substitutions between lineages should be no greater than the mean number. The dispersion index, $R(t)$, tests whether this is the case: if the data fit a Poisson process then $R(t) = 1.0$, if not then $R(t) > 1.0$ and the clock is said to be **overdispersed** (although recent studies indicate that $R(t)$ is a poor statistic for determining whether the data fit a Poisson process).

Initially $R(t)$ was calculated by estimating the mean and variance in the number of substitutions on the lineages of a star phylogeny, where all sequences are taken from species that diverged within a short time period, such as the orders of placental mammals. A star phylogeny is used because it removes any phylogenetic structure which would complicate the calculations. However,

continued

Box 7.2 *continued*

whether the mammalian radiation is a star phylogeny is questionable as this may simply reflect a lack of knowledge about their branching order. For this reason, later work using $R(t)$ is more often based on the unrooted tree of just three mammalian orders—primates, rodents and artiodactyls.

2 *The relative rate test*

The simplest way to test the accuracy of the molecular clock is to estimate the difference in the number of substitutions between two closely related taxa in comparison with a third, more distantly related outgroup species. The advantage of this test is that it does not require any knowledge of the divergence times of the taxa in question. An example is presented in Fig. B7.1. Here Old World monkeys (species 1) and humans (species 2) are compared with a more distantly related New World monkey (species 3). If the human and Old World monkey sequences have evolved according to a molecular clock, both should be equally distant to the New World monkey, so that $d_{13} = d_{23}$. However, the data analysed here show that humans have in fact evolved at significantly lower rates than Old World monkeys.

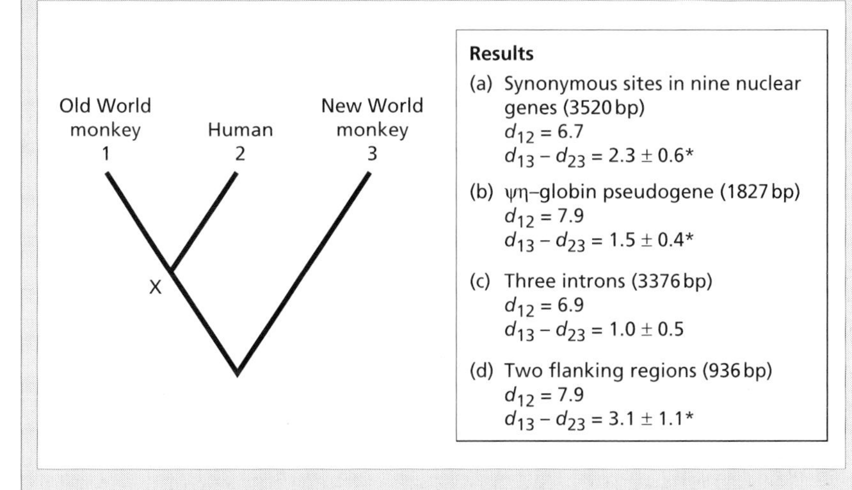

Fig. B7.1 The relative rate test. Because Old World monkeys (species 1) and humans (species 2) share a more recent common ancestor (node x) than either does to New World monkeys (species 3), any difference in rate between them must have occurred since they split from this common ancestor. This can be determined by measuring their respective distances to the outgroup (species 3) and testing whether any difference in these distances ($d_{13} - d_{23}$) is significantly greater than zero. In the test shown here the human lineage appears to be evolving slower than the Old World monkey lineage because it has a smaller distance to the New World monkey. Distances given as percentage substitutions. *Significant at the 1% level. Data from Li *et al.* (1987) and Li (1993).

continued on p. 256

Box 7.2 *continued*

Despite their simplicity there are problems with relative rate tests. First, it is critical that species 1 and 2 are the most closely related, so that some (albeit limited) knowledge of phylogeny is required. In cases such as the orders of placental mammals phylogenetic relationships are uncertain, which may lead to errors. Second, it is important not to use an outgroup that is too distantly related to species 1 and 2 because the more divergent it is, the greater the problem of multiple substitution and the smaller the impact any difference in rate will have on the distance measurements. Finally, relative rate tests are often simultaneously applied to the same sequences for multiple sets of taxa so that they are not independent. This will make the assignment of significance values more difficult.

7.4.2 Can lineage effects explain variation in the molecular clock?

Another way to explain rate variation, and one which gained momentum during the 1970s, was based on the realisation that if the underlying neutral mutation rate varies, so will the rate of substitution (because $k = \mu$; equation (7.4)). There are three ways for mutation rates to vary between species, depending on differences in (i) generation time, (ii) metabolic rate and (iii) the efficiency of DNA repair. Because these are ways in which the mutation rate can vary between lineages they are collectively known as **lineage effects**. Uncertainties in the phylogenetic relationships of the species under analysis can also be put into this category. The neutralist argument is that lineage effects can explain *all* variation in the tick of the molecular clock while selectionists believe that even if these differences are taken into account genes still show rate variation, a residue which is due to the action of natural selection (**residue effects**).

The **generation time hypothesis** has attracted most attention. This simply states that if the mutation rate is set to generation rather than real (chronological) time, so is the rate of substitution. At the molecular level generation time (g) can be defined as the time it takes for germ-line DNA to reproduce (replicate) itself: for a gamete to produce another gamete (Fig. 7.13). If most mutations are the result of errors in this process and if species have similar numbers of cell divisions per generation, then we should expect that species with longer generation times to accumulate fewer substitutions within a given period of real time than those with shorter generations because there will be fewer opportunities for replication errors in the former. This means that the rate of substitution under neutral theory should be a function of both the mutation rate and generation time:

$$k = \mu / g \qquad (7.7)$$

so that the molecular clock will be set to generation time.

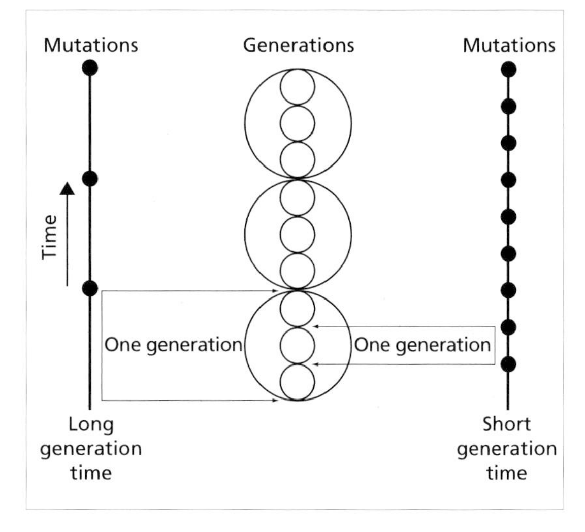

Fig. 7.13 The generation time molecular clock. Each open circle represents a generation from gamete to gamete and each closed circle a mutation (with one occurring in each generation). If a species has a short generation time (small circles) more generations will take place over a given time period, so that there are more opportunities for germ-line mutations and hence a faster molecular clock.

There have been many studies testing whether the molecular clock is better correlated with generation or real time, particularly for mammalian genes. A general conclusion is that the clock at silent sites (and non-coding DNA) is generation time dependent; or rather, that the observed variation in rates of substitution correlates with differences in generation time. For example, Ohta (1995) shows that silent rates are 2.6-fold higher in rodents than primates (Fig. 7.14). Rodents have shorter generation times than primates—those in mouse and rat may be 40-fold (or more) shorter than those in humans—so that mutation rates should be higher in these species. Another example of how generation time affects the molecular clock is the slowdown in rate which

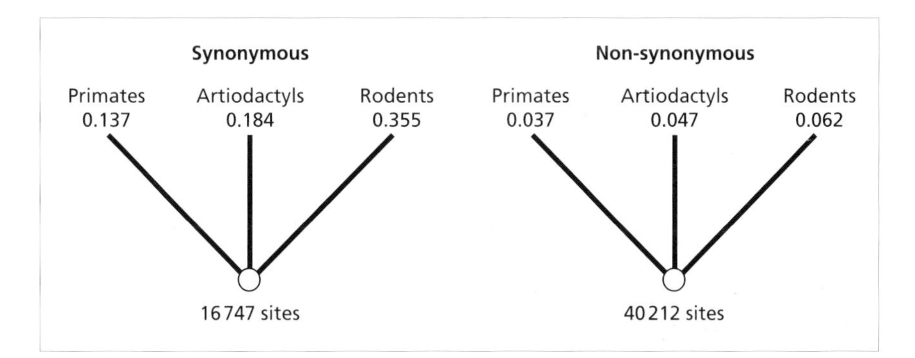

Fig. 7.14 Numbers of synonymous and non-synonymous substitutions for 49 genes from three mammalian orders: primates, rodents and artiodactyls, the phylogenetic relationships of which approximate a 'star phylogeny'. Note that, in both cases, rodents have accumulated more substitutions than primates or artiodactyls. Adapted from Ohta (1995).

has occurred along the human lineage. Li *et al.* (1987) found the silent rates in the orang-utan, gorilla and chimpanzee lineages to be 1.3-, 2.2- and 1.2-fold faster, respectively, than those in humans, which matches the differences in generation times between these species. A similar observation has been made in non-coding regions (see Box 7.2). A generation time molecular clock at silent sites and non-coding DNA supports the neutral theory because these parts of the genome are often regarded as strictly neutral (see section 7.2), so that their rate of substitution will only depend on the mutation rate, itself set to generation time.

More evidence for a generation time clock comes from plants. Although generation time is a difficult concept to apply to plant species, because there is no clear distinction between the germ-line and the soma, an analysis of chloroplast *rbcL* gene sequences revealed that rates of substitution in grasses are five-fold higher than those in palms. This is compatible with their different ages at first flowering, a possible measure of generation time. Other studies of *rbcL* sequences have uncovered similar patterns: for example, annual species of angiosperms evolve more rapidly than perennial species, whilst woody bamboos evolve slowly and have long times to first flowering.

However, the relationship between generation time and substitution rate is not always so simple. One obvious complication is in the assumption that the number of germ-line cell divisions per generation is constant among diverse taxa. If this number were to differ between species then those with similar generation times may actually experience different numbers of replications and hence have different mutation rates. This may in part explain why differences in substitution rate are often not as great as expected given differences in generation time (i.e. rodents do not evolve 40-fold faster than primates). For example, although human and rat females undergo similar numbers of germ-cell divisions per generation (estimated to be 33 and 28, respectively), human males experience many more divisions than their murine counterparts (estimated to be 205 compared with 57 in the rat), which may reduce the difference in substitution rate between these species. The same hypothesis predicts that mutation rates should be higher in males than females because more cell divisions take place in spermatogenesis than oogenesis. This appears to be the case: silent rates are 2.3-fold higher in the human Y-linked zinc-finger-protein gene (*ZFY*) than its X-linked homologue (*ZFX*). Similar results have been found for other mammalian genes which possess copies on both the X and Y chromosomes (and also in birds where the females are heterogametic—ZW). This has been taken as evidence that much of mammalian molecular evolution is male-driven and, crucially, that it is the number of germ-line DNA replications which ultimately determines the rate of substitution, as predicted by the generation time hypothesis. However, an alternative explanation is that the mammalian X chromosome has a selectively lower rate of mutation compared with the Y chromosome and the autosomes because deleterious recessive mutations on the X will be expressed when hemizygous—that is, when an X chromosome is matched with a Y chromosome—but not when heterozygous (i.e. XX, but with different alleles on each).

A more serious problem with the generation time hypothesis is that the expected correlation between generation time and substitution rate is not always found. For example, anomochlooid grass species evolve more slowly than woody bamboos but flower earlier. Furthermore, back in mammals, the molecular clock at non-synonymous sites (or in protein sequences) seems to correlate better with real than generation time. For example, while Ohta shows that rodents evolve 2.6-fold faster than primates at synonymous sites, they are only 1.6-fold faster at non-synonymous sites (Fig. 7.14). Why the clocks at synonymous and non-synonymous sites are set to different time-scales has proven a difficult problem for the neutral theory and one which has been a major impetus in the development of the nearly neutral theory (see section 7.5).

Even more dramatic departures from the generation time clock are found in mtDNA where some mammals, reptiles and fish with relatively short generation times have low rates of substitution. A notable case is sharks in which the rate of silent change is about five- to sevenfold lower than in primates and ungulates despite the fact that these species have similar generation times. Similarly, although whales have shorter generation times than primates, they appear to evolve at lower rates. These results have led to the hypothesis that differences in the metabolic rate of species are a better explanation of the variation in rates of molecular evolution than differences in generation time. The basis of this **metabolic rate hypothesis**, proposed by Martin and Palumbi (1993b), is that organisms with high metabolic rates have increased rates of DNA synthesis and consequently higher rates of mutation than species with lower metabolic rates. Furthermore, free-oxygen radicals, which are produced during aerobic respiration, have mutagenic affects which increase the mutation rate in species with high metabolic rates.

There are two pieces of mtDNA evidence which support this hypothesis. First, small-bodied animals, which have fast metabolisms, tend to have higher rates of substitution than large-bodied animals with slower metabolisms. Second, warm-blooded vertebrates, which might be expected to have high metabolic rates, have higher rates of substitution than cold-blooded vertebrates (Fig. 7.15).

However, both metabolic rate and generation time (as well as other life-history variables) are correlated with body size (see Fig. 7.15) which makes it difficult to discriminate their individual contributions to the rate of sequence evolution. In fact, when the affects of both factors are analysed using phylogenetically independent comparisons, as has been done for mammals, generation time appears as a better predictor of rate of sequence change than metabolic rate.

A final problem with the generation time hypothesis is the assumption that the main source of mutation is the miscopying of bases during DNA replication. However, it is also possible that most substitutions are in fact the

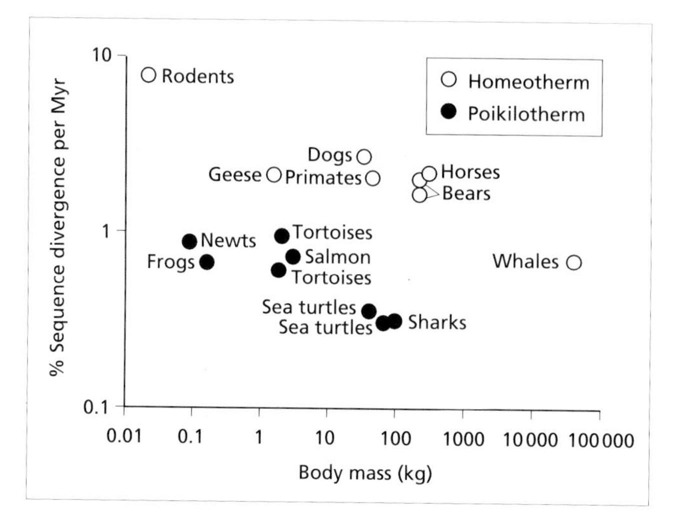

Fig. 7.15 Relationship between the rate of sequence evolution in mtDNA (percentage divergence per million years) and body mass (in kg) for various vertebrates. Mean values are used in all cases and both axes are drawn on a logarithmic scale. Poikilotherms (cold-blooded vertebrates) have lower rates of change than homeotherms (warm-blooded vertebrates) and in both groups larger species (with slower metabolisms) evolve slower than smaller species. Data from Martin and Palumbi (1993b).

result of incorrectly repaired damage to DNA (Fig. 7.16). If species differ in their ability to repair this damage, and rodents are claimed to be less efficient than primates in this respect, then they may also differ in rates of substitution, with species which repair their DNA in the most efficient manner evolving at the lowest rates.

There has been considerable debate about whether differences in **DNA repair** efficiency can produce the observed differences in rates (and patterns) of nucleotide substitution. Unfortunately, repair mechanisms appear to be extremely complex and our knowledge of them is far from complete: not only are many genes involved but there are different repair pathways depending on the type of damage, when it occurs and its location in the genome (see Chapter 3). There is, however, some support for the hypothesis that DNA repair influences mutation rate. First, there is growing evidence for a connection between the machinery of DNA transcription and that of DNA repair such that highly transcribed genes are repaired more efficiently than genes transcribed at low levels. Repair also seems to be directed to the transcribed strand of DNA. This connection could arise if genes are preferentially repaired when there is an open chromatin structure, as is found in transcriptionally active regions, or because the same gene is involved in both processes, which has been found with the basal transcriptor factor 2 (*TFIIH*) gene of humans. Second, base composition and substitution rate at silent sites in mammalian genes tends to be gene rather than species specific, perhaps signifying that homologous genes are transcribed and repaired in a similar manner.

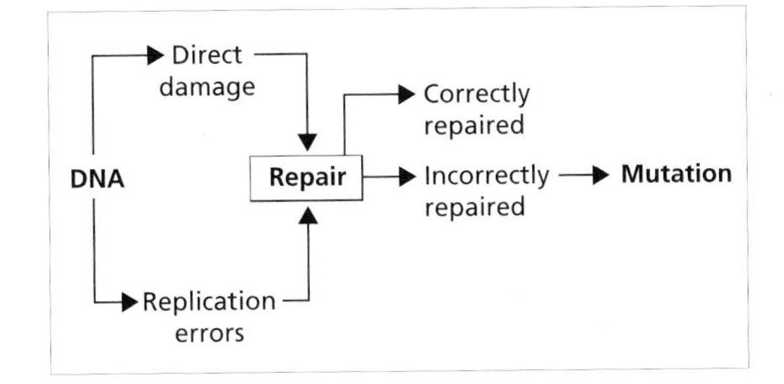

Fig. 7.16 Schematic representation of the relationship between DNA repair and mutation. Damage to DNA can occur either directly, for example through the action of mutagens, or because of miscopying during replication. The DNA repair hypothesis states that the ability to correctly repair DNA varies between species and genes so that the mutation rate will also vary.

Taken together these observations imply that differences in the efficiency of DNA repair might partially determine rates of substitution, although it is unlikely that they can explain all instances of rate variation because closely related species which differ in substitution rate, such as the hominoid primates, are unlikely to differ greatly in the efficiency of DNA repair.

7.5 The nearly neutral theory

So far we have described the ways in which the neutral school has tried to explain variation in the tick of the molecular clock by examining how mutation rates can vary among species and genes. Another way in which variation in substitution rates can be incorporated within the framework of the neutral theory is by incorporating mutations with very small selective coefficients (Fig. 7.17). This is the **nearly neutral theory** of molecular evolution.

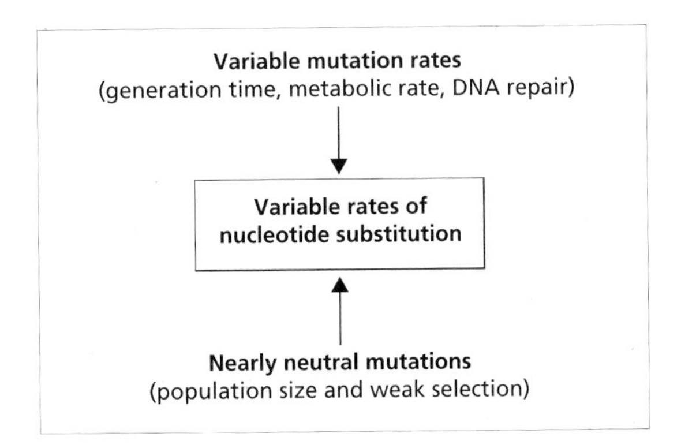

Fig. 7.17 Summary of the possible causes of rate variation under neutral theory.

7.5.1 Development of the nearly neutral theory

By the early 1970s it was becoming clear that the variation in the rate of amino acid substitution in a number of proteins was inconsistent with a Poisson clock (Table 7.3) and that levels of heterozygosity observed for many species were often not as high as expected under the neutral theory (this is discussed in more detail in the next section). These two problems led Tomoko Ohta to propose that most non-synonymous substitutions (or amino acid changes) are not perfectly neutral but slightly deleterious or nearly neutral. This, she thought, was a more biologically realistic view of how mutations would affect the complex interactions between the amino acids which make up a fully functional protein (Fig. 7.18).

Mathematically, nearly neutral mutations are defined as those where the product of the population size and the selective coefficient is *near* zero (for completely neutral mutants, $Ns = 0$). This means that although the selective coefficients against these changes are very small, s, as well as N, will influence their probability of fixation so that they are subject to both (weak) natural selection as well as genetic drift. The inclusion of weak selection together with genetic drift is how the nearly neutral theory differs from its older brother. The rate of substitution (per year) for nearly neutral mutations will therefore depend on population size and the selective coefficient as well as the mutation rate, itself set to *generation time*. Mutations in non-coding DNA and at synonymous sites are still regarded as strictly neutral so that their rate of substitution will depend only on the mutation rate and the species generation time.

Since the nearly neutral model was first proposed in the early 1970s, a number of different versions have been developed which differ in the type of

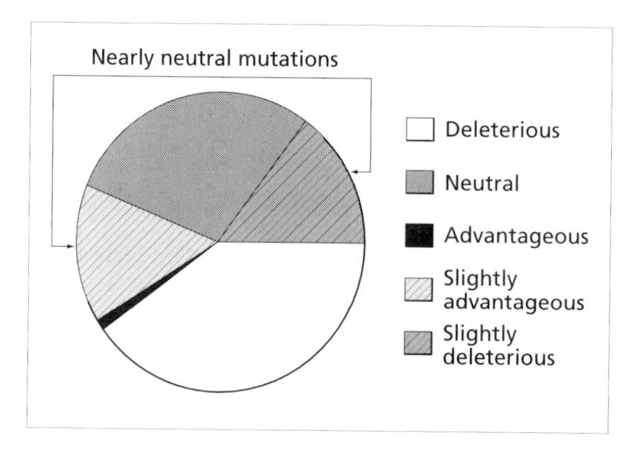

Fig. 7.18 The nearly neutral model of molecular evolution. The pie represents the total number of mutations arising in a gene. When it was originally proposed nearly neutral theory only considered slightly deleterious mutations but more recently it has been modified to incorporate slightly advantageous changes. The proportions assigned to each class of mutation are for graphical purposes only.

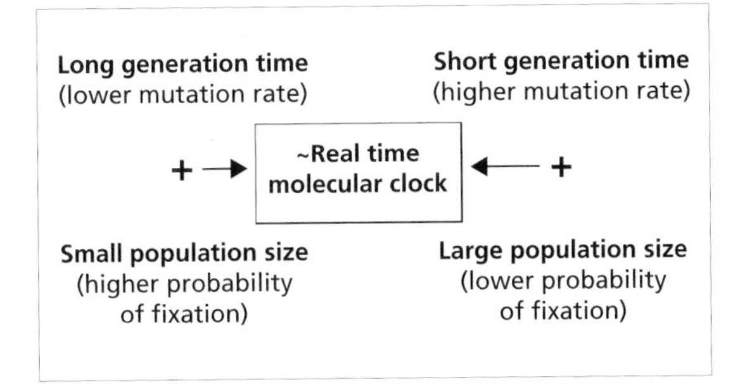

Fig. 7.19 How the nearly neutral theory produces a real time molecular clock at non-synonymous sites. Mutations that occur in species with small populations have a higher probability of fixation, but because these species tend to have long generation times, the rate at which new mutations arise each year is low. Conversely, the probability of fixation will be lower in species with large population sizes but their generation times will be short so that new mutations will arise more frequently.

statistical distribution used to describe the selection coefficients of nearly neutral mutations. The most significant development occurred when the theory was modified to accommodate slightly advantageous mutants, whereas in its original form it was assumed that all nearly neutral mutants were slightly deleterious.

If most nearly neutral mutants are slightly deleterious, then Ohta's theory predicts that the relative importance of (negative) selection or genetic drift, and hence the probability of allele fixation, will depend on the size of the population: negative selection is more important in large populations, so that the probability of fixation of any mutant allele is low, whereas genetic drift is stronger in small populations so that the probability of fixation is higher. This can be used to explain why the molecular clock at non-synonymous sites and in proteins fits real time better than generation time. The mechanism for this is the inverse relationship between population size and generation time which balances out rates of mutation and probabilities of fixation. This works as follows. Species which live in large populations tend to have shorter generation times (and smaller body sizes) than species which have small population sizes. This means that although the nearly neutral mutations which arise in species with small population sizes have a higher probability of being fixed by genetic drift, there will be few such mutations each year because the generation times of these species are long. Conversely, mutations in species with large populations have a lower probability of fixation but more will arise each year because their generation times are shorter and as a consequence mutation rates are higher (Fig. 7.19).

7.5.2 Can nearly neutral theory explain variation in the molecular clock?

As we have already noted, synonymous rates in mammals are more generation

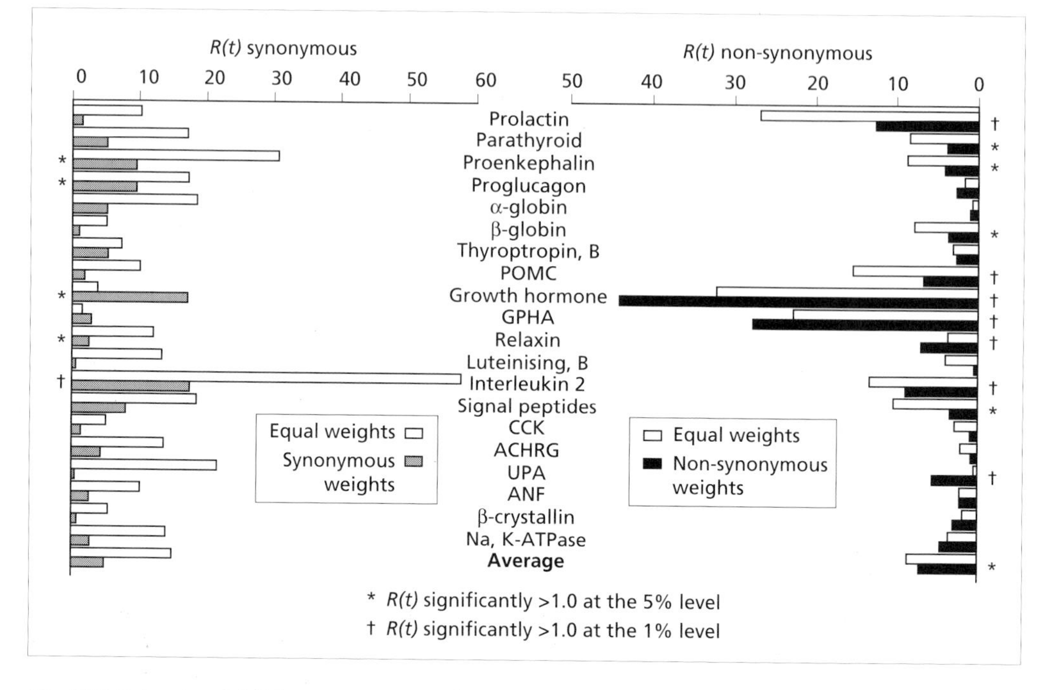

Fig. 7.20 Estimates of *R(t)* for synonymous and non-synonymous substitutions for 20 loci from primates, rodents and artiodactyls. Weights take into account lineage effects and are calculated as the average number of synonymous or non-synonymous substitutions divided by the number of synonymous or non-synonymous sites. No correction is made for lineage effects in the case of equal weights. Values of *R(t)* still significantly greater than 1.0 after weighting are indicated. Lineage effects have less effect at non-synonymous sites but *R(t)* is more often >1.0. Data from Gillespie (1989b).

time dependent than non-synonymous rates, which in turn show a better fit to real time (Fig. 7.14). This could mean that synonymous substitutions are strictly neutral and non-synonymous ones nearly neutral, in keeping with Ohta's theory. This can also be seen in analyses of *R(t)*.

At face value, the *R(t)* values for both synonymous and non-synonymous substitutions from 20 mammalian genes are well over 1.0, thereby rejecting the Poisson clock (Fig. 7.20). However, if a correction ('weighting') is made for generation time (and other lineage effects) then the *R(t)* for synonymous sites falls from 14.41 to 4.64. This means that most rate variation in synonymous rates is indeed due to lineage effects, although there are still some cases where *R(t)* is significantly greater than 1.0 after the correction. After correcting for lineage effects at non-synonymous sites values of *R(t)* only decline from 8.26 to 6.95, meaning that lineage effects have little impact on rates of non-synonymous substitution. Therefore Ohta is right in saying that non-synonymous rates are less generation time dependent than synonymous rates. However, values of *R(t)* for non-synonymous sites are often significantly greater than

1.0 so that there is *not* an overall real time molecular clock of non-synonymous evolution. Ohta suggests that this residue rate variation is due to fluctuations in population size: when populations are small, and particularly during bottlenecks, more nearly neutral substitutions will be fixed because genetic drift will be stronger. Evidence for this comes from *Drosophila* and particularly the observation that Hawaiian species, which have smaller population sizes than their continental cousins, are very diverse at the amino acid level: the ratio of non-synonymous (nearly neutral) to synonymous substitutions is 40–50% higher in Hawaiian *Drosophila* compared with continental species. However, it is also possible that the residue left in $R(t)$ after removing lineage effects is due to natural selection, which is more apparent in the case of non-synonymous substitutions because these have larger effects on protein structure and function.

7.6 Explaining genetic variation within species

7.6.1 Testing the neutral theory within species

So far we have concentrated on the molecular evolution of genes sampled from different species. Yet for a theory to adequately explain molecular evolution it must be able to link change at this level to that which takes place within a single population. To many, the beauty of the neutral theory is that it makes this connection with great simplicity. It is therefore important that we assess whether this is really the case.

The neutral theory makes two very important predictions about levels of genetic variation within species. The first is that the extent of polymorphism (usually measured as average heterozygosity) is a function of only the population size (N) and the mutation rate (μ), so that the larger the population and the higher the mutation rate, the greater the amount of variation seen within species (see Chapter 4). The second prediction is that levels of polymorphism are correlated with the amount of variation between species, so that genes which evolve slowly between species will also exhibit low levels of sequence diversity within them (and vice versa), assuming that functional constraints and mutation rates remain constant. This correlation arises because under neutral theory, genetic variation, whether it occurs within or between species, is the outcome of the *same* evolutionary process—the genetic drift of neutral mutations. The only difference is that this process will have run for a longer period of time between species than within them. For neutralists, polymorphism is therefore a transient 'phase' of molecular evolution. For selectionists, on the other hand, polymorphisms are thought to be actively maintained by selection and not transient phenomena.

During the 1970s the results from electrophoretic studies seemed to cast doubt on the neutralist prediction that there is an association between heterozygosity, population size and mutation rate. Specifically, and rather ironically given that the neutral theory was formulated to explain high levels

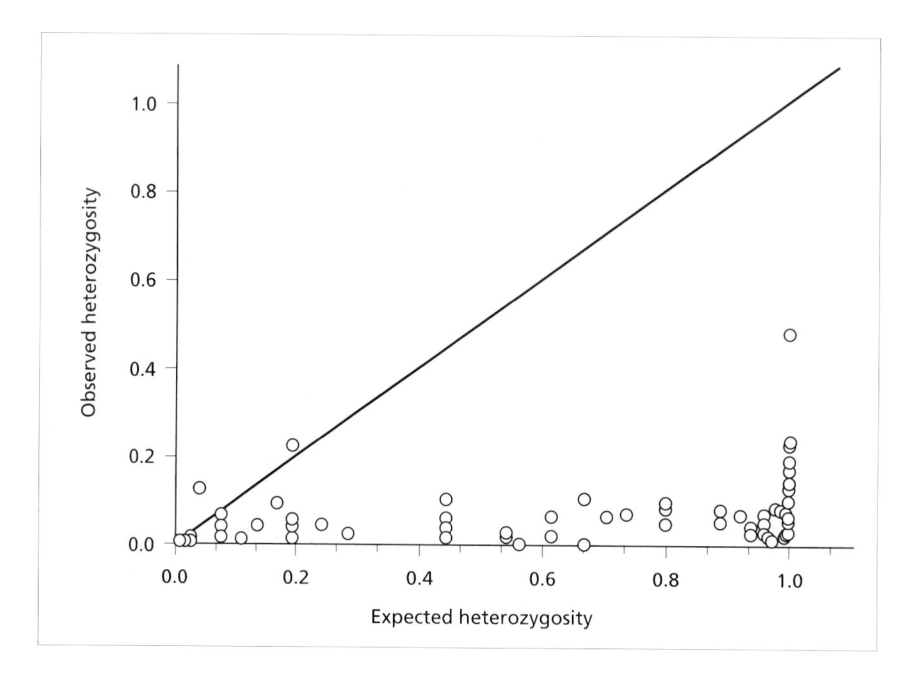

Fig. 7.21 The observed heterozygosities for 77 species (points) compared with those expected under the neutral theory (diagonal line). The expected heterozygosities were calculated using estimates of population size and generation time for each species and assuming a mutation rate per generation of $\mu = 10^{-7}$. Data from Nei and Graur (1984). From Gillespie (1987), with permission of Oxford University Press.

of genetic variation, levels of heterozygosity were found to be *too low* and occupied too narrow a range for the neutralists (Fig. 7.21): for example, humans and *Drosophila* show similar levels of allozyme polymorphism but have very different population sizes. Although these observations suggested that there were problems with the original neutral model and were a key factor in the development of the nearly neutral theory, firm conclusions were difficult to draw because N and μ are hard to estimate and electrophoresis underestimates the extent of genetic variation. More recent studies of allozyme polymorphism based on cross-species comparisons have revealed a general association between heterozygosity and mutation rate. However, analyses of this type are not the best way to test evolutionary models because they lack discriminatory power. We therefore focus on studies of DNA variation.

An important way of assessing the neutral theory in recent years has been to test the proposed correlation between the levels of within species polymorphism and between species divergence (Box 7.1). For example, if synonymous and non-synonymous substitutions are neutral then the ratio of these two types of change will be the *same* within and between species because both are outcomes of the same neutral mutation process. Positive natural selection, on the other hand, would alter this ratio because an advantageous non-synonymous substitution will be fixed quicker by natural selection (i.e. will be a polymorphism for less time) than one subject to only genetic drift, so that there will be less within species non-synonymous variation than expected given what is seen between species.

Over a number of years Martin Kreitman and colleagues have analysed patterns of synonymous and non-synonymous substitution at the alcohol dehydrogenase (*Adh*) locus in *D. melanogaster* and its close relatives and found lower proportions of non-synonymous variation within than between species, as expected if positive natural selection has acted on this gene: 29% of the fixed nucleotide differences between species are non-synonymous compared with only 5% of the polymorphisms within species (Fig. 7.22). This is also true of some other genes in *Drosophila*. In the glucose-6-phosphate dehydrogenase (*G6pd*) gene of *D. melanogaster* and *D. simulans* for example, the level of non-synonymous variation within these species is far less than would be expected given the observed levels of divergence between them—only two non-synonymous polymorphisms compared with 21 fixed non-synonymous changes. Although these patterns of substitution suggest the past action of natural selection, they can also be explained if the non-synonymous substitutions are nearly neutral: if *Drosophila* populations have only recently increased in size (which is uncertain) then little non-synonymous variation will have accumulated within them (because nearly neutral mutations are fixed more

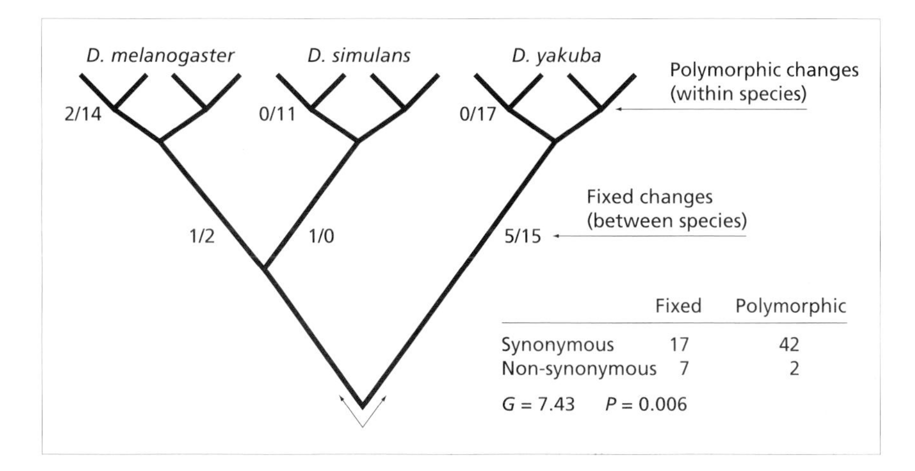

	Fixed	Polymorphic
Synonymous	17	42
Non-synonymous	7	2

$G = 7.43 \quad P = 0.006$

Fig. 7.22 Numbers of non-synonymous and synonymous substitutions in the *Adh* locus within (polymorphic) and between (fixed) three species of *Drosophila*. The results of a *G*-test show that there are significantly fewer non-synonymous polymorphisms than expected by the neutral theory given the number of fixed non-synonymous changes observed between species. The double-headed arrow at the bottom of the tree signifies that the fixed substitutions leading to *D. yakuba* could have occurred at any time since this species separated from *D. melanogaster* and *D. simulans*. Adapted from McDonald and Kreitman (1991).

slowly in large populations) and the observed pattern of high interspecific divergence but low intraspecific variation will be produced.

7.6.2 Recombination and levels of DNA polymorphism in *Drosophila*

The most dramatic incompatibility between levels of genetic variation within and between species occurs in those regions of the *Drosophila* genome where rates of recombination are low, such as near the distal tip of the X chromosome and on the small fourth chromosome, and which also show greatly reduced levels of polymorphism. In contrast, chromosomal regions with higher levels of recombination tend to have higher levels of genetic variation. The extent of this correlation in *D. melanogaster* is given in Fig. 7.23. The neutralist explanation for this pattern is that genomic regions exhibiting low diversity have either lower mutation rates or greater selective constraints and hence evolve at lower rates than other regions. But if this is the case then these regions should also show reduced levels of variation *between* species. However, this is not the pattern observed. For example, the *yellow-achaete* (*y, ac*) region on the X chromosome has a reduced level of polymorphism but the extent of divergence in this region between *D. melanogaster* and *D. simulans* (5.4%) is similar to that observed in a number of other genes (an average of 4.7% for nine genes). Consequently the neutral theory cannot explain the data.

Natural selection, however, can explain this discrepancy through the dual action of **selective sweeps** and **genetic hitchhiking**. Hitchhiking was used in the early 1970s to explain why levels of allozyme polymorphism were often lower than the neutralists predicted. As its name implies, it occurs when a neutral mutation receives a ride with a mutation that selection is driving to fixation. To put it another way, as selection 'sweeps' an advantageous mutation to fixation in a population, all those nucleotides linked to it—those parts

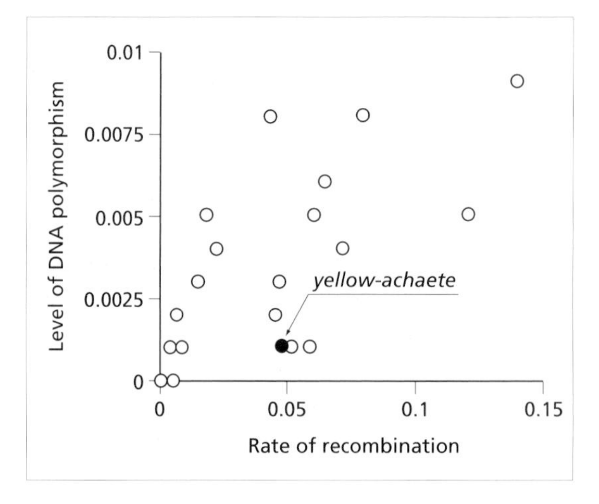

Fig. 7.23 Relationship between the rate of recombination and levels of DNA polymorphism for 20 genes from *D. melanogaster*. The position of *yellow-achaete* (*y, ac*), a gene on the X chromosome with low rates of recombination and low levels of polymorphism, is shown. The correlation coefficient (r^2) is 0.42. Data from Begun and Aquadro (1992).

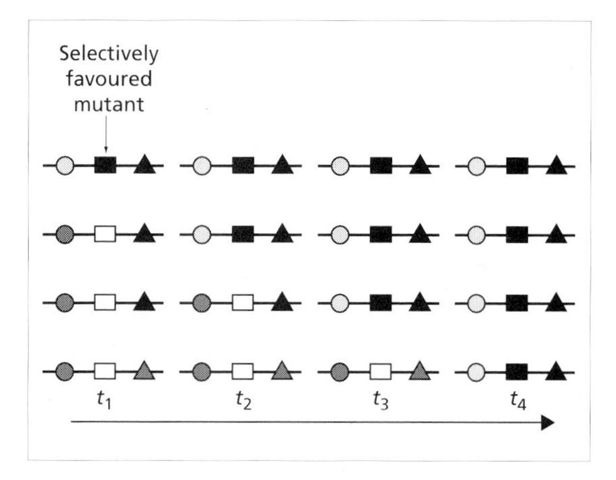

Fig. 7.24 The hitchhiking of neutral sites linked to a selectively favoured mutation. The different combinations of symbols and shades represent different nucleotide combinations in a population of four sequences. If a mutation is selectively advantageous (black square at time t_1) it will sweep through the population at the expense of the other variants until it achieves fixation (at time t_4). Because of the low rate of recombination nucleotides linked to this favoured variant will hitchhike with it and also be swept to fixation.

of the genome that are not routinely separated from the selected site by recombination—are also swept to fixation (Fig. 7.24). Because only one sequence type will be lucky enough to be linked to the selected site, polymorphisms (i.e. other sequence types in the population) near to this site are eliminated and the level of genetic variation is reduced within species, but *not* between them. In parts of the *Drosophila* genome where recombination rates are low many nucleotides will be linked together so that genetic diversity will be reduced over a large area.

Unfortunately, hitchhiking causes more complex effects than a simple reduction in the amount of genetic variation within species. It also means that any variation that does occur in regions subject to hitchhiking will mainly be in the form of alleles at very low frequencies in the population. This is so because these alleles must have arisen *since* the sweep, will not have existed for very long (they have very recent coalescence times—see Chapter 4) and so will not have had time to reach high frequencies in the population. Therefore, not only does hitchhiking reduce the *number* of variable nucleotide positions—called the **segregating sites**—but it also biases the *frequency* distribution, or spectrum, of alleles at these sites towards those that are rare in the population. In contrast, fewer rare alleles are expected under neutral theory because they will, on average, have persisted for longer periods without being swept away, and consequently will be present in a greater proportion of individuals in the population (Box 7.1). Although in some cases where hitchhiking has been proposed in *Drosophila*, such as *yellow-achaete*, this **frequency spectrum of segregating sites** fits the hitchhiking expectation in having an excess of rare variants, in other regions it bears more resemblance to the neutral model! Clearly, a more complex model than selective sweeps and hitchhiking is required for these regions.

One such model is based on **background selection** against deleterious mutations. This was first proposed by Charlesworth *et al.* (1993) who claimed that negative selection against newly arising deleterious mutations may also reduce the amount of genetic variability at linked neutral sites (which will also be removed by negative selection) as long as these deleterious mutations arise at a high enough rate and within a restricted area. One class of mutations that have these characteristics are insertions of transposable elements, such as *P* elements, which are common in the *Drosophila* genome (see Chapter 3). As with hitchhiking, the tighter the linkage between the neutral sites and those subject to selection, the easier background selection will be to detect. The difference between the two models is that hitchhiking considers the outcomes of positive natural selection on the evolution of linked neutral changes, whereas the background selection model looks at the affects of negative selection. Because it deals with negative selection, the background selection model can also be thought of as a neutralist explanation and it predicts a frequency spectrum of segregating sites more like that expected under neutral theory.

Although the DNA sequence data currently available rule out the strict neutralist explanation for the reduced levels of within species variation seen in many *Drosophila* genes where levels of recombination are also reduced, they do not allow us to clearly choose between the background selection and hitchhiking models (although new tests are currently being developed to do just this). It is also possible that a combination of different evolutionary processes, both neutral and selective, have all left their mark on genomes.

7.7 Natural selection at the molecular level

When the neutral theory was first proposed in the late 1960s, there was an immediate response from those who believed that natural selection was the more important force in molecular evolution. As we have seen, part of this response was to question the observations which the neutralists claimed supported their theory. Others attempted to develop models which could explain the observed patterns of molecular evolution through the action of natural selection. The most notable work in this area has been undertaken by John Gillespie. Through a complex statistical argument based around measurements of $R(t)$, Gillespie has concluded that the molecular clock at non-synonymous sites in mammals is episodic—characterised by periods of stasis interspersed with bursts of substitutions as genes adapt to changing environments—rather than one where substitutions are evenly spaced, as predicted under the neutral theory. This **episodic molecular clock** is therefore taken as evidence that natural selection is the driving force of molecular evolution. Furthermore, some lineages are more 'environmentally challenged' than others, because their environments change more frequently, and as a consequence evolve more rapidly, causing the residue effect in estimations of *R(t)*. However, Ohta believes that an episodic clock can also be explained by the nearly neutral theory: bursts of non-synonymous substitution occur when population sizes are small, because many nearly neutral mutations are fixed by genetic drift, while few take place when population sizes are large. Furthermore, the complex interactions between the amino acids of a protein mean that many substitutions are likely to occur together in order to preserve protein structure and function ('compensatory mutations').

Whether non-synonymous changes are nearly neutral or selectively advantageous is the main focus of the neutralist–selectionist debate today. Although it will be difficult to choose between these explanations, selective models such as Gillespie's predict that rates of substitution will be higher in large populations, whereas under nearly neutral theory small populations will have the highest rates of change.

7.7.1 Examples of molecular natural selection

Another approach used to counter the neutral theory, and one which has gathered momentum in recent years, has been to accumulate examples of the action of natural selection at the molecular level. Although detecting molecular natural selection can be difficult, because selective events may involve just a few base changes and because selection and drift may produce similar outcomes, molecular adaptation can still sometimes be inferred through sequence comparisons. To some, each of these examples represents a vote against the neutral theory. Here we outline a few of the most informative.

In some cases it is possible to infer natural selection by studying the evolution of proteins in great detail. Not surprisingly, one such protein is haemoglobin. Because of greatly varying lifestyles, different species have different oxygen requirements, some of which have been met by specific amino acid replacements in the haemoglobin molecule. An interesting example is provided by the crocodilians (crocodiles, alligators and caiman). Although crocodilians spend long periods under water they do not keep large reserves of myoglobin in their tissues, which is how diving mammals and birds store oxygen. Instead, their haemoglobin has adapted to store oxygen in their lungs and blood. The Nobel laureate Max Perutz claims this was achieved with only three amino acid changes. Another case of adaptation in haemoglobin may have occurred in birds that fly at extremely high altitudes, such as the bar-headed goose (*Anser indicus*) which flies at 9000 metres over the Himalayas. Perutz suggests that this could be due to a single proline to alanine replacement in the α-chain (position 126 in the sequence alignment shown in Fig. 7.6) which produces a higher oxygen affinity.

Natural selection has also been detected in proteins that have experienced large-scale convergent or parallel evolution because this is unlikely to have occurred through the chance action of genetic drift. One of the best examples of parallel evolution are the stomach lysozymes of ruminant artiodactyls (such as the cow) and leaf-eating monkeys (such as the langur). The lysozymes of these species have acquired the new function of digesting the bacteria that pass from the fermentative forgut to the true stomach and have independently evolved amino acids which allow this enzyme to function in stomach fluid (see Chapter 5). Convergent evolution is a common occurrence in the envelope protein of HIV-1 where those amino acid changes which allow the virus to escape from immune surveillance evolve regularly in patients infected with unrelated viruses. Despite these examples, there are few well-documented cases of natural selection at the amino acid sequence level.

Traditionally, population geneticists have attempted to detect natural selection by determining the factors which shape heterozygosity either in large multiple allozyme studies across diverse species groups or at a single locus within a much more closely related set of taxa. While the multiple allozyme studies have often been inconclusive, the single-locus work has provided some good evidence for natural selection. This is particularly true of the *Adh* gene of *Drosophila* where there is evidence for a selectively maintained polymorphism at the site of a threonine to lysine amino acid substitution that distinguishes the 'fast' (Adh^f) and 'slow' (Adh^s) alleles, which differ in their ability to process alcohol. Part of the evidence that balancing selection occurs in this gene is that there is more silent polymorphism than expected under neutral theory, especially at sites close to the fast–slow polymorphism. This is suggestive of selection because any silent (neutral) variation linked to a balanced polymorphism that has been maintained for a long period of time would itself persist for longer than if genetic drift were the only process determining its fate (see the coalescent theory section of Chapter 4).

The fast–slow polymorphism in *Adh* has been studied in more detail by Berry and Kreitman (1993) who found a cline in the frequency of the polymorphism in populations of *D. melanogaster* along the east coast of the USA. The frequency of the Adh^f allele ranged from 40% in the south (Florida City, Florida) to 80% in the north (Cherryfield, Maine), whereas other sites in the *Adh* region did not show clinal variation. Although the reasons why populations at different latitudes have different requirements for this polymorphism are unclear, this result is not expected under neutral theory which predicts that all loci (or regions within the same locus) will show the same pattern of geographical differentiation. The reverse pattern was seen in populations of the American oyster, *Crassostrea virginica*. Here DNA markers revealed a marked geographical structuring, with populations from the Atlantic coast and the Gulf of Mexico very distinct. In contrast, no spatial differentiation was observed in various allozymes suggesting that they are under strong selection to maintain particular allele frequencies (Fig. 7.25).

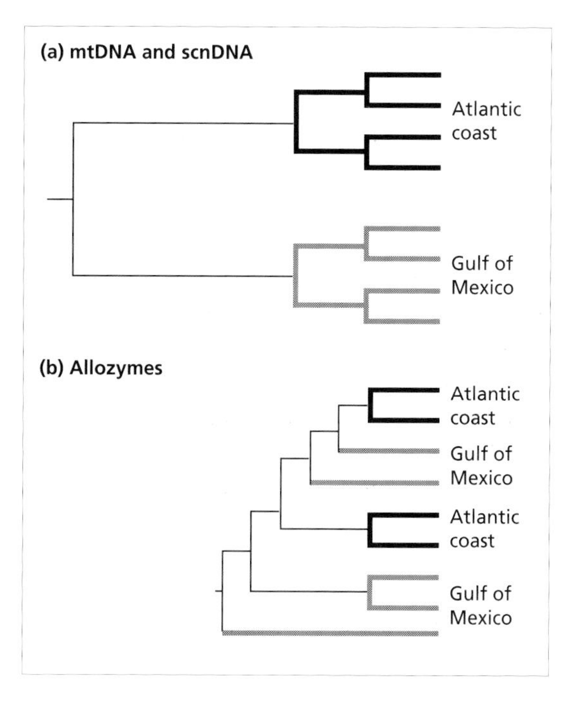

Fig. 7.25 Geographical differentiation and balancing selection in the American oyster, *Crassostrea virginica*. Trees based on mtDNA and scnDNA (single-copy nuclear DNA) show a strong difference between oyster populations sampled from the Atlantic coast compared with those from the Gulf of Mexico (a). In contrast, less geographical structure is seen in trees based on allozyme data from the same populations (b). This indicates that allozymes are subject to balancing selection which maintains homogeneous allele frequencies across large distances. Trees are schematic only; see Karl and Avise (1992) for more details.

7.7.2 Relative rates of non-synonymous and synonymous substitution

For those analysing nucleotide sequence data, the most fruitful way to detect natural selection has perhaps been to document cases where there are greater numbers of non-synonymous (d_N) to synonymous (d_S) substitutions within a gene. This is suggestive of positive selection because it means that the rate of substitution is higher than the rate of mutation, which cannot occur if mutations are neutral (in which the substitution rate is equivalent to the mutation rate). In the case of negative selection, greater numbers of synonymous than non-synonymous substitutions are fixed because the latter are often deleterious and therefore selectively removed (Box 7.1).

To date, most examples of $d_N > d_S$ have been described in genes involved in the interaction between the mammalian immune system and the antigens of intracellular parasites (bacteria, plasmodia and viruses). This can perhaps best be seen at the **major histocompatibility complex (MHC)**—a set of about 100 genes which present parasite antigens for clearance by antibodies and cytotoxic T-lymphocytes (CTLs). In humans these genes are called the **human leukocyte antigens** (HLA) and are located on chromosome 6 (Fig. 7.26). A great deal of research has been directed towards understanding the genetics of the MHC because these genes determine our susceptibility to infectious and autoimmune diseases and are important in tissue transplantation.

Fig. 7.26 Map of the polymorphic human leukocyte antigen (HLA) class I and class II genes on chromosome 6. The number of alleles known for each gene is also shown. From Hill *et al.* (1991), with permission.

Some of the most important HLA genes are those classified as class I and class II, which produce glycoproteins that are expressed on cell surfaces. Class I antigens are expressed on all nucleated cells, whereas class II antigens occur on the antigen-presenting cells of the immune system. Because mammals are exposed to a great number of genetically different parasites, which can impose large selective costs, it is important for the MHC to be as diverse as possible so that the immune system can recognise and control as many of these as possible. Consequently these genes exhibit enormous allelic diversity. For example, there are three class I genes in the human MHC (labelled A, B and C) which have many different alleles (Fig. 7.26). The selective importance of the MHC can be seen in its response to *Plasmodium falciparum* malaria which causes up to 2 million deaths each year in sub-Saharan Africa. Both class I and class II MHC alleles are associated with protection from malaria and most notably the class I allele Bw-53, which is common in West African populations where malaria is endemic but rare elsewhere.

More evidence for natural selection is that some MHC alleles persist for a very long time, so long in fact that polymorphisms are maintained across species boundaries. For example, a deletion polymorphism, denoted $A^{\beta}_{\beta}/A^{v}_{\beta}$, is found in both mice and rats which means that it must be at least as old as this species split, perhaps 15 million years. This is too long for a polymorphism to be maintained by genetic drift because it would be expected that one allele would drift to fixation within this time period (see Chapter 4) so that it must be preserved by some form of balancing selection.

The region of MHC genes that is most intimately involved with the recognition of parasites is called the antigen recognition site (ARS) and there is a great deal of amino acid variation in this region both within and between species. Detailed analyses of the ARS has provided strong evidence for positive selection because values of d_N are greater than d_S, unlike other regions in MHC genes (Fig. 7.27).

Although there is good evidence for natural selection in the MHC it is unclear exactly what form this selection takes. One model is overdominant selection where the fitness of alleles in the heterozygous state is greater than

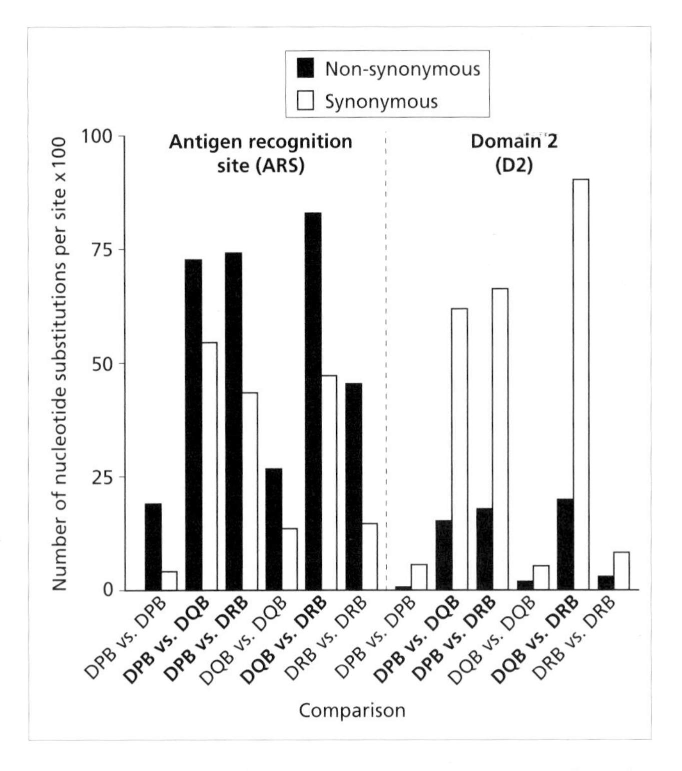

Fig. 7.27 Numbers of nucleotide substitutions per non-synonymous (d_N) and synonymous (d_S) site between various HLA class II genes (comparisons shown in bold) and alleles of these genes. Positive selection dominates in the ARS because the d_N/d_S ratio is usually >1.0, whereas negative selection is more important in D2 because synonymous substitutions are more frequent ($d_N/d_S < 1.0$). Data from Hughes and Nei (1989).

those that are homozygous. An alternative explanation is frequency-dependent selection. Here individuals carrying a new, and hence rare, mutant MHC allele have a selective advantage because parasites will not have had enough time to evolve the ability to evade them. However, this advantage will decrease with time as the MHC allele increases in frequency and is recognised by more parasites.

Host immune systems also act as selection pressures for parasites because those that can avoid being cleared will be at a selective advantage and so will propagate. This often involves surface antigens. For example, in HIV-1 strong selection pressures ($d_N > d_S$) have been detected on those parts of the surface envelope glycoprotein gp120 that are implicated in the escape from immune clearance, whereas the proteins involved in construction of the viral capsid (*gag*) and in replication (*pol*) seem to be under negative selection. Likewise, in *P. falciparum* positive selection has been detected at the major surface antigen (MSA-1) and at the T-cell epitopes of the cell surface circumsporozoite (CS) protein, while in the haemagglutinin 1 (HA 1) gene of human influenza A virus selection occurs at amino acid residues thought to be antigenically important. Positive selection by hosts on parasites has also led to the resurgence of bacterial drug resistance, most notably in tuberculosis (*Mycobacterium tuberculosis*) and *Streptococcus* spp.

While good examples of natural selection at the molecular level are still relatively few in number, it is certain that many more will be uncovered as our ability to distinguish the signatures of natural selection and genetic drift improves and as more comparative DNA sequence data become available.

7.8 Conclusions: can we resolve the neutralist–selectionist debate?

Most support for the neutral theory of molecular evolution has come from comparisons of gene sequences from relatively distantly related species, particularly the different orders of placental mammals, where patterns of substitution have built up over millions of years. Natural selection, on the other hand, is more apparent in sequence comparisons made over shorter time periods, as in the interaction between hosts and parasites or within populations of *Drosophila*. It may be that as time proceeds the evidence for natural selection, the fixed amino acid changes, is obscured by successive and generally neutral changes. For example, in most of the mammalian genes analysed it is likely that the majority of adaptive events occurred in the distant past as these animals radiated and initially adapted to their new environments. Adaptations that have taken place since this time may have required fewer and more subtle changes in sequence (as shown by the haemoglobins), which might have been masked by later neutral substitutions. It is only when we compare genes where adaptation has clearly occurred in the recent past, as in the arms' race between hosts and parasites or in the selective sweeps of *Drosophila*, that we can still detect the footprint of natural selection in the shape of amino acid substitutions.

An example of how the evidence for positive selection can be obscured in sequence comparisons comes from the *jingwei (jgw)* gene found in two African *Drosophila* species, *D. yakuba* and *D. teissieri*. Although *jgw* arose through a retrotransposition event from the adjacent *Adh* gene, it differs from *Adh* in that it lacks introns and has accumulated a large number of non-synonymous substitutions, which initially led to reports that it was in fact a pseudogene. However, closer inspection shows that reports of its death have been greatly exaggerated: *jgw* produces a viable mRNA, is not characterised by the stop mutations and insertions and deletions so often found in pseudogenes, and is in fact a chimera between the processed mRNA of *Adh* and another gene, unrelated and as yet unidentified. Although the overall d_N/d_S ratio is suggestive of neutrality (compatible with *jgw* being a pseudogene), more detailed phylogenetic analysis reveals that this ratio has changed through time, with a burst of non-synonymous substitutions occurring after the initial insertion event, perhaps as the gene acquired a new function (Fig. 7.28). The case of *jgw* also

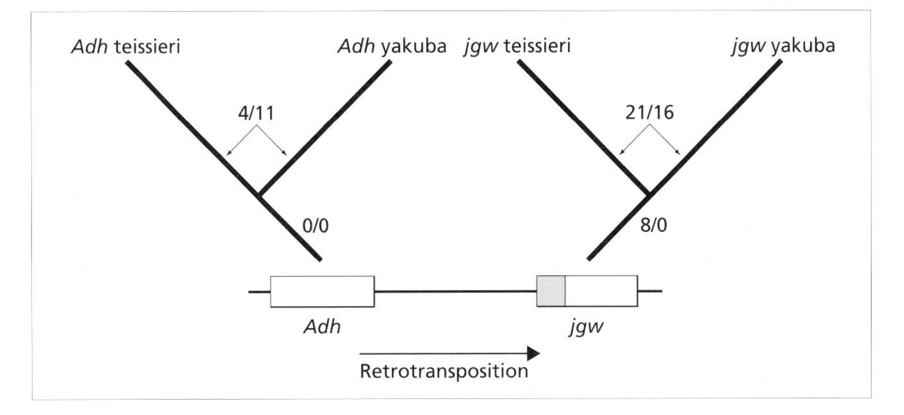

Fig. 7.28 Evolutionary history of the *jingwei (jgw)* gene in two African species of *Drosophila, D. yakuba* and *D. teissieri. jgw* arose through the retrotransposition of an mRNA sequence from the parental *Adh* gene. *jgw* then captured the 5′ region of an unrelated gene (shaded region). The numbers of non-synonymous to synonymous substitutions are shown next to each lineage and indicate that the initial evolution of *jgw* was dominated by positive natural selection (8/0 non-synonymous to synonymous changes) compared with *Adh* (0/0). Adapted from Long and Langley (1993).

illustrates how important it is to consider the phylogenetic relationships of sequences when trying to judge which evolutionary processes have shaped them.

Overall it seems reasonable to conclude that *both* natural selection and genetic drift determine the evolutionary fate of mutations. Neutralists are probably correct in believing that the *majority* of nucleotide substitutions do not affect fitness, and most notably in non-coding DNA and at synonymous sites (in mammals at least). Most selectionists would also agree with this. However, if we look at the dynamics of molecular evolution within a smaller window of time, and particularly at non-synonymous sites, we can clearly make out the footprints of natural selection. To fully resolve the neutralist–selectionist debate will undoubtedly require many more gene sequences from many more species as well as better statistical methods for distinguishing between the two processes, particularly those that incorporate more information about the phylogenetic relationships of the sequences in question.

7.9 Summary

1 A central task of molecular evolution is to determine the processes by which gene sequences change through time.

2 During the 1950s and 1960s extensive genetic variation was discovered within and between species. This posed a problem for theories of natural selection which were thought to impose too high a selective cost on populations.

3 The neutral theory, first proposed by Kimura and King and Jukes, overcame this problem by claiming that molecular evolution is dominated by the genetic drift of neutral mutations which have no selective cost.

4 The best evidence for the neutral theory is the relationship between functional constraint and rate of substitution, so that functionally constrained genes (or regions within genes) evolve at the lowest rates (and vice versa). Other supporting evidence is that rates of silent substitution and patterns of codon usage in mammals are determined mainly by mutation pressure.

5 The opposing argument is that the natural selection of advantageous mutations is the more important force in molecular evolution. Evidence for this comes from within *Drosophila* populations, where there is support for selective sweeps and hitchhiking, and from genes involved in host–parasite interactions where the high numbers of non-synonymous to synonymous substitutions suggest the action of positive selection. Patterns of codon choice in unicellular organisms and *Drosophila* also appear to be selected.

6 The most important debate concerns the molecular clock. The constancy of the clock at synonymous sites in mammals is determined by the constancy (or not) of the underlying neutral mutation rate, itself affected by generation time, metabolic rate and DNA repair, collectively known as lineage effects. In contrast, the clock at non-synonymous sites is less subject to lineage effects but is instead characterised by episodes of substitutions, which has been taken as evidence for natural selection.

7 To overcome the problems of the original neutral theory, Ohta proposed that most non-synonymous substitutions are nearly neutral and so subject to both genetic drift (especially in small populations) and weak negative selection (more important in large populations). The episodic molecular clock at non-synonymous sites can then be explained by fluctuating population sizes.

8 Although it is clear that both natural selection and genetic drift determine the substitution dynamics of gene sequences, the neutralist–selectionist debate continues, particularly with respect to the processes governing non-synonymous evolution.

7.10 Further reading

Good accounts of the early work on genetic variation within and between species and of the development of the neutral theory are given in Avise (1994), Lewontin (1991) and Ridley (1996). Although highly personal, the two most important texts in the neutralist–selectionist debate are Kimura (1983) for the neutralist perspective and Gillespie (1991) for a selectionist argument, whilst Ohta (1992) and Chao and Carr (1993) present the case for the nearly neutral school.

Examples of the correlation between functional constraint and rates of substitution are provided by Dickerson and Geis (1983) (haemoglobin), Shpaer and Mullins (1993) (HIV) and Li *et al.* (1981) (pseudogenes). The neutralist–selectionist debate as it relates to the evolution of base composition and codon usage is well summarised in the papers of Bernardi (1993), Casane *et al.* (1997), Eyre-Walker (1993), Holmquist and Filipski (1994) and Sharp and Matassi (1994), with the situation in *Drosophila* receiving a great deal of recent attention (see Powell and Moriyama, 1997). Papers discussing the molecular clock and the possible causes of rate variation are common, but Li (1993) presents an excellent overview, while more detailed descriptions of the generation time, metabolic rate and DNA repair hypotheses are given in Li *et al.* (1987), Martin and Palumbi (1993) and Filipski (1988), respectively. An important recent analysis is that of Ohta (1995), but also see Gillespie's (1995) companion paper. The evidence for and against the male-driven evolution of mammalian genomes is presented by Shimmin *et al.* (1993) and McVean and Hurst (1997), respectively. The more complex problem of the molecular clock in plants is tackled by Gaut *et al.* (1992, 1997).

Some key papers dealing with genetic variation within species, and particularly the evidence for selective sweeps and hitchhiking are Begun and Aquadro (1992), Eanes *et al.* (1993), Kreitman and Hudson (1991), Langley *et al.* (1993) and McDonald and Kreitman (1991), whilst Charlesworth *et al.* (1993) discuss the evidence for background selection (although also see Braverman *et al.*, 1995).

For those interested in natural selection at the molecular level, Kreitman and Akashi (1995) present an excellent review of the available evidence, while more specific examples are provided by Perutz (1983) (haemoglobin), Stewart *et al.* (1987), (lysozymes), Berry and Kreitman (1993) (*Adh*), Karl and Avise (1992) (oysters), and Hughes and Nei (1989) and Hill *et al.* (1991, 1992) (MHC). The fascinating evolution of *jingwei* is discussed by Long and Langley (1993). Finally, methods for testing the neutrality of mutations are given in McDonald and Kreitman (1991), Hudson *et al.* (1987) and Tajima (1989) (although also see Fu and Li, 1993), and for testing the molecular clock in Gillespie (1989b) and Li (1993).

Chapter 8
Applications of Molecular Phylogenetics

Biologists are making increasing use of phylogenies to address questions across a broad range of scales, from virologists tracking the diversification of viral lineages within a single patient, to biogeographers investigating whether biological diversification is correlated with major geological activity such as continental drift. Previous chapters have emphasised the importance of phylogenies for studying molecular evolution. In this chapter we focus instead on the implications of molecular phylogenies for evolutionary biology. It is not our intention to review all the applications of phylogenies (see the reading list for recent overviews); rather we focus on a sample that highlights the impact molecular phylogenies are having on the study of evolutionary biology.

8.1 Organismal phylogeny

The oldest use for phylogenies of genes is inferring organismal phylogeny. Prior to the molecular revolution, data for phylogenetic inference came from sources such as morphology, ontogeny, behaviour, and geographic distribution. In the 20th century molecular data ranging from immunological distances, allozyme frequencies, restriction sites, DNA hybridisation, and protein and nucleotide sequences have provided much additional data. With depressing regularity, the advent of each new molecular technique has led to claims that *this* technique is the best source of phylogenetic data, and that all previous sources are irredeemably defective. In particular, advocates of molecular techniques have regularly claimed that morphological data is obsolete, or at best too prone to homoplasy to be of much use, whereas molecular data alone yields the true phylogenetic history. In practice, unconditional claims of the superiority of any one type of data have foundered, and it is increasingly appreciated that many different types of organismal attributes retain the trace of those organisms' evolutionary history. Different sources of data each have their strengths and weaknesses, and these may vary depending on the taxonomic group being studied and the kinds of phylogenetic questions being addressed. Molecular data, particularly nucleotide sequences, offer potentially huge data sets that are comparable across a wide taxonomic range. Whereas disparate taxa may share few morphological traits (think of a human and a bacterium), there are numerous homologous genes shared by all the major groups of living organisms. Because different genes (or parts of the same gene) evolve at different rates, molecular data can also be tailored to different time scales, be it the tree of life on Earth unfolding over billions of years, or the evolution of HIV viruses within a single patient over a period of months. Genes that show approximate rate constancy also allow us to make inferences about times of divergence using molecular clocks.

Morphological data have considerable strengths as well. In many groups we have morphological information for many more taxa than we have molecular data; this is particularly true where extinct taxa are concerned. Given that DNA is unlikely to be sequenced from fossil or subfossil remains older than about 40 000 years (Austin *et al.*, 1997), the fossil record is largely the province of morphology. This has implications for extant groups as well, because fossil members of a clade may contain important phylogenetic information that is not present in the surviving taxa. Taxonomic sampling can have a major impact on phylogenetic analyses; results based on a few taxa, or restricted to just the extant members of a clade may give misleading results. In many clades the morphological data base far outweighs the molecular data base.

8.1.1 Combined or separate analysis?

Given that there are numerous sources of phylogenetic data, even if we were to restrict ourselves to molecular data, inevitably the question arises of how to combine these different sources. Three approaches have been proposed. The first, often called **total evidence** (also called 'simultaneous analysis') advocates always combining different data sets into a single data set for analysis. This view is contrasted with 'separate analysis' (also called the 'congruence' or 'consensus' approach) in which the data sets are analysed separately and the resulting trees then combined using consensus trees (see Chapter 2 for a definition of consensus trees). The third approach is 'conditional combination': the data sets are combined unless there is evidence for significant conflict among them.

The primary argument for combining the data in one analysis is that it utilises all the information available to the systematist in a single analysis. Different kinds of data—which may provide information about different parts of the tree—are allowed to jointly affect the result. Conflicts are resolved by weight of evidence, irrespective of the source of data.

Separate analyses seem attractive because they allow for independent estimates of the phylogeny using different data sets. If these independent sources of information yield similar trees, then our confidence in our ability to recover the actual phylogenetic history of a group or organisms is reinforced. Merging all the data together eliminates the opportunity for independent testing. Furthermore, separate analysis allows us to discover whether each data set supports essentially the same (or a compatible set of) relationships, or whether some data sets yield quite different trees. The latter case immediately alerts the investigator to potential problems, and may yield insights into the evolution of different character systems.

The relative merits of combined and separate analyses have been much debated in the literature (see suggested reading at the end of this chapter). Combining data would be a good strategy if the phylogenetic method used was consistent (Chapter 6), such that by adding more data the method would always converge on the correct tree. However, if a method was inconsistent the additional data would not improve the accuracy of the phylogenetic estimate. Indeed, 'good' phylogenetic signal may be swamped by other more numerous—but misleading—characters. Combining all the data into a single analysis assumes that all the data reflect the same evolutionary history. Different genes may have different, conflicting histories, in which case combining their nucleotide or protein sequences into a single data set may be inappropriate. Techniques for dealing with complex, contradictory gene trees are discussed in section 8.2 below.

While separate analysis facilitates detecting conflicts among different data sets, it is not without its own problems. Among the objections raised against separate analysis are concerns about the objectivity of the criteria used to delimit the different data sets, and the limitations of combining trees using consensus methods. In some cases partitioning organisms (and their genomes) into separate categories of data seems arbitrary. Different ontogenetic stages may be discrete (for example, in insects), yet form part of the continuum that is an organism's life cycle. Distinguishing between behavioural and morphological characters may be difficult as a behaviour may be intimately linked with particular morphological features of the organism. Sequencing different fragments of the genome may yield different data sets, but these may reflect the availability of suitable primers and other historical contingencies, rather than 'real' divisions of the genome. While these objections have some validity, a case can be made for the existence of different partitions that are not arbitrary. Unlinked nuclear genes may have different evolutionary histories, which may in turn differ from the histories of organellar genomes such as the mitochondrion. Similarly, in the (almost complete) absence of recombination, animal mitochondrial genes all share the same evolutionary history and hence could be combined and treated as a single data set.

Given that separate analyses will produce one or more trees for each data set, the question naturally arises of how to combine these into a single tree that best summarises all the data sets. One obvious approach is to construct a consensus of the trees from the different data sets, so that the resulting tree contains only those groups found in, say, all of the separate analyses. Such an approach has the apparent virtue of being conservative because groups supported by some but not all the data will not be included. Hence, one might argue that while the price of consensus may be a lack of resolution relative to a combined analysis, at least we can be sure that groups in the final consensus tree are supported by all the data. However, this need not be the case: Barrett *et al.* (1991) constructed a clever counter-example of a case where two data sets analysed separately both yielded support for a clade which was not in the best tree when the two data sets were combined and analysed together (see Box 8.1).

Box 8.1 Why consensus and combination can give different results

Barrett *et al.* (1991) constructed an example where combining data sets and combining trees yield different results. Given these two data sets, each comprising four weighted binary characters (in the diagram below these weights are shown in **boldface**), the best trees for the two data sets treated separately both contain the group BCD, which is therefore in the consensus tree. However, when analysed together the best tree for the combined data does not contain this group.

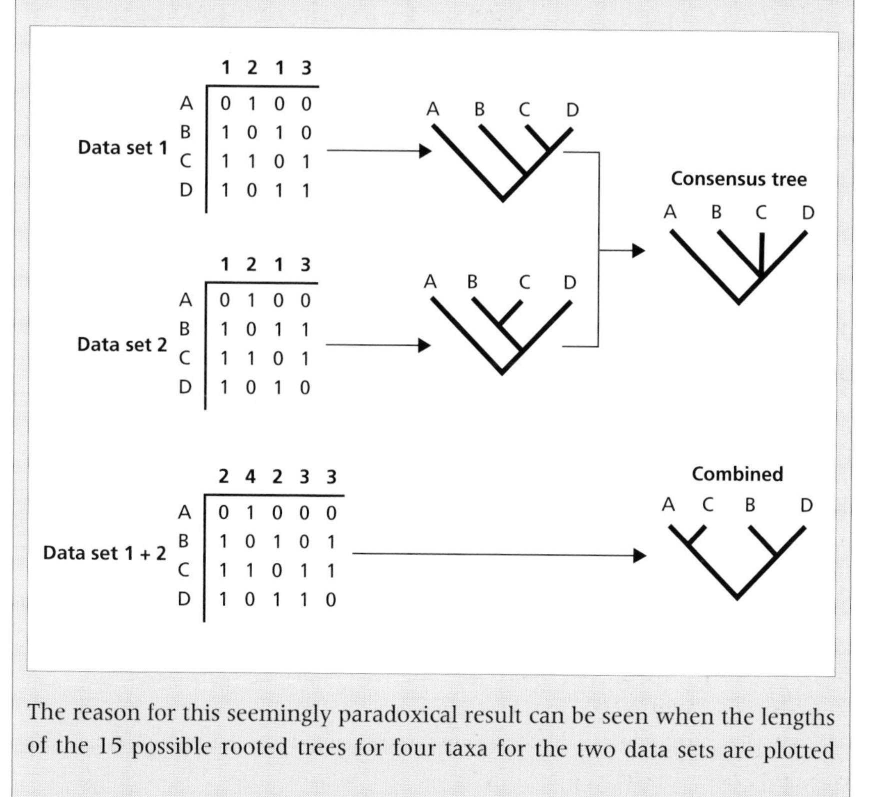

The reason for this seemingly paradoxical result can be seen when the lengths of the 15 possible rooted trees for four taxa for the two data sets are plotted

continued on p. 284

Box 8.1 *continued*

against each other (diagram below). The tinted zone encompasses the possible combination of lengths for trees that are suboptimal for either data set taken separately, but are more parsimonious for the combined data than either of the best trees for the data sets considered separately. Trees falling on the same dashed line (sloping down from left to right) will have the same length for the combined data sets. This length is indicated at the bottom of each line.

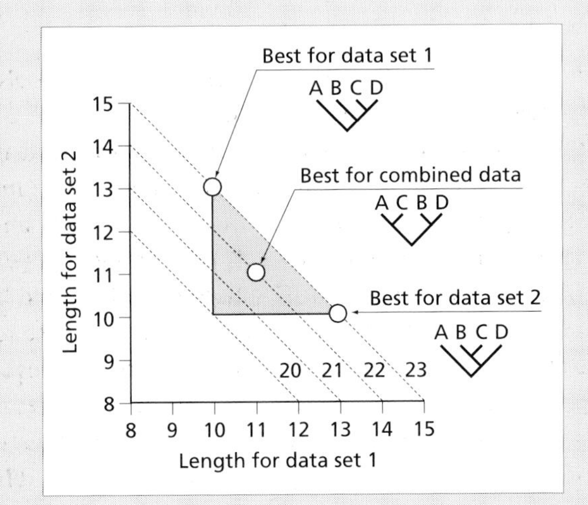

The best trees for the two data sets considered separately have lengths of 10 steps for data sets 1 and 2, respectively. However, each tree requires an additional three steps for the other data set, hence each tree has a total length of 23 steps for the combined data. The tree that is best for the combined data has a total length of 22 steps (11 for each data set). Because it is not the optimal tree for either data set, this tree is not considered in the separate analyses. After Page (1996a).

8.1.2 When to combine data and when not to

An intermediate strategy that incorporates aspects of the 'combined' and 'separate' approaches is to combine different data sets unless there is evidence that one or more data sets markedly conflicts with the remainder. If data sets are telling essentially the same story (within the bounds of sampling error), then combining the data would make the most use of the available information. If the data sets conflict, then separate analyses may be preferable.

Given that our decision to combine or not combine data sets is based on the degree of conflict among those data sets, how can we test whether it is reasonable to combine the data? Several approaches have been suggested, the one we shall describe here is the likelihood heterogeneity test of Huelsenbeck and Bull (1996). This test is a likelihood ratio test like those described in Chapter 6 (section 6.5.2). The null hypothesis (H_0) is that the same tree underlies the different data sets—even though the relative rates of evolution and the values of parameters such as base composition and transition/transversion ratio may vary among the genes, all the genes have the same maximum likelihood tree topology. The alternative hypothesis (H_1) is that different trees underlie the different data sets. As in Chapter 6, the likelihood ratio test statistic is

$$2\Delta = \log L_1 - \log L_2$$

where L_1 is the maximum likelihood of the alternative hypothesis H_1 and L_0 is the maximum likelihood of the null hypothesis H_0.

Applying this test to the bird–mammal problem discussed in Chapter 6 (section 6.8.3), there is significant heterogeneity between the mitochondrial 12S and 16S rRNA genes, valine tRNA, and the nuclear 18S and 28S rRNA genes (Fig. 8.1). Under two different models of DNA evolution, the classical tree grouping birds and crocodiles was the maximum likelihood tree for three of the five genes; under both models, the 18S rRNA gene consistently favoured grouping birds and mammals, whereas most of the other genes favoured the classical grouping. Removing this gene and performing the test on the four remaining genes failed to reject the null hypothesis of homogeneity among the genes, hence the 18S rRNA data appear to provide quite a different signal

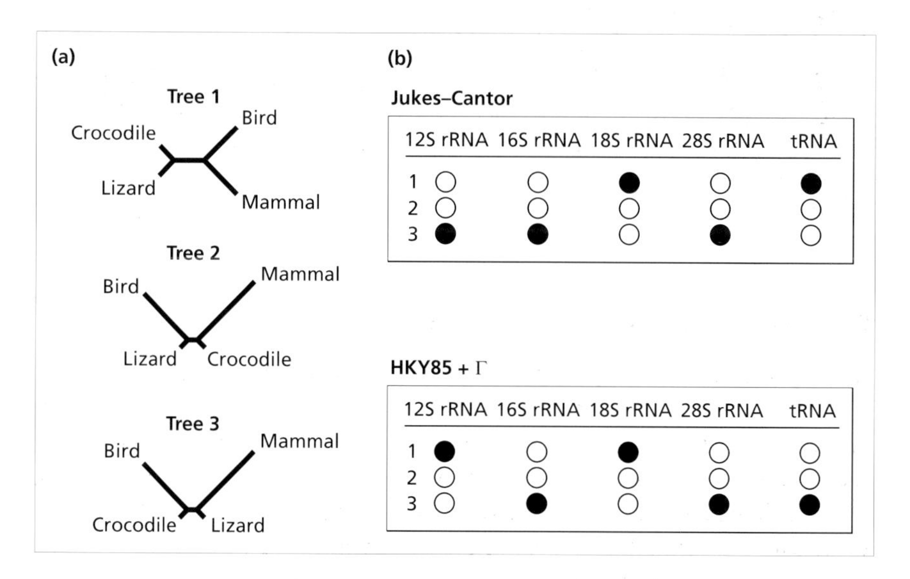

Fig. 8.1 (a) The three unrooted trees for bird, mammal, lizard, and crocodile. (b) For two different models of sequence evolution (Jukes–Cantor and HKY85 + Γ) the table indicates which of the three trees in (a) is the maximum likelihood tree (●) for each of five genes. Adapted from Huelsenbeck and Bull (1996).

from the other genes. Based on this test, the 12S, 16S, 28S and tRNA genes could be combined into a single analysis, and the 18S data should be analysed separately.

8.2 Gene trees and species trees

The implicit assumption made when we use molecular phylogenies to infer organismal relationships (section 8.1) is that gene trees are isomorphic with species trees—the former can be converted into the latter merely by substituting the name of the sequence with the name of the organism from which the sequence was obtained. As sequence data have accumulated it has become increasingly clear that the relationship between gene trees and species trees may be more complex than a simple one-to-one correspondence.

8.2.1 Reconciled trees

One of the first attempts to deal with this problem is the concept of a **reconciled tree**, first introduced by Morris Goodman and colleagues in 1979 to account for discordance between mammalian haemoglobin gene trees and previously accepted notions of mammal phylogeny (similar ideas were proposed independently by workers in parasitology and biogeography; see Box 8.2).

Box 8.2 Historical associations

The association between two or more lineages over evolutionary time is a recurrent theme that spans several different fields within biology, from the molecular level through to macro-evolutionary patterns. Across these disparate levels, there occur many instances of 'historical associations', where genetic or organismal lineages are associated with another, and one lineage may be thought of as tracking the other over evolutionary time with a greater or lesser degree of fidelity.

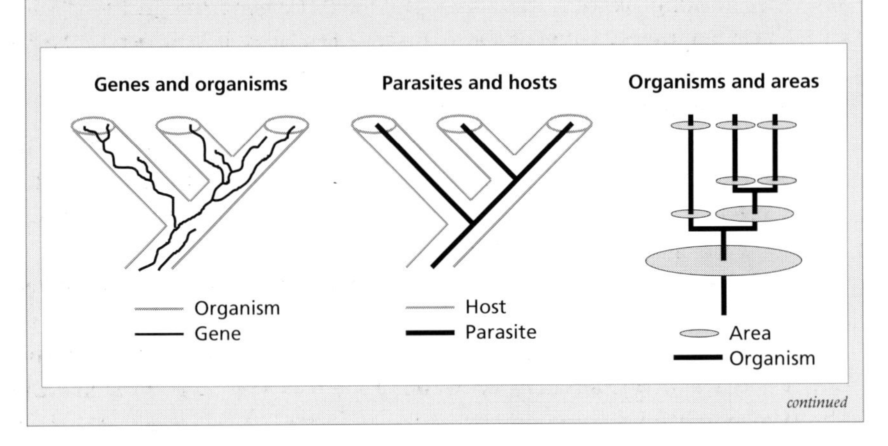

continued

Box 8.2 *continued*

In each case, one entity may be thought of as 'tracking' the other over time. At the molecular level, each gene family has a phylogenetic history that is intimately connected with, but not necessarily identical to, the history of the organisms in which those genes reside. Processes such as gene duplication, lineage sorting and horizontal transfer can produce complex gene trees that differ from organismal phylogeny. Some hosts and their parasites (including viruses) may have a long evolutionary history of close association, reflected in similarities between their evolutionary trees. Similar patterns can be found between other ecological associations, such as insects and their host plants, and the bacterial endosymbionts of insects. At a larger scale still, organisms may track geological history such that sequences of geological events such as continental break-up are directly reflected in the phylogenies of those organisms.

Associations among genes, organisms and areas have traditionally been studied by different biologists from different disciplines, with little interaction between them. Consequently, recognition of the fundamental similarity of the problem faced by molecular systematists, parasitologists and biogeographers has been slow in coming. Despite the relative lack of interaction between these different disciplines, it is striking how similar concepts have arisen independently in these different contexts. Parasitologists (Hopkins, 1948; Clay, 1949) clearly formulated the problem of multiple parasite lineages decades before Fitch's (1970) analogous distinction between paralogous and orthologous genes (Page *et al.*, 1996). Molecular systematists (Goodman *et al.*, 1979) and cladistic biogeographers (Nelson and Platnick, 1981) independently developed very similar methods for interpreting the history of gene trees and biogeograhic patterns, respectively.

If we have a species tree and a gene tree which are incongruent (Fig. 8.2), and we are confident that both are correct for the species and genes, respectively, then we might ask under what circumstances could both be true. If we regard genes as 'tracking' species, then we can embed the gene tree in the species tree. The incongruence between these two trees can be explained by postulating a gene duplication that gave rise to two sets of paralogous genes, of which only four have survived to the present day (Fig. 8.3).

Genes *a* and *b* are orthologous, as are *c* and *d*. Given the duplication δ at the base of the gene tree we would have expected to find two copies of this gene in each taxon a–d. The presence of only a single copy in each requires at least three independent gene losses.

Figure 8.4 shows the reconciled tree computed for the trees shown in Fig. 8.2. This tree can be thought of as the tree obtained by 'unfolding' the gene tree embedded in the species tree in Fig. 8.3 and laying it flat on the page. This tree reconciles the incongruent gene and species trees by postulating that the observed gene tree is a relict of the larger gene tree, resulting from the gene

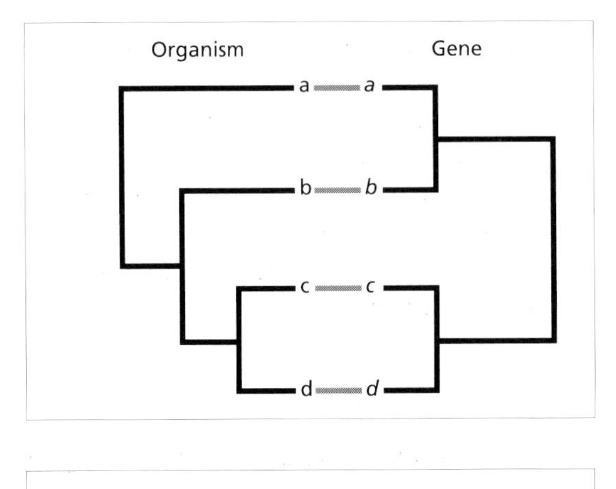

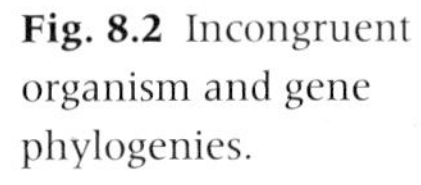
Fig. 8.2 Incongruent organism and gene phylogenies.

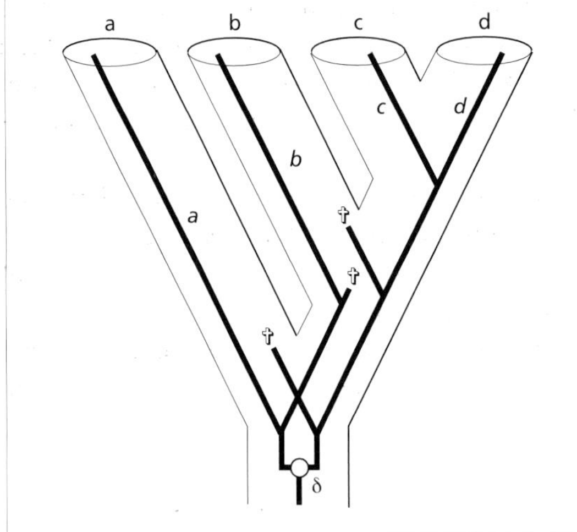

Fig. 8.3 The incongruence between the trees shown in Fig. 8.2 can be explained by hypothesising a gene duplication (δ) at the base of the gene tree, with genes *a* and *b* being paralogous with genes *c* and *d*. The presence of only a single gene in each present day species requires us to postulate three gene losses (†).

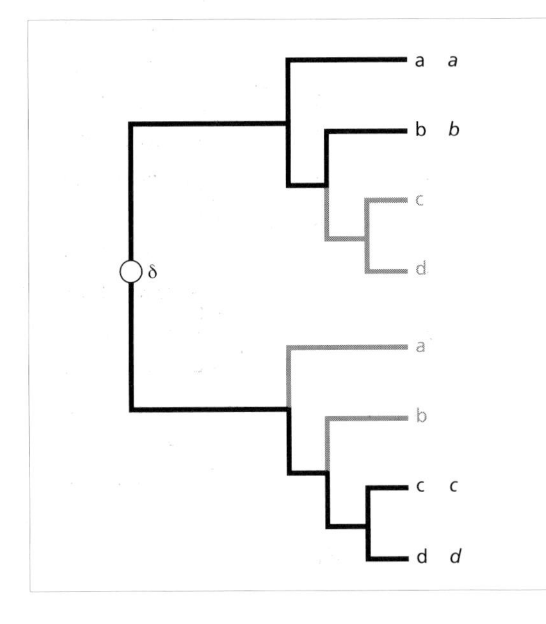

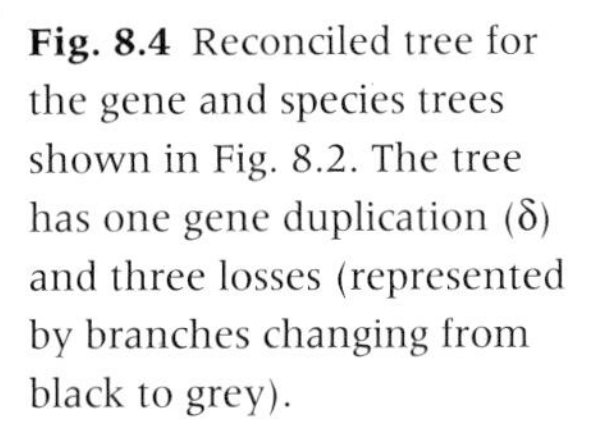
Fig. 8.4 Reconciled tree for the gene and species trees shown in Fig. 8.2. The tree has one gene duplication (δ) and three losses (represented by branches changing from black to grey).

duplication δ. This larger tree is the reconciled tree, and is the tree we would obtain if no gene loss or extinction occurred. As before, given the duplication δ we would have expected two copies of the gene in each species. That we do not see these copies requires us to postulate gene losses in species a and b, and in the ancestor of species c and d. Note that these genes may be present but as yet undetected. The total number of events the reconciled tree postulates (one duplication plus three losses) is the 'cost' of the tree, and can be written $c(G, T)$, where G is the gene tree and T is the species tree.

8.2.2 Reconstructing the history of a gene family

A straightforward use of reconciled trees is to visualise the history of a gene embedded within an organismal phylogeny. From this we can determine how many gene duplications took place, and where in evolutionary time they occurred. This may yield insights into organismal evolution, particularly if episodes of gene duplication are concentrated in particular parts of the organismal phylogeny.

In relatively simple cases the gene tree itself reveals much about the evolution of that gene. However, comparison with the organismal tree may provide evidence of hitherto unsuspected gene duplications. An example of this is the mammalian interleukin-1 (IL) genes.

The gene tree for these genes (Fig. 8.5) shows clear evidence for gene duplications as indicated by the presence of multiple copies of the gene in the

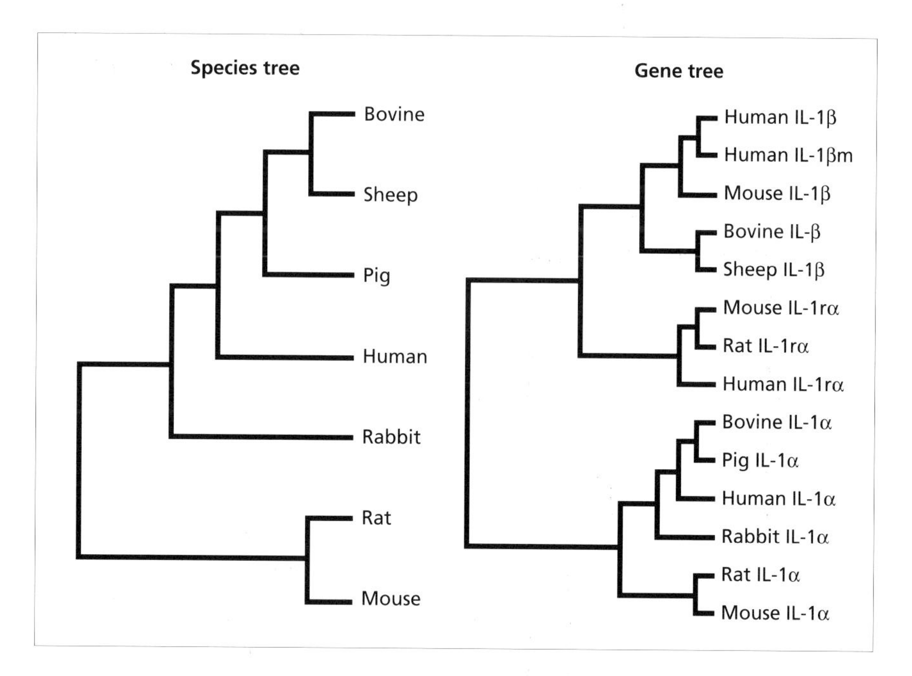

Fig. 8.5 Tree for mammals and for mammalian interleukin-1 genes.

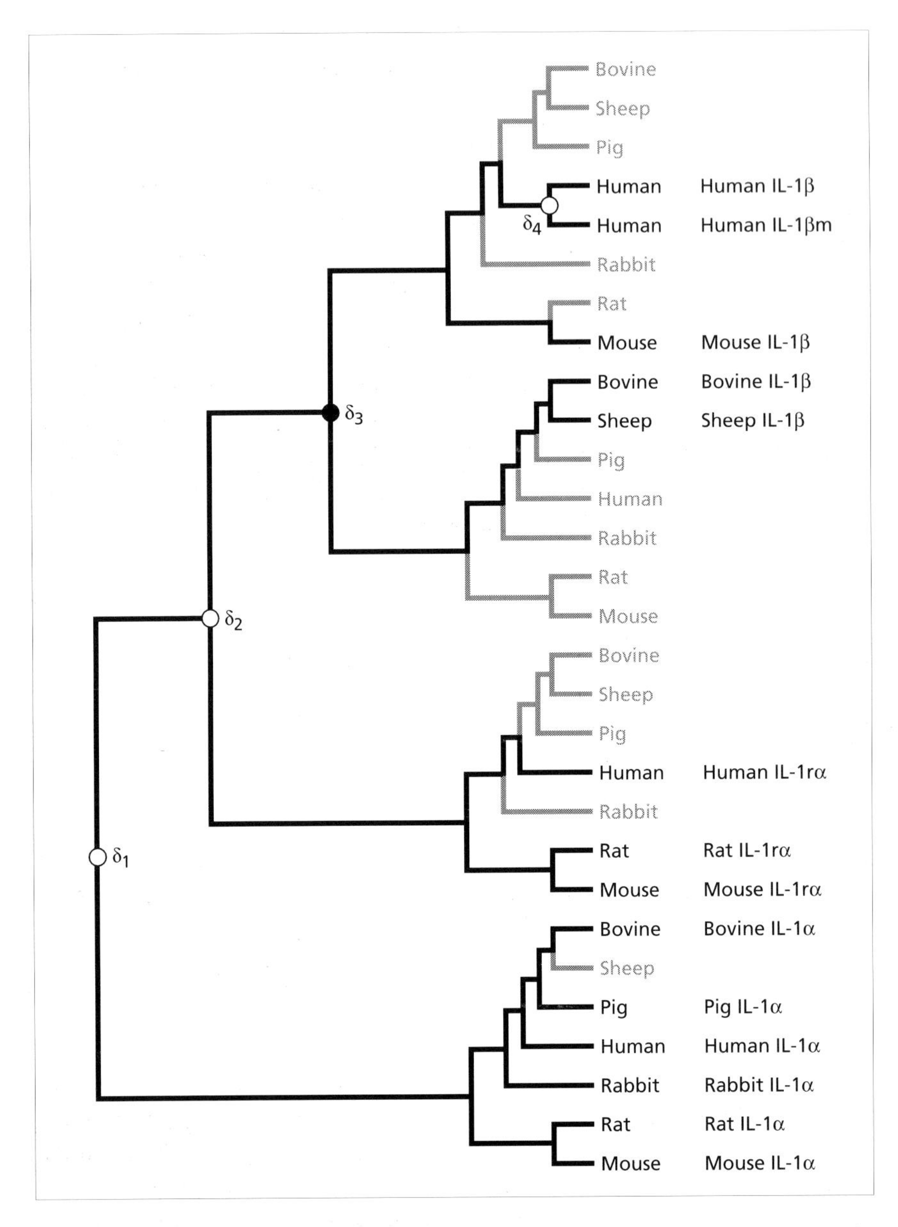

Fig. 8.6 Reconciled tree for the mammalian interleukin-1 gene tree shown in Fig. 8.5. Of the four duplications required, three are supported by the presence of multiple copies of IL in the same mammal species, and one (δ_3) is required to explain the incongruence between IL-1β and mammalian phylogeny.

same species. Given that there are four copies in humans, there must have been at least three duplications in the history of the interleukins. Reconciling the interleukin tree with a tree for the mammals (Fig. 8.6) uncovers evidence for a fourth duplication—mouse IL-1β is more closely related to human IL-β than are bovine and sheep IL-β, which contradicts the species tree.

Box 8.3 Rooting trees using gene duplications

While gene duplications can complicate the task of reconstructing organismal phylogeny, they can also be a useful source of information about the location of the root of the tree. Most methods of tree building produce unrooted trees which are subsequently rooted using an outgroup (a sequence or taxon that does not belong to the group of interest—for example, a chimpanzee sequence could be used to root a tree of human DNA sequences). However, in some cases a suitable outgroup may be lacking; for example, we cannot use an outgroup to root the tree of life. Gene duplications provide an alternative means of locating the root of a species tree. In the example below, we have a gene tree for six sequences from three species (1–3). The genes form two groups (α and β) within which the species relationships are identical.

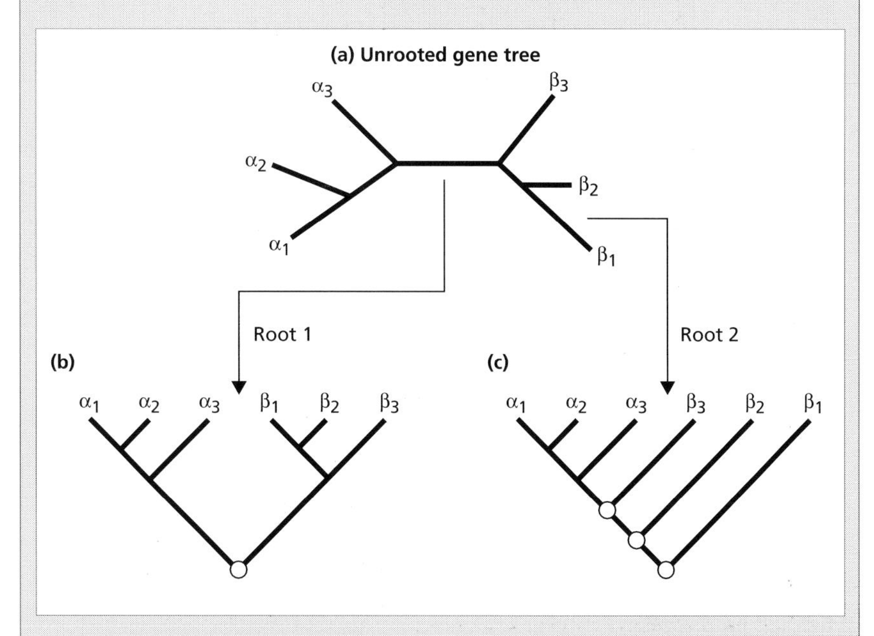

Although we could root the gene tree on any of its nine edges, placing the root between the α and β genes (root 1) requires only a single gene duplication, whereas rooting on any other edge requires additional duplications (two in the case of root 2). Hence the tree is most parsimoniously rooted between the α and β genes.

8.2.3 Inferring organismal phylogeny from complex gene trees

Given that the relationship between gene and organismal phylogeny can be complex, how can we infer species trees when faced with gene duplications? One approach is to use reconciled trees. Like those optimality methods (such

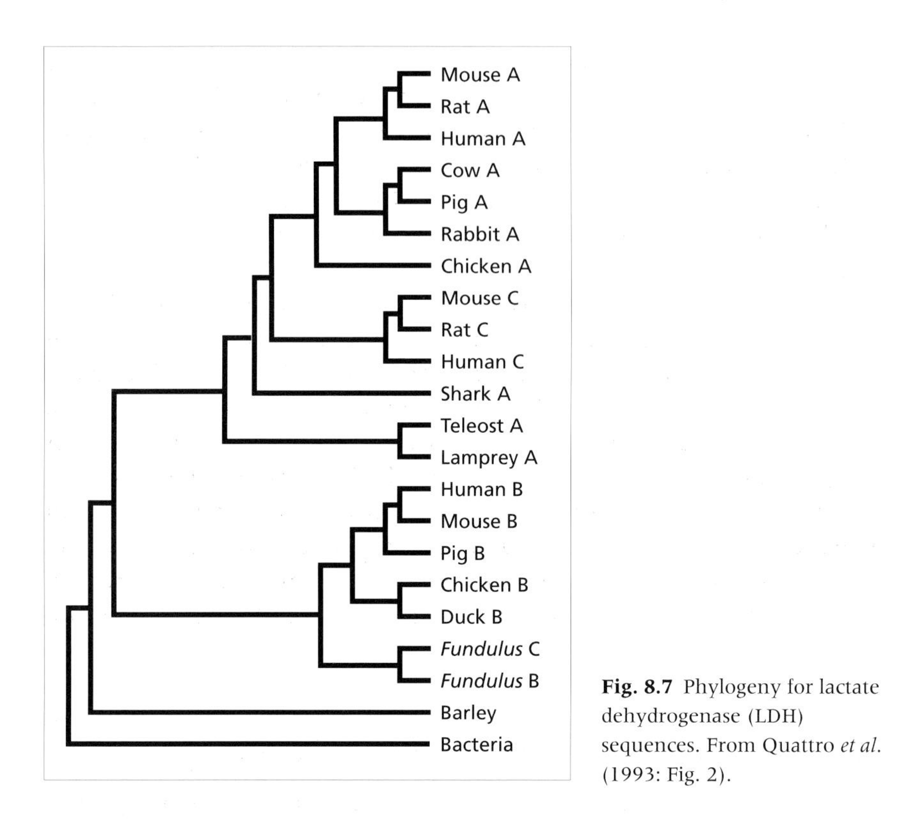

Fig. 8.7 Phylogeny for lactate dehydrogenase (LDH) sequences. From Quattro *et al.* (1993: Fig. 2).

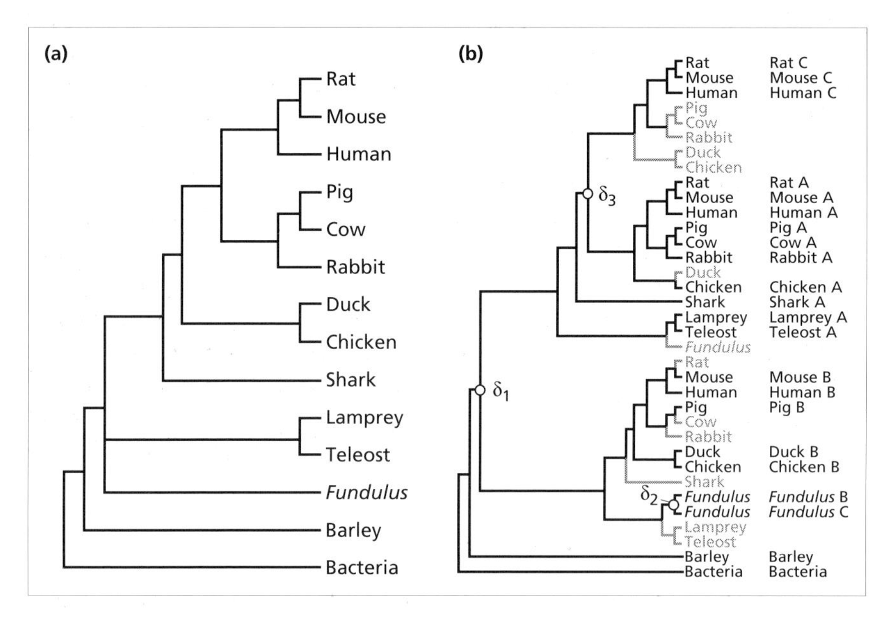

Fig. 8.8 (a) Consensus tree for the five optimal species trees for the LDH gene tree shown in Fig. 8.7; and (b) reconciled tree for the LDH gene tree in Fig. 8.7 and one of the five species trees.

as parsimony and maximum likelihood) that we discussed in Chapter 6, we are trying to solve two problems: (i) for any given tree, what is its 'score' under the chosen optimality criterion, and (ii) which tree has the best score? In this case, the criterion is the number of gene duplications and losses required to embed the gene tree into a species tree. Note that in this instance, we are searching for the best **species tree**, not the best gene tree, as we were in Chapter 6.

As an example of this approach, Fig. 8.7 shows a phylogeny for 22 lactate dehydrogenase (LDH) sequences from organisms as diverse as humans and bacteria (but with a decided bias towards vertebrates). The tree clearly contains information about species relationships, but it is not a species tree as several species occur more than once. There are five species trees that can account for this gene tree with the least cost—three duplications and nine losses. A consensus of these is shown in Fig. 8.8(a). This is in reasonable agreement with accepted relationships among these taxa, with the notable exception of the teleosts, shark and lamprey. This suggests that unless the LDH tree is in error, more duplications have occurred in this gene family than the three shown in Fig. 8.8(b).

Box 8.4 Inferring species trees from multiple gene trees

Reconciled trees can be used to find the best species tree for more than one gene. Given n gene trees $G_1, G_2, \ldots, G_n$, and species tree T, the cost of reconciling all n gene trees is simply the sum of the costs of reconciling the individual gene trees, $c(G_1, T) + c(G_2, T) + \ldots + c(G_n, T)$. Note that this is analogous to parsimony analysis of individual characters (e.g. nucleotide sites) where the total length of the tree is the sum of the minimum number of changes required for each individual character to evolve on the tree (Chapter 6). Guigó *et al.* (1996) used this approach to compute a tree for major eukaryote groups based on trees for 53 different nuclear genes. Disturbingly for molecular systematists, they found only 17 of the 53 genes were perfectly consistent with the preferred species tree, implying that two-thirds of the genes were at least partially misleading about organismal relationships. Reanalysis of their data (Page and Charleston, 1997b) suggests that this figure is nearer one half, but it is still a sobering result. There is considerable scope for evaluating how reliable different genes are as indicators of organismal relationships.

8.3 Host–parasite cospeciation

Host–parasite systems are intrinsically interesting to evolutionary biologists because they potentially signal a long and intimate association between two or more groups of organisms that are often distantly related and quite dissimilar biologically. This long history of association often leads to reciprocal adaptations in the hosts and parasites (classical **coevolution** or **coadaptation**) as well as contemporaneous cladogenetic events in the two lineages (**cospeciation**). The phenomenon of cospeciation is of particular interest to comparative biologists because it allows us to identify events of the same age in the host and parasite phylogenies, and thus provide an internal time calibration for comparative studies of rates of evolution in the two groups. Evidence of cospeciation also can be used to test hypotheses of coadaptation in the hosts and parasites.

A fundamental problem in the study of host–parasite coevolution is the reconstruction of the history of the association between host and parasite. We want to know to what extent hosts and parasites have cospeciated (**association by descent**) and to what extent parasites have switched hosts (**association by colonisation**). These processes, combined with independent speciation and extinction of either member of the association, will leave their trace on the phylogenies of the hosts and parasites.

8.3.1 Molecular phylogenies and cospeciation

Molecular phylogenies are powerful tools for the study of cospeciation, not because molecular data are inherently better than morphological data, but because molecular data offer the prospect of being able to compare host and parasite divergence using comparable units such as nucleotide substitutions per site, especially if homologous genes are studied in both host and parasite.

Molecular divergence may also permit us to decide between the two primary explanations of incongruent host–parasite phylogenies: host switching and multiple lineages. Figure 8.9(a) depicts a pair of incongruent host and parasite cladograms (see Chapter 2 for a definition of a cladogram). The incongruence may be due to host switching or the presence of multiple lineages. By themselves the cladograms do not allow us to decide between these explanations, although the relative likelihood of either 'host switching' or 'multiple lineages and extinction' scenarios may be assessed. A host-switching explanation for Fig. 8.9(a) suggests that the ancestor of parasite P2 colonised host H2 from host H1 and displaced the parasite species (P4) already present on H2 (Fig. 8.9b). This scenario requires two evolutionary events, one host-switching event and the extinction of parasite P4. A multiple lineage explanation suggests that, initially, two parasite lineages were present on the ancestral hosts, and as those parasites cospeciated with their hosts, three parasite lineages went extinct (Fig. 8.9c). Which scenario we decide is most likely may be determined by the biology of the host–parasite system, i.e. how common are host switches relative to extinctions?

If we have information on relative time of divergence between the host and parasite species then we may be able to use that information to choose between these explanations. In Fig. 8.9(d) the parasite lineage P2 is younger than its host H2, consistent with host switching, whereas in Fig. 8.9(e) the greater antiquity of P2 supports a multiple lineages explanation (the persistence

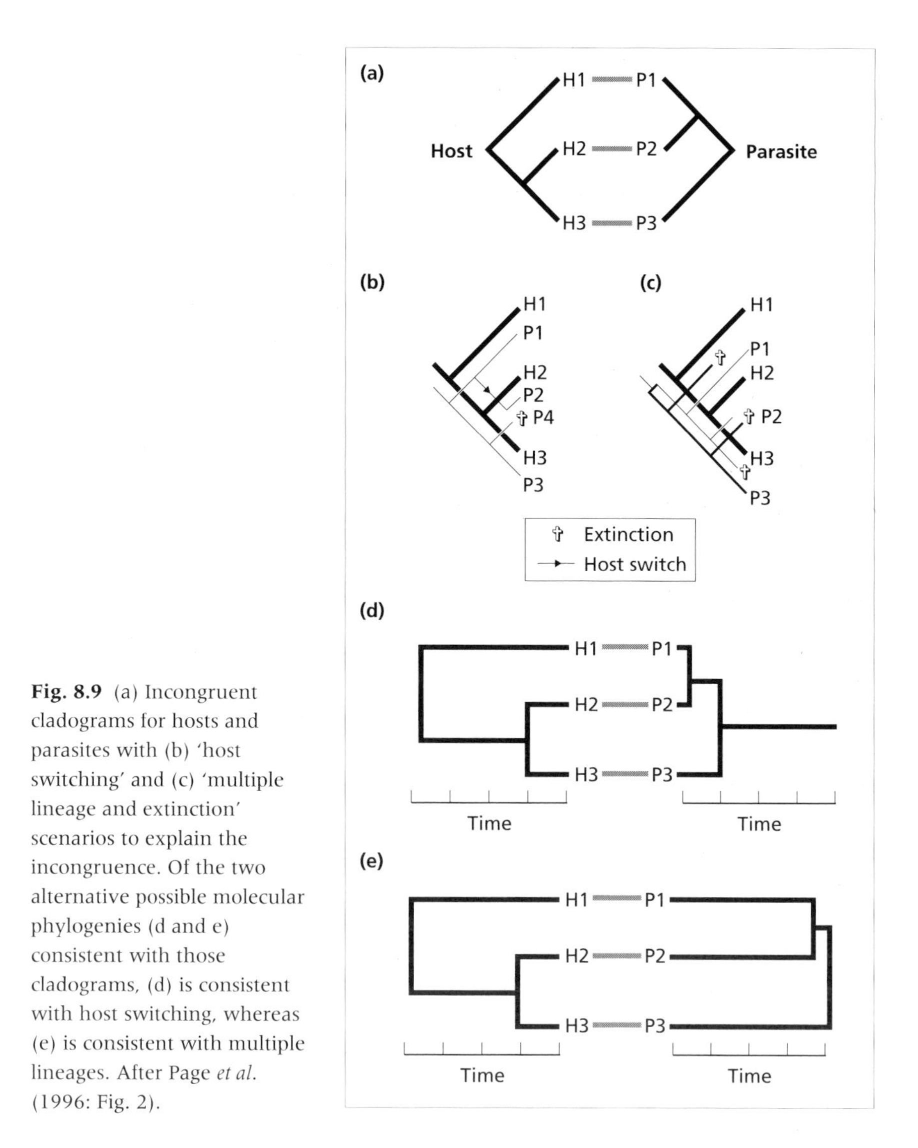

Fig. 8.9 (a) Incongruent cladograms for hosts and parasites with (b) 'host switching' and (c) 'multiple lineage and extinction' scenarios to explain the incongruence. Of the two alternative possible molecular phylogenies (d and e) consistent with those cladograms, (d) is consistent with host switching, whereas (e) is consistent with multiple lineages. After Page *et al.* (1996: Fig. 2).

of relict parasites on their original hosts). Data on timing may come from molecular clocks (see Chapter 7); hence estimates of times of divergence from molecular data may help discriminate between competing explanations of incongruence between host and parasite phylogenies.

8.3.2 Calibrating rates of evolution in cospeciating clades

Given estimates of the amount of divergence between two pairs of taxa, say two insects and two mammals, and estimates of the age of these taxa we can readily compare rates of evolution in the two taxa. Estimating sequence divergence is relatively straightforward (Chapter 5). However, estimating taxon ages can be problematic as it relies on having a sufficiently complete fossil record. Incompleteness, as revealed by a poor correlation between cladistic rank and taxon age, is a widespread problem in the fossil record, and seriously hampers calibration of rates of evolution.

In order to reduce reliance on the fossil record, some workers have calibrated molecular evolution in one lineage using the fossil record of a lineage with which it is closely associated. Examples include endosymbiotic bacteria of aphids and cockroaches, and vertebrate alpha-herpes viruses. If two clades are cospeciating then equivalent events in the two lineages can be identified. Nodes in the lineage lacking a fossil record can be dated using the ages of the corresponding nodes in the lineage having a fossil record (Fig. 8.10).

If we can go a step further and compare sequence divergence in both host and associates then we are freed from any reliance on the fossil record, providing we are content to measure relative rather then absolute rates of evolution. Alternatively, we can maximise the usefulness of the fossil record by effectively combining evidence from the fossil records of both host and parasite.

Fitting a line to a plot of parasite divergence against host divergence (Fig. 8.11) allows us to describe two aspects of host–parasite divergence. The slope of the line is the relative rate of host and parasite evolution; the intercept measures the divergence of parasites when their hosts speciate. An intercept of zero implies synchronous cospeciation: host and parasite speciate together. A negative intercept implies delayed cospeciation: the parasites tend to speciate

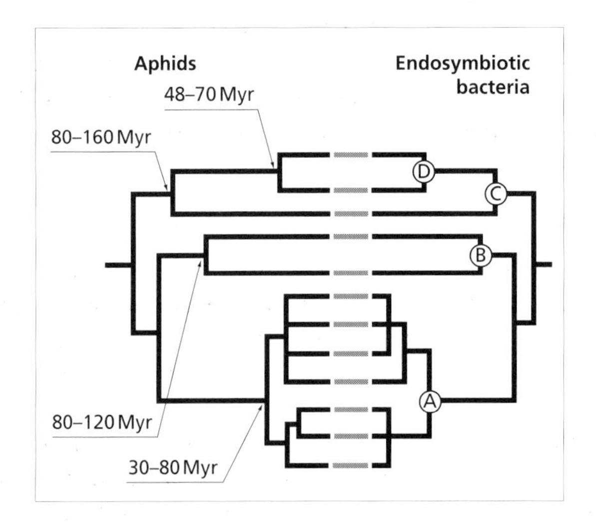

Fig. 8.10 Phylogeny of 12 aphid species and their endosymbiotic bacteria. The two phylogenies are nearly identical, indicating that the bacteria have cospeciated with their insect hosts. Four nodes in the bacterial phylogeny (labelled A–D) can be approximately dated using ages for the equivalent nodes in the aphid tree derived from the fossil record and biogeography. These dates can then be used to calibrate the rate of evolution of the endosymbionts, which have no fossil record themselves. After Moran *et al.* (1993: Fig. 1).

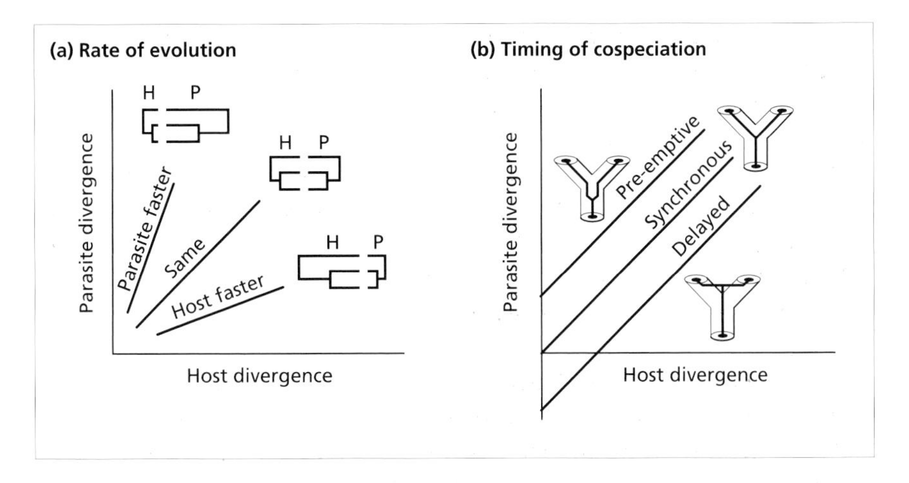

Fig. 8.11 Bivariate plots of the relationship between parasite divergence and host divergence. The slope of the relationship (a) indicates the relative rates of evolution in the two clades, the intercept (b) indicates the relative timing of speciation. From Hafner and Page (1995).

after their hosts; a positive intercept indicates that the parasites diverge prior to their hosts.

Returning to the analogy with gene trees, if the bivariate plots shown in Fig. 8.11 were instead plotting sequence divergence for a given gene against taxon divergence time, then a positive intercept would reflect the average sequence divergence within a single species, i.e. heterozygosity (Lynch and Jarrell, 1993). By analogy, a positive intercept for parasite sequence divergence reflects differentiation among parasites of a single host lineage. We would expect the intercept to be positive in situations where the parasites of different host populations (or different individual hosts) are genetically divergent.

8.3.3 Maximum likelihood tests of cospeciation

Maximum likelihood methods similar to those encountered in Chapter 6 can be employed to make detailed comparisons of host and parasite phylogenies. Recently, Huelsenbeck and colleagues (1997) outlined a series of nested hypotheses concerning cospeciation that can be tested using likelihood ratio tests (Fig. 8.12).

The first, most general question is whether host and parasite phylogenies have the same branching order, irrespective of details of branch length or speciation time (Fig. 8.12a). In other words, the question is whether the cladograms for host and parasite are identical. Huelsenbeck *et al.* point out that this is equivalent to the problem of whether different data partitions of a data set are congruent (see section 8.1.2); in this instance the two partitions are the host and parasite nucleotide sequences, respectively, and the null hypothesis is that they are estimating the same tree.

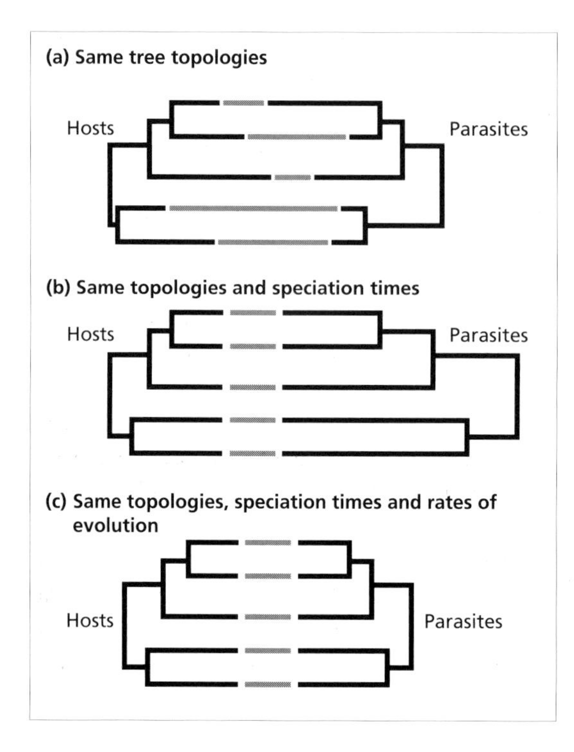

Fig. 8.12 Three null hypotheses concerning host–parasite cospeciation. The first (a) is that the topologies of the host and parasite trees are identical, although the rates of evolution and speciation times may differ. The second (b) is that, given that the topologies are the same, the relative speciation times are identical, even if the rates are different in the two clades. The third (c) is that topologies, speciation times, and rates are identical in host and parasite. After Huelsenbeck *et al.* (1997: Fig. 3).

Even if host and parasite phylogenies have the same cladistic topology, host and parasite could have speciated at different times (Fig. 8.12b). This is because there are typically several different orderings in which speciation events can occur but still give rise to the same tree topology (Fig. 8.13). In the terminology used in Chapter 2, a cladogram may correspond to one or more ultrametric trees.

If we have a molecular clock, the null hypothesis of identical speciation times ('temporal cospeciation') can be tested using ultrametric trees for host

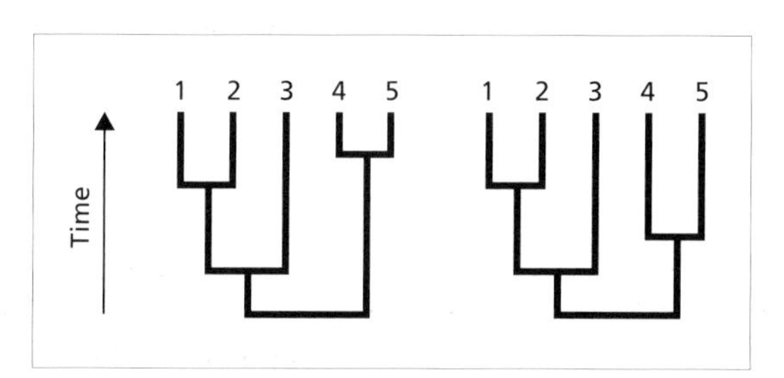

Fig. 8.13 Two phylogenies that have the same topology but different speciation times. In the tree on the left, taxa 4 and 5 speciated after taxa 1 and 2, whereas in the tree on the right, the split between 4 and 5 predates the split between 1 and 2. After Huelsenbeck *et al.* (1997: Fig. 2).

and parasite that have both the same order of speciation and the same relative ratio of time between speciation events (Fig. 8.12b)—although the host and parasite clades may be evolving at different rates.

The final null hypothesis is that host and parasite are cospeciating at the same time *and* are evolving at an identical rate (Fig. 8.12c). This is equivalent to the previous hypothesis but with the additional constraint of equality of rates in host and parasite.

Huelsenbeck and colleagues applied these tests to the five *Orthogeomys* gophers and their lice (Box 8.5). For these five pairs of hosts and parasites Huelsenbeck *et al.* could not reject the hypothesis that the COI sequences were clock like, nor could they reject the null hypotheses that the tree topologies were identical (Fig. 8.12a) and had the same speciation times (Fig. 8.12b). However, the null hypothesis of equal rates of nucleotide substitution (Fig. 8.12c) was rejected; the lice are evolving 3.02 ± 0.53 times as rapidly as their hosts.

At present, these tests all suffer the limitation that they require each parasite to have a single host, and each host to have a single parasite. Furthermore, the tests of temporal cospeciation and equal rates of evolution require identical host and parasite phylogenies. Future developments may broaden the applicability of these methods.

8.4 Age and rates of diversification

One of the great challenges in reconstructing the history of life on earth is to establish the age of origin of different lineages. The only direct evidence for the age of a lineage is the fossil record, which is frequently rather poor. Furthermore, fossils provide minimum estimates of taxon age; the oldest fossil for a taxon tells us that lineage must be at least that old—there is still the possibility that older fossils remain to be discovered. These limitations can to some extent be addressed by interpreting fossils in light of phylogeny. A good illustration of this is the 'sister group rule'. This rule follows from the fact that the two immediate descendants of an ancestral species are of the same age. Hence, if A and B are two sister species (i.e. each other's closest relative) then A and B are at least as old as the oldest fossil of either A or B (Fig. 8.14).

8.4.1 Ages of taxa

Perhaps the most powerful approach to combining phylogenies and the fossil record is to use the fossil record to calibrate rates of molecular evolution, then use those calibrated rates to extrapolate the ages of taxa for which fossils are not available. A recent example of this is provided by Cooper and Penny's (1997) investigation of the age of modern birds. There is considerable controversy about the extent and nature of the extinctions at the Cretaceous/ Tertiary (K/T) boundary some 65 million years ago. A popular hypothesis is

Box 8.5 Pocket gophers and their lice

Pocket gophers are small rodents found in the western and central United States, Mexico, and Central America. They are host to two genera of ectoparasitic lice, *Geomydoecus* and *Thomomydoecus*. These flightless insects are highly host specific and spend their entire life history on their mammalian hosts. Because of the similarity of the gopher and louse phylogenies these taxa have played a prominent role in recent studies of host–parasite cospeciation. Hafner *et al.* (1994) obtained DNA sequences for a homologous 379-bp region of the mitochondrial cytochrome oxidase subunit I (COI) gene for both gophers and their lice. These data have been used by a number of authors to illustrate methods for comparing rates of evolution in hosts and parasites. Hafner *et al.* originally suggested that the lice show a 10-fold higher rate of substitution than gophers; subsequent reanalyses suggest the actual disparity is nearer 2–3 (Huelsenbeck *et al.*, 1997; Page, 1996b).

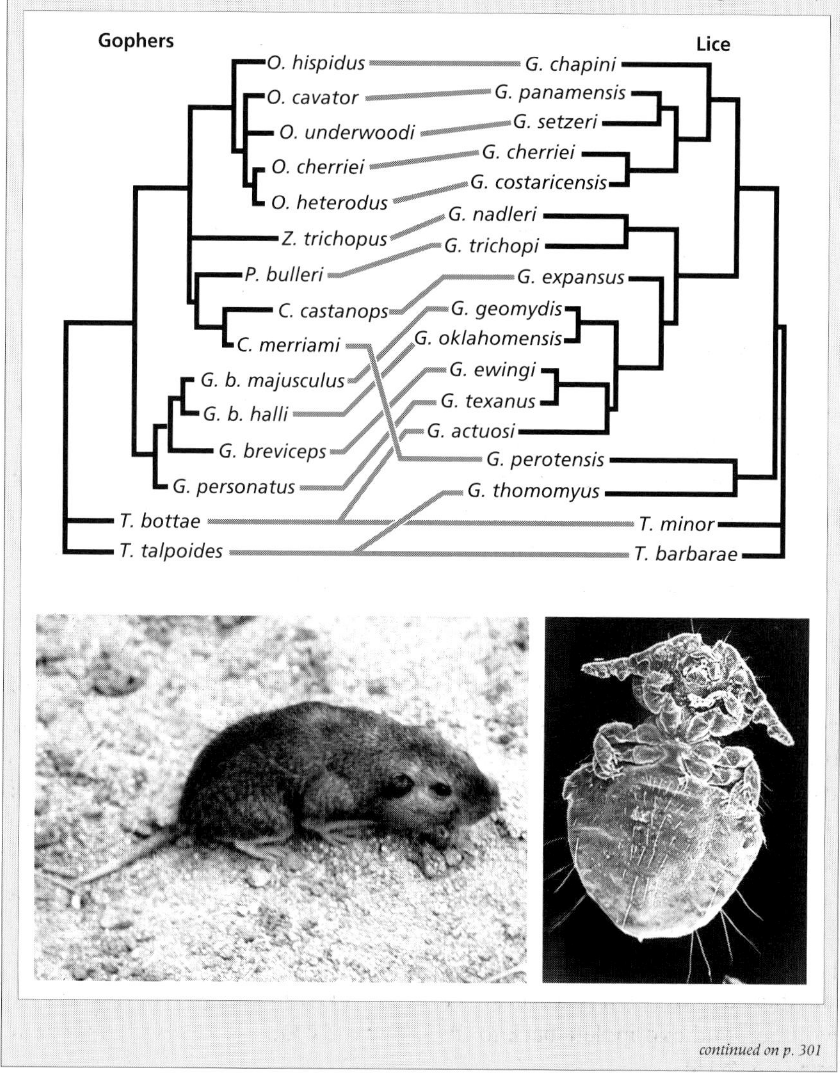

continued on p. 301

Box 8.5 *continued*

The figure above shows phylogenies for pocket gophers and their chewing lice parasites (the organisms are shown in inset) with branch lengths proportional to numbers of substitutions at the third codon position in the COI gene. Pocket gopher genera are *Orthogeomys, Pappogeomys, Cratogeomys, Geomys, Thomomys* and *Zygogeomys*; louse genera are *Geomydoecus* and *Thomomydoecus*. Grey lines connect gophers with their species specific parasite. After Hafner *et al.* (1994).

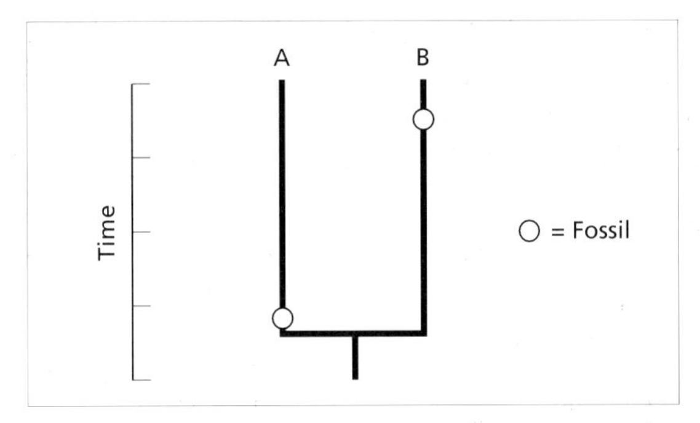

Fig. 8.14 The sister group rule. Lineages A and B both have fossils, although of different ages. Because they are sister taxa, lineage B is at least as old as A, even though B's fossil record does not extend back nearly as far as that of A. This rule means that even groups with no fossil record can be assigned a minimum age using the fossil record of their nearest relatives.

that modern bird (and mammal) orders only radiated after the K/T boundary; prior to that time, dinosaurs were the dominant terrestrial vertebrates. The bird fossil record is patchy as most birds have lightweight skeletons and (at least in modern birds) no teeth, which does not favour their preservation. Although some fossils of putatively modern bird orders have been found in Cretaceous fossil beds, the identity of these fossils has been disputed. Hence, a conservative interpretation of the fossil record is consistent with a post-Cretaceous radiation of modern birds.

Cooper and Penny set out to test this hypothesis by calibrating the rates of evolution in two genes, mitochondrial 12S rRNA and the nuclear *c-mos* gene, in 16 orders of birds using the avian fossil record. They then extrapolated back to estimate the divergence times for modern bird orders. Because the details of bird phylogeny are controversial, Cooper and Penny chose to use quartets of birds about which there is little argument concerning their relationships. For example, given the ostrich, rhea, loon, and shearwater, there is little doubt that ostrich and rhea go together, as do the loon and the shearwater (each quartet was chosen such that the root lay between the two pairs of taxa). We can then use the oldest fossil for each pair of birds to calibrate the rate of evolution and extrapolate back to the age of the common ancestor of the four taxa (Fig. 8.15).

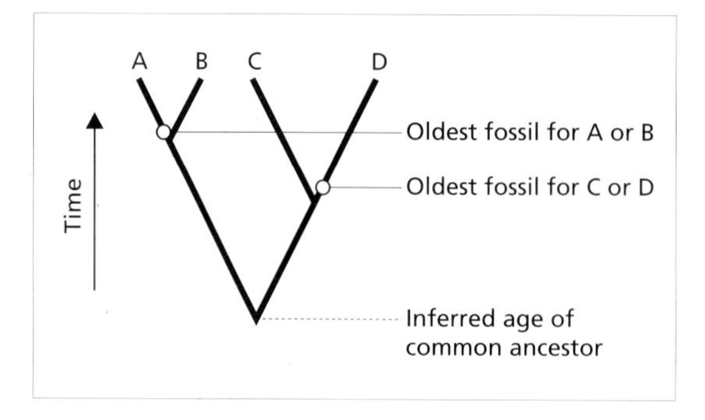

Fig. 8.15 The quartet method used by Cooper and Penny (1997) to infer the age of the common ancestor of four taxa A–D. The rate of molecular evolution in the taxon pairs (A, B) and (C, D) is calibrated using the oldest fossil for each pair of taxa. Given this calibration, and the divergence between the two pairs of taxa, the age of their common ancestor can be inferred.

The other advantage of using quartets is that they reduce the dependence on there being a molecular clock ticking uniformly across all birds because each non-overlapping quartet of taxa is calibrated separately. Cooper and Penny found that different quartets yielded similar rate estimates. It should be noted, however, that these estimates are not completely independent due to overlap among the subtrees for the quartets that span the root of bird phylogeny.

Applying this method to the 12S rRNA and *c-mos* data strongly suggests that modern birds as a whole originated in the Early Cretaceous and that at least 21 lineages survived the K/T extinction, leading Cooper and Penny to talk of 'mass survival' of birds rather than mass extinction (Fig. 8.16). However, it should be noted that survival of lineages does not mean that modern birds were unaffected by events at the K/T boundary; the lineages that survived may be only a subset of those extant in the Cretaceous, with each surviving lineage having undergone a reduction in diversity at the time of the K/T boundary.

The dates for the radiation of birds (Cooper and Penny, 1997), mammals (Hedges *et al.*, 1996) and animals in general (Wray *et al.*, 1996) obtained using molecular data have caused some controversy, given that they suggest many lineages are older than previously thought. However, this discrepancy between fossil and molecular evidence may be due to a too literal reading of the fossil record. Palaeontologists using phylogenetic methods have argued for some time that lineages may be older than a naïve reading of the fossil record would suggest (Novacek and Norell, 1982).

8.4.2 Rates of diversification

The fossil record is imperfect, but seemingly is the only source of information

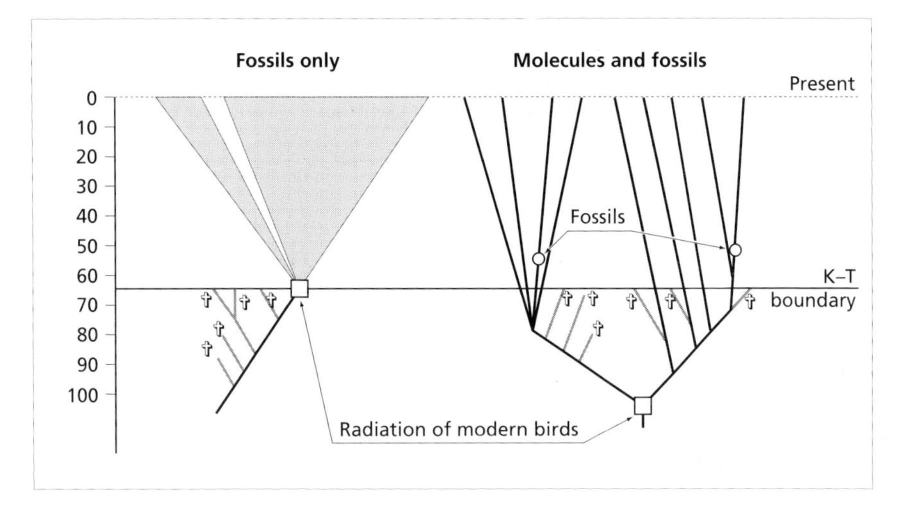

Fig. 8.16 Two alternative views of the evolution of birds. One interpretation of the fossil record suggests that modern birds radiated after the K/T boundary extinction event; Cretaceous birds were present but did not belong to the modern orders and went extinct at the end of the Cretaceous. However, calibrating rates of molecular evolution with fossils of modern birds suggests that many modern bird orders were already present in the Cretaceous, and that the post K/T 'radiation' is an artefact of interpreting the fossil record too literally.

about extinction prior to human historical records—the *prima facie* evidence for extinction is that there are organisms found as fossils that are not found alive today. However, recent work suggests that using molecular phylogenies for present-day taxa it is possible to make inferences about extinction rates, even in the absence of fossils. This approach relies on molecular phylogenies that are both consistent with a molecular clock and are largely complete in their sampling of the entities (species, populations, strains, etc.) being studied.

Consider the diagram in Fig. 8.17(a) of a phylogeny that has been sampled at six different intervals; for example, at interval 3 there were six lineages extant, of which three went extinct and three survived and speciated to yield the seven lineages present today. If we plot the actual number of lineages present at each time interval (a 'lineages through time' plot), and the number of lineages that are ancestral to extant species, we obtain the two curves shown in Fig. 8.17(b). If we considered only the phylogeny of the extant species (shown as thick lines in Fig. 8.17a) then we would underestimate the number of lineages actually present in several of the time intervals; this is shown by the gap between the two curves in Fig. 8.17(b).

If we assume a simple model of speciation in which the rates of speciation (b or 'birth rate') and extinction (d or 'death rate') is constant over time then the lineages through time plot for a group of extant species has a characteristic shape: a steep rise at the beginning, a middle phase with a flat slope, then an

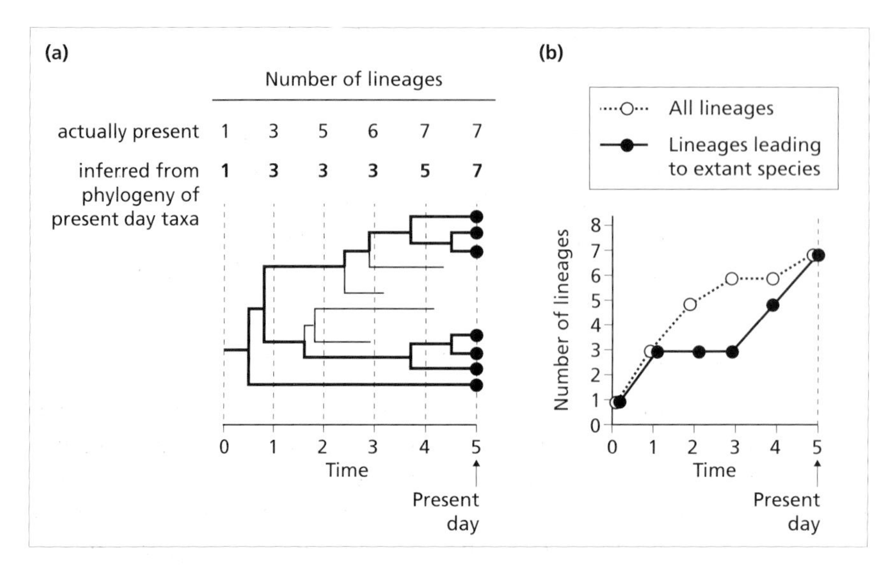

Fig. 8.17 Lineages through time plots. (a) A phylogeny for a group of seven extant species and their ancestors. Four lineages (thin lines) have gone extinct. (b) A plot of the number of lineages actually present at each time interval (○), and the number of lineages that have descendants alive today (●). The gap between the two lines is due to lineage extinction. After Nee *et al.* (1994).

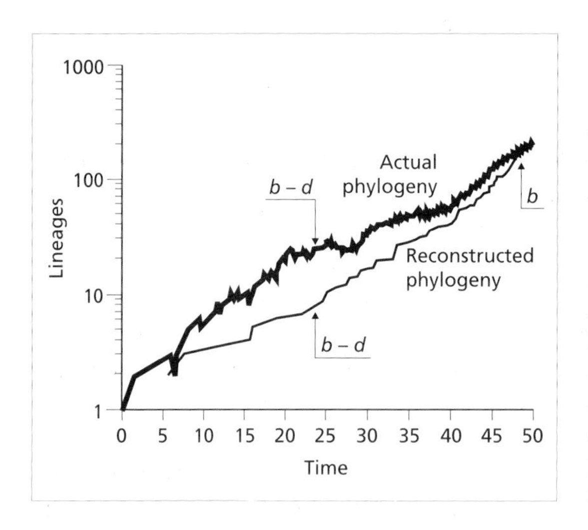

Fig. 8.18 Lineages through time plots for a simulated phylogeny and its reconstruction. The slopes of both curves are the difference between rates of speciation (b) and extinction (d) over most of the clade's history, asymptotically approaching the speciation rate b at the present day. The gap between the actual and reconstructed curves varies with the rate of extinction; the greater the extinction rate the more pronounced the gap. Modified from Nee *et al.* (1994).

upward rise near the present (Fig. 8.18). The slope of the line in the middle is the birth rate minus the death rate, $b - d$, and represents the effective rate at which lineages are added. The upswing as we approach the present is due to the decreasing likelihood of observing a lineage extinction once the time interval between our observation point and the present day becomes short relative to the probability of a lineage going extinct—we are more likely to observe at least one lineage going extinct over a period of 10 million years than over a period of 10 000 years. The slope of the curve as it approaches the present day can be used to estimate the birth rate for the clade, *b*.

Analyses of lineages through time have the potential to provide new insights into rates of organismal differentiation. Rates can be compared among different clades, and within the same clade but over different time periods. Estimates of lineage extinction derived from molecular phylogenies can also be used to assess the completeness of a clade's representation in the fossil record.

8.5 Phylogenies in molecular epidemiology

Since the start of the AIDS epidemic in the early 1980s, there has been a heightened interest in the emergence and evolution of infectious diseases. It is clearly extremely important to know what factors allowed new infections like HIV to appear, or older ones to reappear, and then to track their spread through populations. These tasks form part of the science of **molecular epidemiology**.

Although tracing the spread of infectious diseases was traditionally approached by serology, that is by studying the movement of pathogen strains which differed in immunological (= serological) response, today a great deal of research in this area is undertaken using comparative gene sequence data. This in turn means that phylogenetic trees have become an important analytical tool. In this section we shall illustrate a few of the ways in which phylogenies have been used to study the origin and spread of viral infections.

One of the most basic questions to ask with any infectious disease is where it comes from. As we saw in Chapter 7, there are two types of HIV: a highly virulent global type (HIV-1) and a somewhat less virulent strain which is found most often in West Africa (HIV-2). HIV-like viruses are also commonly found in many African monkeys, and are referred to as simian immunodeficiency viruses (SIV). Because they are infected at such high frequencies, African monkeys are almost certainly the animal reservoir that ultimately gave us the AIDS viruses. However, if we construct phylogenetic trees of gene sequences from these different viral groups, we find that the relationships between them are rather complex. A tree depicting the phylogenetic relationships among the known primate immunodeficiency viruses, summarised from a number of sources, is shown in Fig. 8.19.

The first thing to take in from this tree is that HIV-1 and HIV-2 are most closely related to *different* primate species: HIV-1 to the viruses found in chimpanzees (SIV_{CPZ}) and HIV-2 to those found in sooty mangabey monkeys (SIV_{SM}). Because HIV-2 is predominantly a West African virus, and sooty mangabey are West African monkeys, it is possible that the sooty mangabey is the direct source of HIV-2. However, whether chimpanzees are the ultimate source of HIV-1 is less clear as chimpanzees are very rarely infected in the wild. Furthermore, because other monkey species are known to carry immunodeficiency viruses which have not yet been sequenced, it may be that

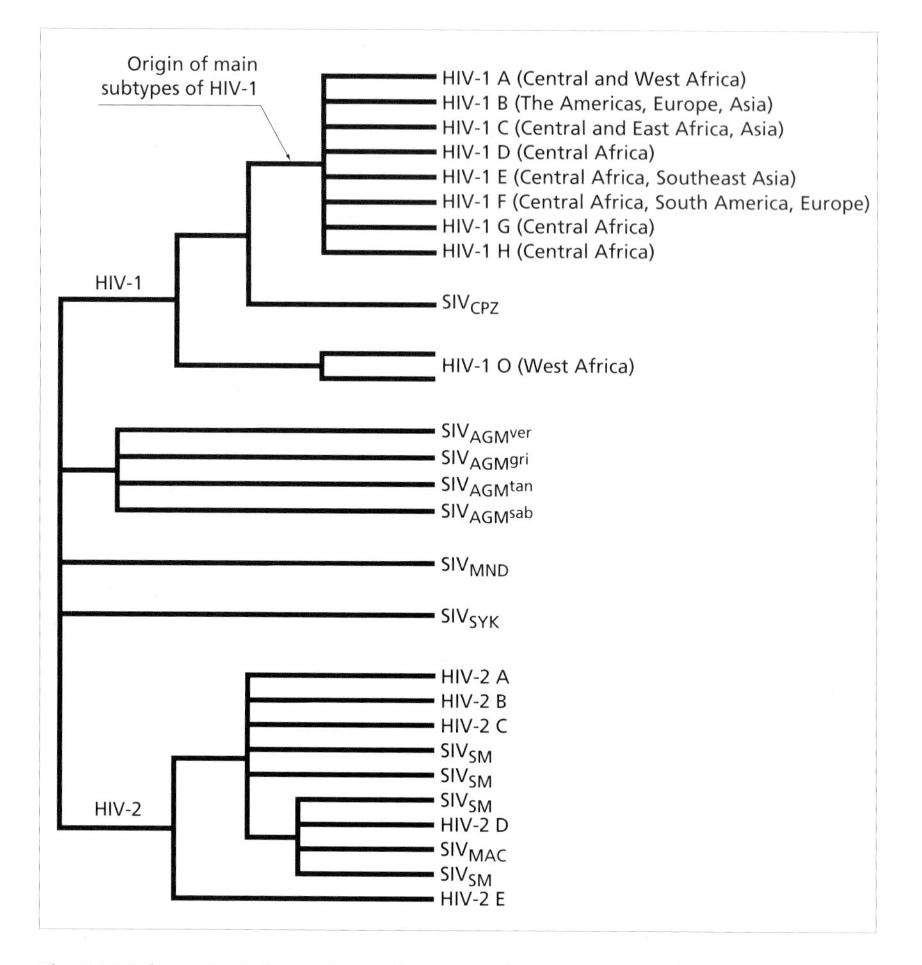

Fig. 8.19 Schematic phylogenetic tree illustrating the evolutionary relationships between different primate immunodeficiency viruses and the different subtypes of HIV-1 and HIV-2 (denoted by letters). Branch lengths are not drawn to scale. Abbreviations for simian immunodeficiency viruses are as follows: SIV_{AGM} = African green monkeys (ver = vervet, gri = grivet, tan = tantalus, sab = sabaeus); SIV_{MAC} = macaque; SIV_{MND} = mandrill; SIV_{SM} = sooty mangabey; SIV_{SYK} = Sykes' monkey. Adapted from Holmes (1998).

both HIV-1 and SIV_{CPZ} were derived from another, as yet undescribed, monkey source. Whatever their precise origin, the separation of HIV-1 and HIV-2 on the tree means that they must have entered human populations on *different* occasions.

Even more interesting patterns arise if we look at the phylogenetic relationships of the main viral lineages within HIV-1 and HIV-2. These lineages are called 'subtypes' and ten (denoted A to J) have so far been identified in HIV-1 worldwide, although more may be lurking below the surface (and only the most common eight are shown on the tree here). The most interesting features of these subtypes are that they may have spread through different populations at different times and sometimes by different routes. Subtype A, for example, is commonly found in sub-Saharan Africa where it is predominantly transmitted through heterosexual intercourse. It may also be one of the oldest of all subtypes, although it still probably emerged within the last 50 years or so. In contrast, subtype B is associated with the HIV epidemic in homosexual men and injecting drug users from North America, Europe and Japan and probably did not emerge until the late 1970s or early 1980s. Most striking of all are the subtype O viruses which infect some people in parts of West Africa. These are separated from the other HIV-1 sequences by those viruses present in the chimpanzee. This means that HIV-1 must also have entered humans from other primates at least twice. Multiple entry of viruses into humans has also taken place in HIV-2, where there is more of a mix of human and monkey lineages. Overall, this phylogenetic analysis tells us that rather than being very rare events, the emergence of AIDS viruses in humans has been depressingly common.

Another interesting virus on the tree is found in rhesus macaque monkeys (SIV_{MAC}). The puzzle here is that rhesus macaques are Asian monkeys, yet every other SIV collected to date comes from an African monkey species. How could an African virus have got to an Asian monkey? The answer seems to be that rhesus macaques were kept in primate colonies in the United States along with sooty mangabeys and that the virus somehow managed to transfer between the two species. Proof that rhesus macaques are not normally SIV infected is that wild populations do not harbour the virus.

Similar studies have been performed on many other emerging viruses. One example which generated a great deal of interest concerned the appearance of a mysterious and highly virulent respiratory (pulmonary) illness in the 'Four Corners' region of the Southwestern United States (the shared border region between Arizona, Colorado, New Mexico and Utah). This infection killed 26 people in 1993 and prompted a large-scale epidemiological investigation about its likely cause. In the first part of the study, blood samples from infected individuals were tested, using standard serological techniques, against a wide variety of pathogens to see which, if any, were responsible for the disease. The only positive results were with the Puumula strain of the hantaviruses. These RNA viruses have long been known in Europe (like Puumula), and more commonly in Asia, where they are responsible for renal and haemorrhagic diseases, killing from 4000 to 20 000 people each year. However, hantaviruses primarily infect rodents rather than humans, were not known to cause human disease in the Americas and were certainly not associated with pulmonary symptoms. But although a hantavirus was an unusual suspect for the Four Corners outbreak, a subsequent molecular epidemiological analysis revealed that this cluster of cases was indeed caused by a new strain of hantavirus, and one which had evidently adapted to infect the lungs instead of the kidneys (Fig. 8.20). However, rodents were still a very big factor in its spread: this new

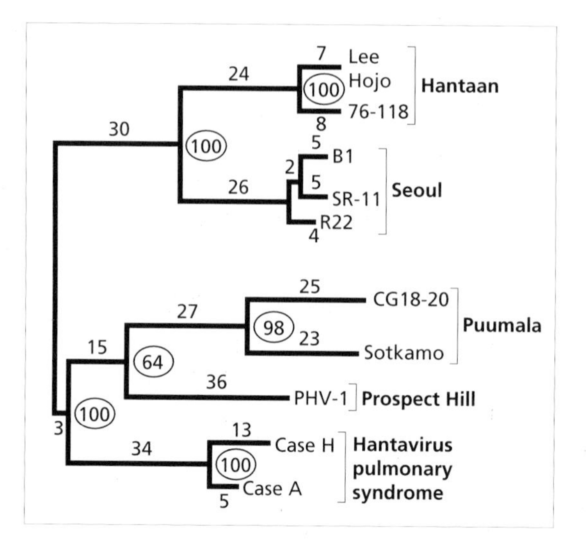

Fig. 8.20 Phylogenetic relationships between the Hantavirus pulmonary syndrome virus (now called 'Sin Nombre' virus) which led to the Four Corners outbreak and viral sequences from other hantaviruses. The tree was reconstructed on a 241 bp region of the M segment of the viral genome using parsimony. Values of bootstrap support >50% are shown in ovals next to the appropriate nodes and branch lengths (numbers of substitutions) are also indicated. The details of the other viruses included in the tree are as follows. The Hantaan strain is from Asia, causes a serious haemorrhagic fever with renal syndrome (HFRS), and circulates in field mice (*Apodemus agrarius*), while the Seoul strain causes a milder form of HFRS and is found worldwide, using the rat (*Rattus norvegicus*) as a reservoir. The Puumula strain is European in origin and again causes a mild HFRS, this time with the bank vole (*Clethrionomys glareolus*) as its host population. Finally, the Prospect Hill virus is North American in origin and only known to infect bank voles (*Microtus pennsylvanicus*). From Nichol *et al.* (1993).

strain of hantavirus—originally called 'Hantavirus pulmonary syndrome', and now 'Sin Nombre virus'—was found to be circulating at high frequencies in deer mice (*Peromyscus maniculatus*) from the Four Corners region. Furthermore, these rodents had recently experienced a massive expansion in population size, evidently bringing with it increased opportunities for disease transmission to humans.

As we have seen with both HIV and hantaviruses, the movement of pathogens from animals into humans is often associated with the outbreak of

---

Fig. 8.21 (*Opposite*) Phylogenetic tree (reconstructed using parsimony) of 89 nucleoprotein (NP) gene sequences from different isolates of influenza A virus. Isolates are designated by their strain name and their HN type. The species from which the strains were isolated are indicated by the animal symbols and roman numerals denote the five viral groups (also based on NP analysis) that have been proposed. In some cases, dates of appearance have also been estimated and placed next to the appropriate nodes. Branch lengths are drawn to scale. From Webster *et al.* (1995), with permission from Cambridge University Press.

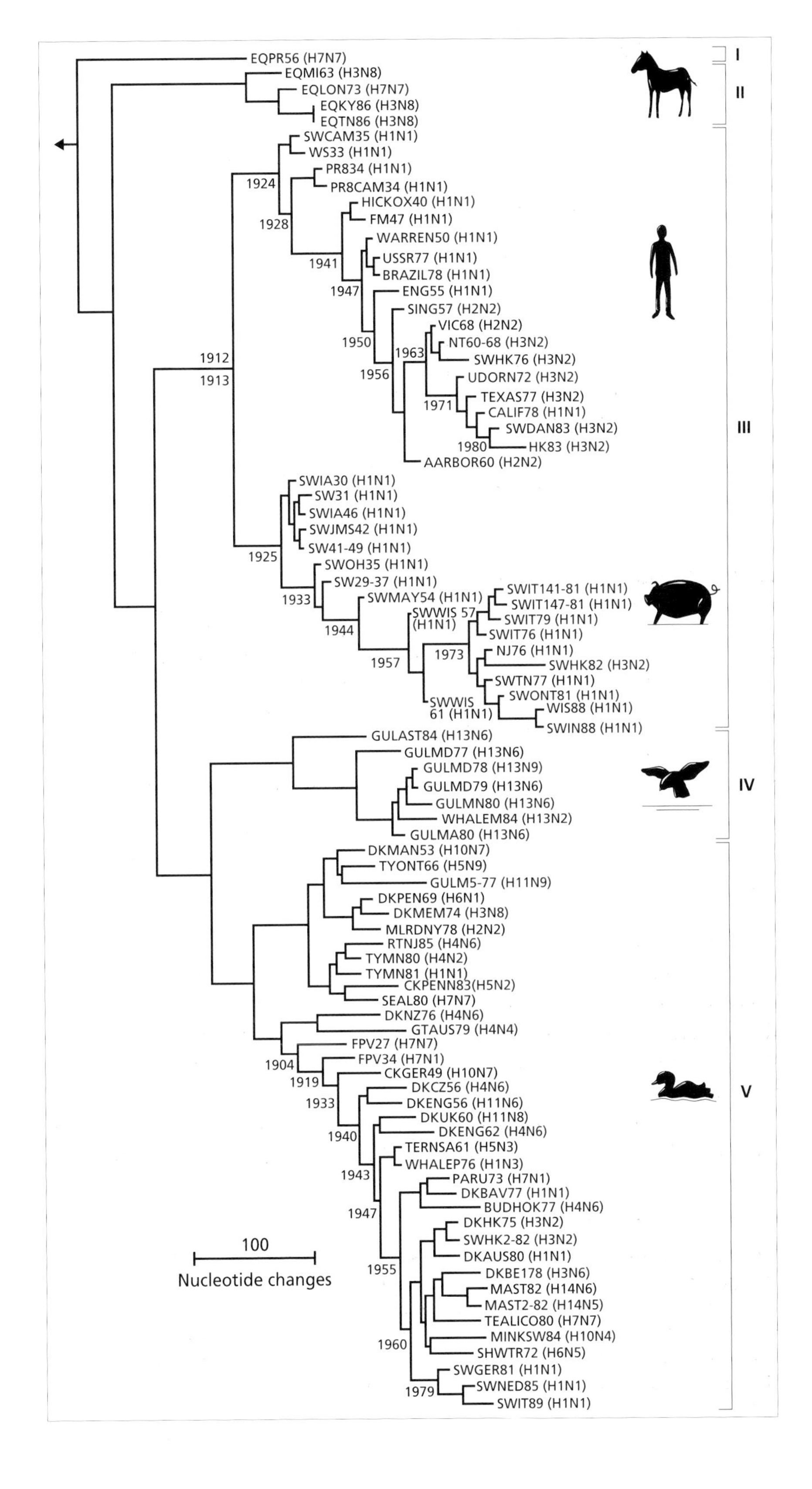
EQPR56 (H7N7)
EQMI63 (H3N8)
EQLON73 (H7N7)
EQKY86 (H3N8)
EQTN86 (H3N8)
I
II
SWCAM35 (H1N1)
WS33 (H1N1)
PR834 (H1N1)
PR8CAM34 (H1N1)
HICKOX40 (H1N1)
FM47 (H1N1)
WARREN50 (H1N1)
USSR77 (H1N1)
BRAZIL78 (H1N1)
ENG55 (H1N1)
SING57 (H2N2)
VIC68 (H2N2)
NT60-68 (H3N2)
SWHK76 (H3N2)
UDORN72 (H3N2)
TEXAS77 (H3N2)
CALIF78 (H1N1)
SWDAN83 (H3N2)
HK83 (H3N2)
AARBOR60 (H2N2)
1924
1928
1941
1947
1950
1956
1963
1971
1980
1912
1913
III
SWIA30 (H1N1)
SW31 (H1N1)
SWIA46 (H1N1)
SWJMS42 (H1N1)
SW41-49 (H1N1)
SWOH35 (H1N1)
SW29-37 (H1N1)
SWMAY54 (H1N1)
SWWIS 57 (H1N1)
SWIT141-81 (H1N1)
SWIT147-81 (H1N1)
SWIT79 (H1N1)
SWIT76 (H1N1)
NJ76 (H1N1)
SWHK82 (H3N2)
SWTN77 (H1N1)
SWONT81 (H1N1)
SWWIS 61 (H1N1)
WIS88 (H1N1)
SWIN88 (H1N1)
1925
1933
1944
1957
1973
GULAST84 (H13N6)
GULMD77 (H13N6)
GULMD78 (H13N9)
GULMD79 (H13N6)
GULMN80 (H13N6)
WHALEM84 (H13N2)
GULMA80 (H13N6)
IV
DKMAN53 (H10N7)
TYONT66 (H5N9)
GULM5-77 (H11N9)
DKPEN69 (H6N1)
DKMEM74 (H3N8)
MLRDNY78 (H2N2)
RTNJ85 (H4N6)
TYMN80 (H4N2)
TYMN81 (H1N1)
CKPENN83(H5N2)
SEAL80 (H7N7)
DKNZ76 (H4N6)
GTAUS79 (H4N4)
FPV27 (H7N7)
FPV34 (H7N1)
CKGER49 (H10N7)
DKCZ56 (H4N6)
DKENG56 (H11N6)
DKUK60 (H11N8)
DKENG62 (H4N6)
TERNSA61 (H5N3)
WHALEP76 (H1N3)
PARU73 (H7N1)
DKBAV77 (H1N1)
BUDHOK77 (H4N6)
DKHK75 (H3N2)
SWHK2-82 (H3N2)
DKAUS80 (H1N1)
DKBE178 (H3N6)
MAST82 (H14N6)
MAST2-82 (H14N5)
TEALICO80 (H7N7)
MINKSW84 (H10N4)
SHWTR72 (H6N5)
SWGER81 (H1N1)
SWNED85 (H1N1)
SWIT89 (H1N1)
1904
1919
1933
1940
1943
1947
1955
1960
1979
V
100
Nucleotide changes

disease. Another important example is provided by an infection that has been with us since Greek and Roman times—influenza. It may come as a surprise to learn that the worst epidemic of infectious disease ever recorded was that of 'Spanish flu', which killed at least 20 million people worldwide in 1918–19. The virus also managed to infect a staggering 28% of the US population and had a mortality rate of >2.5%, when rates of <0.1% are the norm. Why was this virus so virulent? Once again, molecular phylogenies can provide us with important clues.

There are three different types of influenza virus: influenza A, the most virulent and mutagenic form and which is associated with the major epidemics and pandemics in humans; influenza B which occasionally causes severe illness; and influenza C which, if it causes a disease at all, is usually very mild. Not surprisingly, most attention has focused on influenza A.

Viruses similar in structure and sequence to influenza A are found in a variety of animal species, such as pigs, horses and, most notably, birds. Within the birds, waterfowl (such as ducks) are the most commonly infected group but suffer no ill effects. Phylogenetic trees constructed on sequences from the nucleoprotein (NP) gene from all these species show that viral lineages are very species-specific (Fig. 8.21). Influenza viruses clearly do not cross species boundaries that often. However, when they do, the results can be devastating. The key to understanding this process is the genome of the virus itself.

Influenza virus has a segmented genome which means that recombination (called **reassortment** in this context) among the haemagglutinin (H) and neuraminidase (N) proteins found on the surface of the virus takes place regularly. This reassortment allows new strains, with different combinations of H and N types, to be formed. Birds naturally carry many different H and N combinations (look at the tips of the tree in Fig. 8.21), some of them potentially highly pathogenic for humans. Fortunately, these virulent combinations are usually unable to infect humans directly because they lack a particular nucleoprotein gene sequence which seems to determine host range. This explains why the NP gene tree is so species-specific. Pigs, however, can be infected with *both* the bird and human viruses and so serve as a 'mixing vessel' for these viruses, allowing new H and N combinations to acquire human NP sequences. These recombinant viruses, with new and possibly dangerous cocktails of H and N proteins, may then enter and spread through human populations, sometimes with terrible consequences. This process of slower evolution of viral strains from year to year, seemingly driven by natural selection, is known as 'antigenic drift'. For example, the 1918–19 flu pandemic was caused by the combination H1N1. This appeared again in 1977 in the guise of 'Russian flu'. In the interim period, two other major pandemics ravaged human populations; the 'Asian flu' of 1957, caused by an H2N2 strain, and the 'Hong Kong flu' of 1968 which was the product of an H3N2 outbreak. Despite the importance of pigs, very occasionally a virulent viral strain can move directly from birds into humans, as occurred in the 'Asian flu' of 1957, and more recently in Hong Kong, where a new and seemingly highly virulent surface protein combination, H5N1, has entered humans from chickens.

We will consider one more use of phylogenetic trees in molecular epidemiology, as a 'forensic' tool in documenting the source of particular infection events. We have already come across one famous case of this in Chapter 1—where trees of HIV sequences were used to show that a dentist in Florida had infected a group of his former patients with this virus. Similar studies have been performed on another important blood-borne human pathogen, and one that also evolves frighteningly quickly—hepatitis C virus (HCV).

As with HIV, HCV (another RNA virus) can also be transmitted through contaminated blood products, such as the clotting factors used by haemophiliacs. Despite this, it came as a surprise to many when it was discovered that the virus had been transmitted in anti-rhesus D immunoglobulin—a blood product which is routinely given to rhesus-negative mothers to prevent them from being immunologically incompatible with their babies if they happen to be rhesus-positive. However, just such a transmission took place in Ireland in 1977 where up to 900 women were infected with HCV following exposure to the same, contaminated, batches of this immunoglobulin. The conclusive

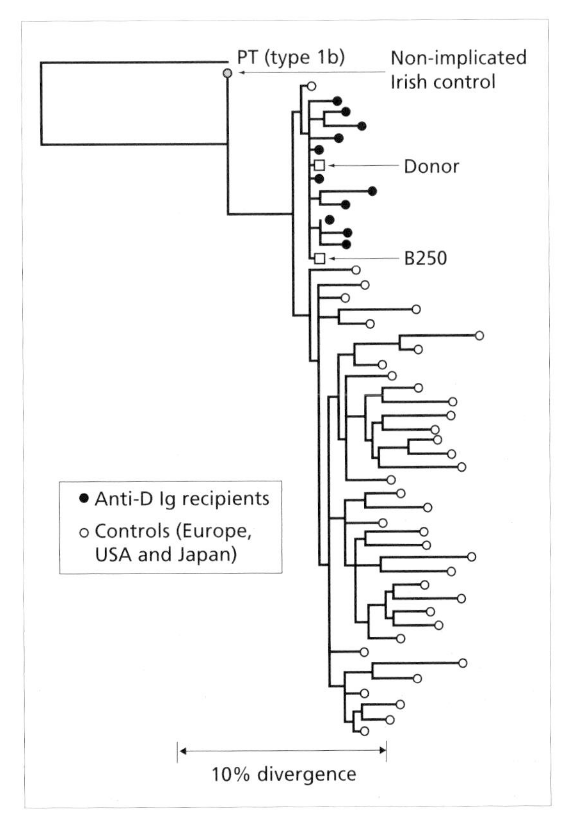

Fig. 8.22 Phylogenetic tree of HCV sequences depicting an anti-rhesus D immunoglobulin (Ig) transmission case in Ireland. Sequences from those HCV infected individuals who received the implicated batches of anti-D Ig are denoted by closed circles and cluster together along with those sequences recovered from one of the original batches (B250) and the suspected HCV infected donor (denoted by open squares). A background population of individuals from different parts of the world infected with the same HCV subtype (1a) are shown as open circles. A sequence from HCV type 1b (PT) was used to root the phylogenetic tree. Data from Power *et al.* (1995).

evidence that infected batches of anti-rhesus D immunoglobulin were responsible for this epidemic came from a phylogenetic analysis of HCV sequences. All women who had received the implicated batches of immunoglobulin were infected with viruses classified as type 1a, whereas others infected in the same area had type 1b, and formed a tight cluster on the phylogenetic tree, indicative of a transmission group. Crucially, viral sequences were also available from one of the suspected batches and from the original HCV-infected blood donor and these grouped closely with the infected women (Fig. 8.22).

We have illustrated only a few of the uses phylogenetic trees have in molecular epidemiology. As pathogens become more routinely studied at the nucleotide sequence level, it is certain that molecular phylogenies will become an increasingly useful tool by which to understand their origins and spread.

8.6 Summary

1 Inferring organismal phylogeny may require combining data from different genes, or from more diverse sources such as morphology and behaviour. These data sets may be combined into a single analysis, analysed separately, or combined on the condition that there is no significant heterogeneity among the data sets.

2 The relationship between gene trees and species trees can be complex, particularly if the gene has undergone several duplications in its history. Reconciled trees can be used to visualise this history.

3 Cospeciating lineages of hosts and associated organisms can be used to calibrate rates of molecular evolution in distantly related taxa, with or without using the fossil record.

4 Combining molecular evidence, phylogenies, and the fossil record can yield new interpretations of the time of diversification of major clades.

5 Rates of lineage diversification and extinction can be inferred from 'lineage through time' plots of molecular phylogenies.

6 Phylogenies of viruses contain important information about their epidemiological history and can be used to determine from where they (and other pathogens) originated and how they have then spread through populations. It is even possible to determine the source of localised transmission events.

8.7 Further reading

The book edited by Harvey *et al.* (1996) is an excellent overview of the range of evolutionary questions that can be addressed using phylogenies. More specific topics, such as host–parasite cospeciation, biogeography, and character evolution are discussed by Brooks and McLennan (1991) and Harvey and Pagel (1991). There is a large, robust literature on the total evidence debate. Recent reviews include de Quieroz *et al.* (1995), Nixon and Carpenter (1996), and

Huelsenbeck *et al.* (1996). The gene tree-species tree problem is discussed by Pamilo and Nei (1988), Doyle (1992), and Brower *et al.* (1996), among others. The reconciled tree method was first proposed by Goodman *et al.* (1979), and has recently received renewed attention (Guigó *et al.*, 1996; Mirkin *et al.*, 1995; Page, 1994). Our treatment here is based on Page and Charleston (1997a; 1997b).

Brooks and McLennan (1991; 1993) review the application of phylogenetics to the study of cospeciation. For recent, opposing views on the methodological details of comparing host and parasite trees see Hoberg *et al.* (1997) and Paterson and Gray (1997). Hafner and Page (1995,1996) describe the use of cospeciating host and parasites to compare rates of evolution. Applications of this approach include Hafner *et al.* (1994), McGeoch *et al.* (1995), Moran *et al.* (1995) and Huelsenbeck *et al.* (1997). The utility of molecular data in host–parasite cospeciation studies is emphasised by Page *et al.* (1996). The evidence that the bird and mammal radiations predate the K/T boundary is presented by Hedges *et al.* (1996) and Cooper and Penny (1997). See Gibbons (1997) for a commentary and Mindell (1997) for the most recent overview of avian molecular phylogenetics. Wray *et al.* (1996) present the molecular evidence that the Burgess Shale is much younger than the radiation of the Metazoa. Smith (1994) is an excellent introduction to the phylogenetic interpretation of the fossil record.

The theory behind lineage through time plots is developed by Nee, Harvey, and colleagues (Harvey *et al.*, 1994; Nee *et al.*, 1992, 1994a,b, 1995).

Applications of phylogenies to epidemiological studies are discussed in Holmes (1998), Holmes and Garnett (1994) and Holmes and Harvey (1998). With respect to the specific viral examples used, the evolution of primate immunodeficiency viruses was approached by Hirsch *et al.* (1995) and Sharp *et al.* (1994, 1995), that of Sin Nombre virus by Nichol *et al.* (1993), of influenza by Scholtissek (1987) and Webster *et al.* (1995), and of HCV by Power *et al.* (1995).

References and Bibliography

Aguadé M, Meyers W, Long AD & Langley CH (1994). Single-stranded conformation polymorphism analysis coupled with stratified DNA sequencing reveals reduced sequence variation in the *su(s)* and *su(w^a)* regions of the *Drosophila melanogaster* X chromosome. *Proceedings of the National Academy of Sciences of the USA* **91**, 4658–4662.

Akashi H (1994). Synonymous codon usage in *Drosophila melanogaster*: natural selection and translational accuracy. *Genetics* **136**, 927–935.

Altschul SF & Lipman DJ (1989). Trees, stars, and multiple biological sequence alignment. *SIAM Journal of Applied Mathematics* **49**, 197–209.

Andrews P & Cronin JE (1982). The relationships of *Sivapithecus* and *Ramapithecus* and the evolution of the orang-utan. *Nature* **297**, 541–545.

Armour JAL, Anttinen T, May CA, Vega EE, Sajantila A, Kidd JR, Kidd KK, Bertranpetit J, Pääbo S & Jeffreys AJ (1996). Minisatellite diversity supports a recent African origin for modern humans. *Nature Genetics* **13**, 154–160.

Årnason U & Johnsson E (1992) The complete mitochondrial DNA sequence of the harbor seal, *Phoca vitulina*. *Journal of Molecular Evolution* **34**, 493–505.

Austin JJ, Smith AB & Thomas RJ (1997). Palaeontology in a molecular world: the search for authentic ancient DNA. *Trends in Ecology and Evolution* **12**, 303–306.

Averof M & Akam M (1993). *HOM/Hox* genes of *Artemia*: implications for the origin of insect and crustacean body plans. *Current Biology* **3**, 73–78.

Avise JC (1994). *Molecular Markers, Natural History & Evolution*. Chapman & Hall, New York.

Ayala FJ, Escalante A, O'hUigin C & Klein J (1994). Molecular genetics of speciation and human origins. *Proceedings of the National Academy of Sciences of the USA* **91**, 6787–6794.

Bailey WJ, Kim J, Wagner GP & Ruddle FH (1997). Phylogenetic reconstruction of vertebrate *Hox* cluster duplications. *Molecular Biology and Evolution* **14**, 843–853.

Bandelt H-J & Dress AWM (1992). Split decomposition: a new and useful approach to phylogenetic analysis of distance data. *Molecular Phylogenetics and Evolution* **1**, 242–252.

Barthélemy J-P & Guénoche A (1991). *Trees and Proximity Representations*. John Wiley & Sons, Chichester.

Barrett M, Donoghue MJ & Sober E (1991). Against consensus. *Systematic Zoology* **40**, 486–493.

Barton NH (1996). Natural selection and random genetic drift as causes of evolution on islands. *Philosophical Transactions of the Royal Society of London B* **351**, 785–795.

Barton NH & Turelli M (1989). Evolutionary quantitative genetics: how little do we know? *Annual Review of Genetics* **23**, 337–370.

Bates G & Lehrach H (1994). Trinucleotide repeat expansions and human genetic disease. *BioEssays* **4**, 277–284.

Begun DJ & Aquadro CF (1992). Levels of naturally occurring DNA polymorphism correlate with recombination rates in *D. melanogaster*. *Nature* **356**, 519–520.

Begun DJ & Aquadro CF (1993). African and North American populations of *Drosophila melanogaster* are very different at the DNA level. *Nature* **365**, 548–550.

Begun DJ & Aquadro CF (1995). Evolution at the tip and base of the X chromosome in an African population of *Drosophila melanogaster*. *Molecular Biology and Evolution* **12**, 382–390.

Bernardi G (1993). The vertebrate genome: isochores and evolution. *Molecular Biology and Evolution* **10**, 186–204.

Bernardi G, Mouchiroud D & Gautier C (1993). Silent substitutions in mammalian genomes and their evolutionary implications. *Journal of Molecular Evolution* **37**, 583–589.

Bernardi G, Olofsson B, Filipski J, Zerial M, Salinas J, Cuny G, Meunier-Rotival M & Rodier F (1985). The mosaic genome of warm-blooded vertebrates. *Science* **228**, 953–958.

Berry AJ & Kreitman M (1993). Molecular analysis of an allozyme cline: alcohol dehydrogenase in *Drosophila melanogaster* on the east coast of the United States. *Genetics* **134**, 869–893.

Berry AJ, Ajioka JW & Kreitman M (1991). Lack of polymorphism on the *Drosophila* fourth chromosome resulting from selection. *Genetics* **129**, 1111–1117.

Bickmore WA & Summer AT (1989). Mammalian chromosome banding—an expression of genome organisation. *Trends in Genetics* **5**, 144–149.

Bird AP (1987). CpG islands as gene markers in the vertebrate nucleus. *Trends in Genetics* **3**, 342–347.

Bird AP (1995). Gene number, noise reduction and biological complexity. *Trends in Genetics* **11**, 94–100.

Bohr VA & Wassermann K (1988). DNA repair at the level of the gene. *Trends in Biochemical Sciences* **13**, 429–433.

Bowcock AM, Ruiz-Linares A, Tomfohrde J, Minch E, Kidd JR & Cavalli-Sforza LL (1994). High resolution of human evolutionary trees with polymorphic microsatellites. *Nature* **368**, 455–457.

Braverman JM, Hudson RR, Kaplan NL, Langley CH & Stephan W (1995). The hitchhiking effect on the site frequency spectrum of DNA polymorphisms. *Genetics* **140**, 783–796.

Britten RJ (1986). Rates of DNA sequence evolution differ between taxonomic groups. *Science* **231**, 1393–1398.

Bromham L, Rambaut A & Harvey PH (1996). Determinants of rate variation in mammalian DNA sequence evolution. *Journal of Molecular Evolution* **43**, 610–621.

Brooks DR & McLennan DA (1991). *Phylogeny, Ecology, and Behavior.* Chicago University Press, Chicago.

Brooks DR & McLennan DA (1993). *Parascript: Parasites and the Language of Evolution.* Smithsonian Institution Press, Washington.

Brooks M, Emelianov I, McMillan O, Rose O & Mallet J (1996). A quick guide to genetic and molecular markers used in biodiversity studies. In *Biodiversity. A Biology of Numbers and Differences* (KJ Gaston, ed.), pp. 48–53. Blackwell Science, Oxford.

Brower AVZ, DeSalle R & Vogler A (1996). Gene trees, species trees, and systematics: a cladistic perspective. *Annual Review of Ecology and Systematics* **27**, 423–450.

Brown H, Sanger F & Kitai R (1955). The structure of pig and sheep insulins. *Biochemical Journal* **60**, 556–565.

Brown WM, Prager EM, Wang A & Wilson AC (1982). Mitochondrial DNA sequences of primates: tempo and mode of evolution. *Journal of Molecular Evolution* **18**, 225–239.

Bryson V & Vogel HJ (eds) (1965). *Evolving Genes and Proteins.* Academic Press, New York.

Bruford MW & Wayne RK (1993). Microsatellites and their application to population genetic studies. *Current Opinion in Genetics and Development* **3**, 939–943.

Bull JJ, Huelsenbeck JP, Cunningham CW, Swofford DL & Waddell PJ (1993). Partitioning and combining data in phylogenetic analysis. *Systematic Biology* **42**, 384–397.

Bulmer M (1987). Coevolution of codon usage and transfer RNA abundance. *Nature* **325**, 728–730.

Bulmer M (1991). The selection–mutation–drift theory of synonymous codon usage. *Genetics* **129**, 897–907.

Bulmer M, Wolfe KH & Sharp PM (1991). Synonymous nucleotide substitution rates in mammalian genes: implications for the molecular clock and the relationship of mammalian orders. *Proceedings of the National Academy of Sciences of the USA* **88**, 5974–5978.

Bult CJ, White O, Olsen GJ, Zhou LX, Fleischmann RD, Sutton GG, Blake JA, Fitzgerald LM, Clayton RA, Gocayne JD, Kerlavage AR, Dougherty BA, Tomb JF, Adams MD, Reich CI,

Overbeek R, Kirkness EF, Weinstock KG, Merrick JM, Glodek A, Scott JL, Geoghagen NSM, Weidman JF, Fuhrmann JL, Nguyen D, Utterback TR, Kelley JM, Peterson JD, Sadow PW, Hanna MC, Cotton MD, Roberts KM, Hurst MA, Kaine BP, Borodovsky M, Klenk HP, Fraser CM, Smith HO, Woese CR & Venter JC (1996). Complete genome sequence of the methanogenic archaeon, *Methanococcus jannaschii*. *Science* **273**, 1058–1073.

Cann RL, Stoneking M & Wilson AC (1987). Mitochondrial DNA and human evolution. *Nature* **325**, 31–36.

Cannio R, Rossi M & Bartolucci S (1994). A few amino acid substitutions are responsible for the higher thermostability of a novel NAD^+-dependent bacillar alcohol dehydrogenase. *European Journal of Biochemistry* **222**, 345–352.

Casane D, Boissinot S, Chang BH-J, Shimmin LC & Li W-H (1997). Mutation pattern variation among regions of the primate genome. *Journal of Molecular Evolution* **45**, 216–226.

Chang BH-J & Li W-H (1995). Estimating the intensity of male-driven evolution in rodents by using X-linked and Y-linked *Ube 1* genes and pseudogenes. *Journal of Molecular Evolution* **40**, 70–77.

Chang BH-J, Shimmin LC, Shyue S-K, Hewett-Emmett D & Li W-H (1994). Weak male-driven molecular evolution in rodents. *Proceedings of the National Academy of Sciences of the USA* **91**, 827–831.

Chao L (1990). Fitness of RNA virus decreased by Muller's ratchet. *Nature* **348**, 454–455.

Chao L & Carr DE (1993). The molecular clock and the relationship between population size and generation time. *Evolution* **47**, 688–690.

Charlesworth B, Morgan MT & Charlesworth D (1993). The effect of deleterious mutations on neutral molecular variation. *Genetics* **134**, 1289–1303.

Charlesworth B, Sniegowski P & Stephan W (1994). The evolutionary dynamics of repetitive DNA in eukaryotes. *Nature* **371**, 215–220.

Christopher DA & Hallick RB (1989). *Euglena gracilis* chloroplast ribosomal protein operon: a new chloroplast gene for ribosomal protein L5 and description of a novel organelle intron category designated group III. *Nucleic Acids Research* **17**, 7591–7608.

Clay T (1949). Some problems in the evolution of a group of ectoparasites. *Evolution* **3**, 279–299.

Clegg MT, Gaut BS, Learn GH Jr & Morton BR (1994). Rates and patterns of chloroplast DNA evolution. *Proceedings of the National Academy of Sciences of the USA* **91**, 6795–6801.

Clegg MT, Cummings MP & Durbin ML (1997). The evolution of plant nuclear genes. *Proceedings of the National Academy of Sciences of the USA* **94**, 7791–7798.

Cohan FM (1994). The effects of rare but promiscuous genetic exchange on evolutionary divergence in prokaryotes. *American Naturalist* **143**, 965–986.

Cohen J (1997). The flu pandemic that might have been. *Science* **277**, 1600–1601.

Collins DW & Jukes TH (1993). Relationship between G+C in silent sites of codons and amino acid composition of human proteins. *Journal of Molecular Evolution* **36**, 201–213.

Comuzzie AG, Hixson JE, Almasy L, Mitchell BD, Mahaney MC, Dyer TD, Stern MP, MacCluer JW & Blangero J (1997). A major quantitative trait locus determining serum leptin levels and fat mass is located on human chromosome 2. *Nature Genetics* **15**, 273–275.

Cooper A & Penny D (1997). Mass survival of birds across the Cretaceous–Tertiary boundary: molecular evidence. *Science* **275**, 1109–1113.

Coyne JA (1992). Genetics and speciation. *Nature* **355**, 511–515.

Coyne JA & Orr HA (1989). Patterns of speciation in *Drosophila*. *Evolution* **43**, 362–381.

Coyne JA, Crittenden AP & Mah K (1994). Genetics of a pheromonal difference contributing to reproductive isolation in *Drosophila*. *Science* **265**, 1461–1464.

Cummings MP, Otto SP & Wakely J (1995). Sampling properties of DNA sequence data in phylogenetic analysis. *Molecular Biology and Evolution* **12**, 814–822.

The Cystic Fibrosis Genetic Analysis Consortium (1990). Worldwide survey of the ΔF508 mutation—report from the Cystic Fibrosis Genetic Analysis Consortium. *American Journal of Human Genetics* **47**, 354–359.

Darnell J, Lodish H & Baltimore D (1990). *Molecular Cell Biology* (2nd edn). Scientific American Books. Distributed by WH Freeman & Co., New York.

Darwin C (1859). *On The Origin of Species*. John Murray, London.

Dayhoff MO (1978). *Atlas of Protein Sequence and Structure, Volume 5, Supplement 3*, National Biomedical Research Foundation, Washington, DC.

Dean M, Carrington M, Winkler C, Huttley GA, Smith MW, Allikmets R, Goedert JJ, Buchbinder SP, Vittinghoff E, Gomperts E, Donfield S, Vlahov D, Kaslow R, Saah A, Rinaldo C, Detels R, Hemophilia Growth and Development Study, Multicenter AIDS Cohort Study, Multicenter Hemophilia Cohort Study, San Francisco City Cohort, ALIVE Study & O'Brien SJ (1996). Genetic restriction of HIV-1 infection and progression to AIDS by a deletion allele of the CKR5 structural gene. *Science* **273**, 1856–1862.

de Queiroz A, Donoghue MJ & Kim J (1995). Separate versus combined analysis of phylogenetic evidence. *Annual Review of Ecology and Systematics* **26**, 657–681.

De Robertis EM (1997). The ancestry of segmentation. *Nature* **387**, 25–26.

DeSalle R & Templeton AR (1988). Founder effects and the rate of mitochondrial DNA evolution in Hawaiian *Drosophila*. *Evolution* **42**, 1076–1084.

Dickerson RE (1983). The DNA helix and how it is read. *Scientific American* **249**, 94.

Dickerson RE & Geis I (1983). *Hemoglobin: Structure, Function, Evolution, and Pathology*. The Benjamin/Cummings Publishing Company Inc., Menlo Park, CA.

Donnelly P & Tavaré S (1995). Coalescents and genealogical structure under neutrality. *Annual Review of Genetics* **29**, 401–421.

D'Onofrio G, Mouchiroud D, Aïssani B, Gautier C & Bernardi G (1991). Correlations between the compositional properties of human genes, codon usage, and amino acid composition of proteins. *Journal of Molecular Evolution* **32**, 504–510.

Dopazo J, Dress A & von Haeseler A (1993). Split decomposition: a technique to analyze viral evolution. *Proceedings of the National Academy of Sciences of the USA* **90**, 10 320–10 324.

Dover GA (1982). Molecular drive: a cohesive mode of species evolution. *Nature* **299**, 111–117.

Dover GA (1987). DNA turnover and the molecular clock. *Journal of Molecular Evolution* **26**, 47–58.

Dover GA (1993). Evolution of genetic redundancy for advanced players. *Current Opinion in Genetics and Development* **3**, 902–910.

Doyle JJ (1992). Gene trees and species trees: molecular systematics as one-character taxonomy. *Systematic Botany* **17**, 144–163.

Drake JW (1993). Rates of spontaneous mutation among RNA viruses. *Proceedings of the National Academy Sciences of the USA* **90**, 4171–4175.

Dubrova YE, Nesterov VN, Krouchinsky NG, Ostapenko VA, Neumann R, Neil DL & Jeffreys AJ (1996). Human minisatellite mutation rate after the Chernobyl accident. *Nature* **380**, 683–686.

Eanes WF, Kirchner M & Yoon J (1993). Evidence for adaptive evolution of the *G6pd* gene in the *Drosophila melanogaster* and *Drosophila simulans* lineages. *Proceedings of the National Academy of Sciences of the USA* **90**, 7475–7479.

Engels WR (1992). The origin of *P* elements in *Drosophila melanogaster*. *BioEssays* **14**, 681–686.

Erlich HA & Gyllensten UB (1991). Shared epitopes among HLA class II alleles: gene conversion, common ancestry and balancing selection. *Immunology Today* **12**, 411–414.

Eyre-Walker AC (1991). An analysis of codon usage in mammals: selection or mutation bias? *Journal of Molecular Evolution* **33**, 442–449.

Eyre-Walker AC (1992). The effect of constraint on the rate of evolution in neutral models with biased mutation. *Genetics* **131**, 233–234.

Eyre-Walker AC (1993). Recombination and mammalian genome evolution. *Proceedings of the Royal Society of London B* **252**, 237–243.

Eyre-Walker A (1994). DNA mismatch repair and synonymous codon evolution in mammals. *Molecular Biology and Evolution* **11**, 88–98.

Farris JS (1981). Distance data in phylogenetic analysis. In *Advances in Cladistics: Proceedings of the First Meeting of the Willi Hennig Society* (VA Funk & DR Brooks, eds), pp. 3–23. New York Botanical Garden, New York.

Felsenstein J (1978). Cases in which parsimony or compatibility methods will be positively misleading. *Systematic Zoology* **27**, 401–410.

Felsenstein J (1981). Evolutionary trees from DNA sequences: a maximum likelihood approach. *Journal of Molecular Evolution* **17**, 368–376.

Felsenstein J (1992). Estimating effective population size from samples of sequences: inefficiency of pairwise and segregating sites as compared to phylogenetic estimates. *Genetical Research* **59**, 139–147.

Ferguson RA & Goldberg DM (1997). Genetic markers of alcohol abuse. *Clinica Chimica Acta* **257**, 199–250.

Figueroa F, Günther E & Klein J (1988). MHC polymorphism pre-dating speciation. *Nature* **355**, 265–267.

Filipski J (1988). Why the rate of silent coding substitutions is variable within a vertebrate's genome. *Journal of Theoretical Biology* **134**, 159–164.

Fitch WM (1970). Distinguishing homologous and analogous proteins. *Systematic Zoology* **19**, 99–113.

Fitch WM & Beintema JJ (1990). Correcting parsimonious trees for unseen nucleotide substitutions: the effect of dense branching as exemplified by ribonuclease. *Molecular Biology and Evolution* **7**, 438–443.

Fitch WM, Leiter JME, Li X & Palese P (1991). Positive Darwinian evolution in human influenza A viruses. *Proceedings of the National Academy of Sciences of the USA* **88**, 4270–4274.

Flavell AJ, Pearce SR & Kumar A (1994). Plant transposable elements and the genome. *Current Opinion in Genetics and Development* **4**, 838–844.

Fleischmann RD, Adams MD, White O, Clayton RA, Kirkness EF, Kerlavage AR, Bult CJ, Tomb JF, Dougherty BA, Merrick JM, MCKenney K, Sutton G, Fitzhugh W, Fields C, Gocayne JD, Scott J, Shirley R, Liu LI, Glodek A, Kelley JM, Weidman JF, Phillips CA, Spriggs T, Hedblom E, Cotton MD, Utterback TR, Hanna MC, Nguyen DT, Saudek DM, Brandon RC, Fine LD, Fritchman JL, Fuhrmann JL, Geoghagen NSM, Gnehm CL, McDonald LA, Small KV, Fraser CM, Smith HO & Venter JC (1995). Whole genome random sequencing and assembly of *Haemophilus influenzae*. *Science* **269**, 496–512.

Fu Y-X (1994). A phylogenetic estimator of effective population size or mutation rate. *Genetics* **136**, 685–692.

Fu Y-X & Li W-H (1993). Statistical tests of neutrality of mutations. *Genetics* **133**, 693–709.

Fullerton SM, Harding RM, Boyce AJ & Clegg JB (1994). Molecular and population genetic analysis of allele sequence diversity at the human β-globin locus. *Proceedings of the National Academy of Sciences of the USA* **91**, 1805–1809.

Galtier N & Gouy M (1995). Inferring phylogenies from DNA sequences of unequal base compositions. *Proceedings of the National Academy of Sciences of the USA* **92**, 11 317–11 321.

Gaut BS, Muse SV, Clark D & Clegg MT (1992). Relative rates of nucleotide substitution at the *rbcL* locus of monocotyledonous plants. *Journal of Molecular Evolution* **35**, 292–303.

Gaut BS, Morton BR, McCaig BC & Clegg MT (1996). Substitution rate comparisons between grasses and palms: synonymous rate differences at the nuclear *Adh* parallel rate differences at the plastid gene *rbcL*. *Proceedings of the National Academy of Sciences of the USA* **93**, 10 274–10 279.

Gaut BS, Clark LG, Wendel JF & Muse SV (1997). Comparisons of the molecular evolutionary process at *rbcL* and *ndhF* in the grass family. *Molecular Biology and Evolution* **14**, 769–777.

Garcia-Fernàndez J & Holland PWH (1994). Archetypal organization of the amphioxus *Hox* gene cluster. *Nature* **370**, 563–566.

Gibbons A (1997). Did birds sail through the K-T boundary with flying colors? *Science* **275**, 1068.

Gilbert W (1978). Why genes in pieces? *Nature* **271**, 501.

Gilbert W, de Souza SJ & Long M (1997). Origin of genes. *Proceedings of the National Academy of Sciences of the USA* **94**, 7698–7703.

Gill P, Jeffreys AJ & Werrett DJ (1985). Forensic applications of DNA 'fingerprints'. *Nature* **318**, 577–579.

Gillespie JH (1984). The molecular clock may be episodic. *Proceedings of the National Academy of Sciences of the USA* **81**, 8009–8013.

Gillespie JH (1986a). Rates of molecular evolution. *Annual Review of Ecology and Systematics* **17**, 637–665.

Gillespie JH (1986b). Natural selection and the molecular clock. *Molecular Biology and Evolution* **3**, 138–155.

Gillespie JH (1987). Molecular evolution and the neutral allele theory. *Oxford Surveys in Evolutionary Biology* **4**, 10–37.

Gillespie JH (1989a). Molecular evolution and polymorphism: SAS-CFF meets the mutational landscape. *American Naturalist* **134**, 638–658.

Gillespie JH (1989b). Lineage effects and the index of dispersion of molecular evolution. *Molecular Biology and Evolution* **6**, 636–647.

Gillespie JH (1991). *The Causes of Molecular Evolution*. Oxford University Press, Oxford.

Gillespie JH (1993). Substitution processes in molecular evolution. I. Uniform and clustered substitutions in a haploid model. *Genetics* **134**, 971–981.

Gillespie JH (1994a). Substitution processes in molecular evolution. II. Exchangeable models from population genetics. *Evolution* **48**, 1101–1113.

Gillespie JH (1994b). Substitution processes in molecular evolution. III. Deleterious alleles. *Genetics* **138**, 943–952.

Gillespie JH (1995). On Ohta's hypothesis: most amino acid substitutions are deleterious. *Journal of Molecular Evolution* **40**, 64–69.

Go M (1981). Correlation of DNA exonic regions with protein structural units in haemoglobin. *Nature* **291**, 90–92.

Goedde HW, Agarwal DP, Fritze G, Meier-Tackmann D, Singh S, Beckmann G, Bhatia K, Chen LZ, Fang B, Lisker R, Paik YK, Rothhammer F, Saha N, Segal B, Srivastava LM & Czeizal A (1992). Distribution of ADH_2 and $ALDH_2$ genotypes in different populations. *Human Genetics* **88**, 344–346.

Goldman N (1993). Statistical tests of models of DNA substitution. *Journal of Molecular Evolution* **36**, 182–198.

Goldman N (1994). Variance to mean ratio, *R(t)* for poisson processes on phylogenetic trees. *Molecular Phylogenetics and Evolution* **3**, 230–239.

Goldman N & Yang Z (1994). A codon-based model of nucleotide substitution for protein-coding DNA sequences. *Molecular Biology and Evolution* **11**, 725–736.

Goldstein DB, Ruiz Linares A, Cavalli-Sforza LL & Feldman MW (1995). Genetic absolute dating based on microsatellites and the origin of modern humans. *Proceedings of the National Academy of Sciences of the USA* **92**, 6723–6727.

Goodman M, Czelusniak J, Moore GW, Romero-Herrera AE & Matsuda G (1979). Fitting the gene lineage into its species lineage: a parsimony strategy illustrated by cladograms constructed from globin sequences. *Systematic Zoology* **28**, 132–168.

Gottelli D, Sillero-zubiri C, Applebaum GD, Roy MS, Girman DJ, Garcia-moreno J, Ostrander EA & Wayne RK (1994). Molecular genetics of the most endangered canid—the Ethiopian wolf *Canis simensis*. *Molecular Ecology* **3**, 301–312.

Gottesman II (1997). Twins: en route to QTLs for cognition. *Science* **276**, 1522–1523.

Goudsmit J, De Ronde A, Ho DD & Perelson AS (1996). Human immunodeficiency virus *in vivo*: calculations based on a single zidovudine resistance mutation at codon 215 of reverse transcriptase. *Journal of Virology* **70**, 5662–5664.

Grantham R, Gautier C, Gouy M, Mercier R & Pavé A (1980). Codon catalog usage and the genome hypothesis. *Nucleic Acids Research* **8**, 49–62.

Graur D & Li W-H (1991). Neutral mutation hypothesis test. *Nature* **354**, 114–115.

Graur D, Hide WA & Li W-H (1991). Is the guinea-pig a rodent? *Nature* **351**, 649–652.

Gray MW (1993). Origin and evolution of organelle genomes. *Current Opinion in Genetics and Development* **8**, 9884–9890.

Griffiths AJF, Miller JH, Suzuki DT, Lewontin RC & Gelbart WM (1993). *An Introduction to Genetic Analysis* (5th edn). WH Freeman & Co., New York.

Guigó R, Muchnik I & Smith TF (1996). Reconstruction of ancient molecular phylogeny. *Molecular Phylogenetics and Evolution* **6**, 189–213.

Hafner MS & Page RDM (1995). Molecular phylogenies and host–parasite cospeciation: gophers and lice as a model system. *Philosophical Transactions of the Royal Society of London B* **349**, 77–83.

Hafner MS, Sudman PD, Villablanca FX, Spradling TA, Demastes JW & Nadler SA (1994). Disparate rates of molecular evolution in cospeciating hosts and parasites. *Science* **265**, 1087–1090.

Hagelberg E, Gray IC & Jeffreys AJ (1991). Identification of the skeletal remains of a murder victim by DNA analysis. *Nature* **352**, 427–429.

Haldane JBS (1957). The cost of natural selection. *Journal of Genetics* **55**, 511–524.

Hall BG (1990). Spontaneous point mutations that occur more often when they are advantageous than when they are neutral. *Genetics* **126**, 5–16.

Hamilton WD, Axelrod R & Tanese R (1990). Sexual reproduction as an adaptation to resist parasites (a review). *Proceedings of the National Academy of Sciences of the USA* **87**, 3566–3573.

Hammer MF (1995). A recent common ancestry for human Y chromosomes. *Nature* **378**, 376–378.

Harding RM (1996). New phylogenies: an introductory look at the coalescent. In *New Uses for New Phylogenies* (PH Harvey, AJ Leigh Brown, J Maynard Smith & S Nee, eds), pp. 15–22. Oxford University Press, Oxford.

Harding RM, Fullerton SM, Griffiths RC, Bond J, Cox MJ, Schneider JA, Moulin DS & Clegg JB (1997). Archaic African and Asian lineages in the genetic ancestry of modern humans. *American Journal of Human Genetics* **60**, 772–789.

Hardison RC (1991). Evolution of globin gene families. In *Evolution at the Molecular Level* (RK Selander, AG Clark & TS Whittam, eds), pp. 272–289. Sinuaer Associates, Sunderland MA.

Harpending HC, Sherry ST, Rogers AR & Stoneking M (1993). The genetic structure of ancient human populations. *Current Anthropology* **34**, 483–496.

Harris H (1966). Enzyme polymorphisms in man. *Proceedings of the Royal Society of London B* **164**, 298–310.

Hartl DL & Clark AG (1989). *Principles of Population Genetics* (2nd edn). Sinauer Associates, Sunderland MA.

Hartl DL, Moriyama EN & Sawyer S (1994). Selection intensity for codon bias. *Genetics* **138**, 227–234.

Harvey PH & Pagel MD (1991). *The Comparative Method in Evolutionary Biology*. Oxford University Press, Oxford.

Harvey PH, May RM & Nee S (1994). Phylogenies without fossils. *Evolution* **48**, 523–529.

Harvey PH, Leigh Brown AJ, Maynard Smith J & Nee S (1996). *New Uses for New Phylogenies*. Oxford University Press, Oxford.

Hasegawa M, Kishino H & Yano T-A (1985). Dating of the human–ape splitting by a molecular clock of mitochondrial DNA. *Journal of Molecular Evolution* **22**, 160–174.

Hedges SB, Kumar S & Tamura K (1991). Human origins and analysis of mitochondrial DNA sequences. *Science* **255**, 737–739.

Hedges SB, Parker PH, Sibley CG & Kumar S (1996). Continental breakup and the ordinal diversification of birds and mammals. *Nature* **381**, 226–229.

Hedrick PW (1994). Evolutionary genetics of the major histocompatibility complex. *American Naturalist* **143**, 945–964.

Hein J (1990). Unified approach to alignment and phylogenies. *Methods in Enzymology* **183**, 626–645.

Hendy M (1991). A combinatorial description of the closest tree algorithm for finding evolutionary trees. *Discrete Mathematics* **96**, 51–58.

Hendy MD & Penny D (1993). Spectral analysis of phylogenetic data. *Journal of Classification* **10**, 5–24.

Henikoff S, Greene EA, Pietrokovski S, Bork P, Attwood TK & Hood L (1997). Gene families: the taxonomy of protein paralogs and chimeras. *Science* **278**, 609–614.

Hey J (1997). Mitochondrial and nuclear genes present conflicting portraits of human origins. *Molecular Biology and Evolution* **14**, 166–172.

Hickson RE, Simon C, Cooper A, Spicer GS, Sullivan J & Penny D (1996). Conserved sequence motifs, alignment, and secondary structure for the third domain of animal 12S rRNA. *Molecular Biology and Evolution* **13**, 150–169.

Hill AVS, Allsopp CEM, Kwiatkowski D, Anstey NM, Twumasi P, Rowe PA, Bennett S, Brewster D, McMichael AJ & Greenwood BM (1991). Common West African HLA antigens are associated with protection from severe malaria. *Nature* **352**, 595–600.

Hill AVS, Elvin J, Willis AC, Aidoo M, Allsopp CEM, Gotch FM, Gao M, Takiguchi M, Greenwood BM, Townsend ARM, McMichael AJ & Whittle HC (1992). Molecular analysis of the association of HLA-B53 and resistance to severe malaria. *Nature* **360**, 434–439.

Hillis DM (1996). Inferring complex phylogenies. *Nature* **383**, 130–131.

Hillis DM & Dixon MT (1991). Ribosomal DNA: molecular evolution and phylogenetic inference. *Quarterly Review of Biology* **66**, 410–453.

Hillis DM, Bull JJ, White ME, Badgett MR & Molineux IJ (1992). Experimental phylogenetics: generation of a known phylogeny. *Science* **255**, 589–592.

Hillis DM, Huelsenbeck JP & Cunningham CW (1994). Application and accuracy of molecular phylogenies. *Science* **264**, 671–677.

Hiratsuka J, Shimada H, Whittier R, Ishibashi T, Sakamoto M, Mori M, Kondo C, Honji Y, Sun CR, Meng BY, Li YQ, Kanno A, Nishizawa Y, Hirai A, Shinozaki K & Sugiura M (1989). The complete sequence of the rice (*Oryza sativa*) chloroplast genome: intermolecular recombination between distinct transfer-RNA genes accounts for a major plastid DNA inversion during the evolution of the cereals. *Molecular and General Genetics* **217**, 185–194.

Hirsch VM, Dapolito G, Goeken R & Campbell BJ (1995). Phylogeny and natural history of the primate lentiviruses, SIV and HIV. *Current Opinion in Genetic Development* **5**, 798–806.

Hoberg EP, Brooks DR & Seigel-Causey D (1997). Host–parasite co-speciation: history, principles, and prospects. In *Host–Parasite Evolution: General Principles and Avian Models* (DH Clayton and J Moore, eds), pp. 212–235. Oxford University Press, Oxford.

Holland PWH, Garcia-Fernàndez J, Williams NA & Sidow A (1994). Gene duplications and the origins of vertebrate development. *Development* Suppl. S125–S133.

Holland PWH, Holland LZ, Williams NA & Holland ND (1992). An amphioxus homeobox gene: sequence conservation, spatial expressing during development and insights into vertebrate evolution. *Development* **116**, 653–661.

Holmes EC (1998). Molecular phylogenies and the genetic structure of viral populations. In *Evolution in Health & Disease* (S Stearns, ed.). Oxford University Press, Oxford. In press.

Holmes EC & Harvey PH (1993). Fitting the bill. *Current Biology* **3**, 776–777.

Holmes EC & Garnett GP (1994). Genes, trees and infections: molecular evidence in epidemiology. *Trends in Ecology and Evolution* **9**, 256–260.

Holmes EC, Zhang LQ, Simmonds P, Ludlam CA & Leigh Brown AJ (1992). Convergent and divergent sequence evolution in the surface envelope glycoprotein of human immunodeficiency virus type 1 within a single infected patient. *Proceedings of the National Academy of Sciences of the USA* **89**, 4835–4839.

Holmquist GP & Filipski J (1994). Organisation of mutations along the genome: a prime determinant of genome evolution. *Trends in Ecology and Evolution* **9**, 65–69.

Hopkins GHE (1948). Some factors which have modified the phylogenetic relationship between parasite and host in the Mallophaga. *Proceedings of the Linnean Society, London* **161**, 37–39.

Hubby JL & Lewontin RC (1966). A molecular approach to the study of genic heterozygosity in natural populations. I. The number of alleles at different loci in *Drosophila pseudoobscura*. *Genetics* **54**, 577–594.

Hudson RR (1990). Gene genealogies and the coalescent process. In *Oxford Surveys in Evolutionary Biology*, Volume 7 (D Futuyma & J Antonovics, eds), pp. 1–44. Oxford University Press, Oxford.

Hudson RR (1993). Levels of DNA polymorphism and divergence yield important insights into evolutionary processes. *Proceedings of the National Academy of Sciences of the USA* **90**, 7425–7426.

Hudson RR (1994). How can the low levels of DNA sequence variation in regions of the *Drosophila* genome with low recombination rates be explained? *Proceedings of the National Academy of Sciences of the USA* **91**, 6815–6818.

Hudson RR & Kaplan NL (1995a). The coalescent process in models with selection and recombination. *Genetics* **120**, 831–840.

Hudson RR & Kaplan NL (1995b). Deleterious background selection with recombination. *Genetics* **141**, 1605–1617.

Hudson RR, Kreitman M & Aguadé M (1987). A test of neutral molecular evolution based on nucleotide data. *Genetics* **116**, 153–159.

Hudson RR, Slatkin M & Maddison WP (1992). Estimation of levels of gene flow from DNA sequence data. *Genetics* **132**, 583–589.

Huelsenbeck JP & Hillis DM (1993). Success of phylogenetic methods in the four-taxon case. *Systematic Biology* **42**, 247–264.

Huelsenbeck JP & Bull JJ (1996). A likelihood ratio test to detect conflicting phylogenetic signal. *Systematic Biology* **45**, 92–98.

Huelsenbeck JP & Rannala B (1997). Phylogenetic methods come of age: testing hypotheses in an evolutionary context. *Science* **276**, 227–232.

Huelsenbeck JP, Bull JJ & Cunningham CW (1996a). Combining data in phylogenetic analysis. *Trends in Ecology and Evolution* **11**, 152–157.

Huelsenbeck JP, Hillis DM & Jones R (1996b). Parametric bootstrapping in molecular phylogenetics: applications and performance. In *Molecular Zoology: Advances, Strategies, and Protocols* (JD Ferraris & SR Palumbi, eds), pp. 19–45. Wiley–Liss, New York.

Huelsenbeck JP, Rannala B & Yang Z (1997). Statistical tests of host–parasite cospeciation. *Evolution* **51**, 410–419.

Hughes AL (1992). Positive selection and interallelic recombination at the merozoite surface antigen-1 MSA-1 of *Plasmodium falciparum*. *Molecular Biology and Evolution* **9**, 381–393.

Hughes AL & Nei M (1988). Pattern of nucleotide substitution at major histocompatibility complex class I loci reveals overdominant selection. *Nature* **335**, 367–370.

Hughes AL & Nei M (1989). Nucleotide substitution at major histocompatibility complex class II loci: evidence for overdominant selection. *Proceedings of the National Academy Sciences of the USA* **86**, 958–962.

Hughes MK & Hughes AL (1993). Evolution of duplicate genes in a tetraploid animal, *Xenopus laevis*. *Molecular Biology and Evolution* **10**, 1360–1369.

Hull DL (1988). *Science as a Process*. University of Chicago Press, Chicago.

Hurst LD (1994). The uncertain origin of introns. *Nature* **371**, 381–382.

Ikemura T (1985). Codon usage and tRNA content in unicellular and multicellular organisms. *Molecular Biology and Evolution* **2**, 13–34.

Janecek LL, Honeycutt RL, Adkins RM & Davis SK (1996). Mitochondrial gene sequences and the molecular systematics of the artiodactyl subfamily Bovinae. *Molecular Phylogenetics and Evolution* **6**, 107–119.

Jarne P & Lagoda PJL (1996). Microsatellites from molecules to populations and back. *Trends in Ecology and Evolution* **11**, 424–429.

Jeffreys AJ, Wilson V & Thein SL (1985). Hypervariable 'minisatellite' regions in human DNA. *Nature* **314**, 67–73.

Jermann TM, Opitz JG, Stackhouse J & Benner SA (1995). Reconstructing the evolutionary history of the artiodactyl ribonuclease superfamily. *Nature* **374**, 57–59.

John B & Miklos G (1988). *The Eukaryote Genome in Development and Evolution*. Allen and Unwin, London.

Jukes TH & Cantor CR (1969). Evolution of protein molecules. In *Mammalian Protein Metabolism III* (HN Munro, ed.), pp. 21–132. Academic Press, New York.

Jukes TH & Osawa S (1991). Recent evidence for evolution of the genetic code. In *Evolution of Life: Fossils, Molecules and Culture* (S Osawa & T Honjo, eds), pp. 79–95. Springer-Verlag, Tokyo.

Kaplan NL, Hudson RR & Langley CH (1989). The 'hitchhiking effect' revisited. *Genetics* **123**, 887–899.

Karl SA & Avise JC (1992). Balancing selection on allozyme loci in oysters: implications from nuclear RFLPs. *Science* **256**, 100–102.

Kasper P, Simmonds P, Schneweis KE, Kasier R, Matz B, Oldenburg J, Brackmann H-H & Holmes EC (1995). The genetic diversification of the HIV-1 *gag* p17 gene in patients infected from a common source. *AIDS Research and Human Retroviruses* **11**, 1197–1201.

Keeling PJ, Charlebois RL & Doolittle WF (1994). Archaebacterial genomes: eubacterial form and eukaryotic content. *Current Opinion in Genetics and Development* **4**, 816–822.

Keeling PJ, Charlebois RL & Doolittle WF (1996). A non-canonical genetic code in an early diverging eukaryotic lineage. *EMBO Journal* **15**, 2285–2290.

Kellam P, Boucher CAB, Tijnagel JMGH & Larder BA (1994). Zidovudine treatment results in the selection of human immunodeficiency virus type 1 variants whose genotypes confer increasing levels of drug resistance. *Journal of General Virology* **75**, 341–351.

Kendrew JC (ed.) (1994). *The Encyclopedia of Molecular Biology*. Blackwell Science, Oxford.

Kidwell MG (1993). Lateral transfer in natural populations of eukaryotes. *Annual Review of Genetics* **27**, 235–256.

Kidwell MG & Lisch D (1997). Transposable elements as sources of variation in animals and plants. *Proceedings of the National Academy of Sciences of the USA* **94**, 7704–7711.

Kimura M (1968). Evolutionary rate at the molecular level. *Nature* **217**, 624–626.

Kimura M (1969). The rate of evolution considered from the standpoint of population genetics. *Proceedings of the National Academy of Sciences of the USA* **78**, 454–458.

Kimura M (1980). A simple method for estimating evolutionary rates of base substitutions through comparative studies of nucleotide sequences. *Journal of Molecular Evolution* **16**, 111–120.

Kimura M (1983). *The Neutral Theory of Molecular Evolution*. Cambridge University Press, Cambridge.

Kimura M (1987). Molecular evolutionary clock and the neutral theory. *Journal of Molecular Evolution* **26**, 24–33.

Kimura M (1991). Recent development of the neutral theory viewed from the Wrightian tradition of theoretical population genetics. *Proceedings of the National Academy of Sciences of the USA* **88**, 5969–5973.

Kimura M & Ohta T (1969). The average number of generations until fixation of a mutant gene in a finite population. *Genetics* **61**, 763–771.

Kimura M & Ohta T (1971). On some principles governing molecular evolution. *Proceedings of the National Academy of Sciences of the USA* **71**, 2848–2852.

Kimura M & Ohta T (1974). Protein polymorphism as a phase of molecular evolution. *Nature* **229**, 467–469.

King JL & Jukes TH (1969). Non-Darwinian evolution: random fixation of selectively neutral mutations. *Science* **164**, 788–798.

Kingman JFC (1982). On the genealogy of large populations. *Journal of Applied Probability* **19A**, 27–43.

Kishino H & Hasegawa M (1989). Evaluation of the maximum likelihood estimate of the evolutionary tree topologies from DNA sequence data, and the branching order of the Hominoidea. *Journal of Molecular Evolution* **29**, 170–179.

Klein J, Gutknecht J & Fischer N (1990). The major histocompatibility complex and human evolution. *Trends in Genetics* **6**, 7–11.

Klein J & O'hUigin C (1993). Composite origin of major histocompatibility complex genes. *Current Opinion in Genetic Development* **3**, 923–930.

Kondrashov AS (1988). Deleterious mutations and the evolution of sexual reproduction. *Nature* **336**, 435–440.

Kornegay JR, Schilling JW & Wilson AC (1994). Molecular adaptation of a leaf-eating bird: stomach lysozyme of the hoatzin. *Molecular Biology and Evolution* **11**, 921–928.

Krappen C & Ruddle FH (1993). Evolution of a regulatory gene family: *HOM/HOX* genes. *Current Opinion in Genetics and Development* **3**, 931–938.

Kreitman M & Hudson RR (1991). Inferring the evolutionary histories of the *Adh* and *Adh-dup* loci in *Drosophila melanogaster* from patterns of polymorphism and divergence. *Genetics* **127**, 565–582.

Kreitman M & Akashi H (1995). Molecular evidence for natural selection. *Annual Review of Ecology and Systematics* **26**, 403–422.

Krings M, Stone A, Schmitz RW, Krainitzki H, Stoneking M & Pääbo S (1997). Neanderthal DNA sequences and the origin of modern humans. *Cell* **90**, 19–30.

Kuhner MK, Yamato J & Felsenstein J (1995). Estimating effective population size and mutation rate from sequence data using Metropolis–Hastings sampling. *Genetics* **140**, 1421–1430.

Kunkel TA (1992). Biological asymmetries and fidelity of eukaryotic DNA replication. *BioEssays* **14**, 303–309.

Lai C, Lyman RF, Long AD, Langley CH & Mackay TFC (1994). Naturally occurring variation in bristle number and DNA polymorphisms at the *scabrous* locus of *Drosophila melanogaster*. *Science* **266**, 1697–1702.

Lanave C, Preparata G, Saccone C & Serio G (1984). A new method for calculating evolutionary substitution rates. *Journal of Molecular Evolution* **20**, 86–93.

Langley CH & Fitch WM (1974). An examination of the constancy of the rate of molecular evolution. *Journal of Molecular Evolution* **3**, 161–177.

Langley CH, MacDonald J, Miyashita N & Aguadé M (1993). Lack of correlation between interspecific divergence and intraspecific polymorphism at the suppressor of forked region in *Drosophila melanogaster* and *Drosophila simulans*. *Proceedings of the National Academy of Sciences of the USA* **90**, 1800–1803.

Larder BA & Kemp SD (1989). Multiple mutations in HIV-1 reverse transcriptase confer high-level resistance to zidovudine (AZT). *Science* **246**, 1155–1158.

Lawrence JG & Ochman H (1997). Amelioration of bacteria genomes: rates of change and exchange. *Journal of Molecular Evolution* **44**, 383–397.

LeClerc JE, Li B, Payne WL & Cebula TA (1996). High mutation frequencies among *Escherichia coli* and *Salmonella* pathogens. *Science* **274**, 1208–1211.

Leeds JM, Slabaugh MB & Mathews CK (1985). DNA precursor pools and ribonuclease reductase activity: distribution between the nucleus and cytoplasm of mammalian cells. *Molecular and Cellular Biology* **5**, 3443–3450.

Leigh Brown AJ (1997). Analysis of HIV-1 *env* gene sequences reveals evidence for a low effective number in the viral population. *Proceedings of the National Academy of Sciences of the USA* **94**, 1862–1865.

Leigh Brown AJ & Holmes EC (1994). The evolutionary biology of human immunodeficiency virus. *Annual Review of Ecology and Systematics* **25**, 127–165.

Leitner T, Escanilla D, Franzen C, Uhlen M & Albert J (1996). Accurate reconstruction of a known HIV-1 transmission history. *Proceedings of the National Academy of Sciences of the USA* **93**, 10864–10869.

Lento GM, Hickson RE, Chambers GK & Penny D (1995). Use of spectral analysis to test hypotheses on the origin of pinnipeds. *Molecular Biology and Evolution* **12**, 28–52.

Lewontin RC (1974). *The Genetic Basis of Evolutionary Change*. Columbia University Press, New York.

Lewontin RC (1988). On measures of gametic disequilibrium. *Genetics* **120**, 849–852.

Lewontin RC (1991). Electrophoresis in the development of evolutionary genetics: milestone or millstone? *Genetics* **128**, 657–662.

Lewontin RC & Hubby JL (1966). A molecular approach to the study of genic heterozygosity in natural populations. II. Amount of variation and degree of heterozygosity in natural populations of *Drosophila pseudoobscura*. *Genetics* **54**, 595–609.

Li D & Stevenson KJ (1997). Purification and sequence analysis of a novel NADP(H)-dependent type III alcohol dehydrogenase from *Thermococcus* strain AN1. *Journal of Bacteriology* **179**, 4433–4437.

Li W-H (1993). So, what about the molecular clock hypothesis? *Current Opinion in Genetics and Development* **3**, 896–901.

Li W-H (1997). *Molecular Evolution*. Sinauer Associates, Sunderland MA.

Li W-H & Tanimura M (1987). The molecular clock runs more slowly in man than in apes and monkeys. *Nature* **326**, 93–96.

Li W-H & Graur D (1991). *Fundamentals of Molecular Evolution*. Sinauer Associates, Sunderland MA.

Li W-H & Sadler LA (1991). Low nucleotide diversity in man. *Genetics* **129**, 513–523.

Li W-H, Gojobori T & Nei M (1981). Pseudogenes as a paradigm of neutral evolution. *Nature* **292**, 237–239.

Li W-H, Wu C-I & Luo C-C (1985). A new method for estimating synonymous and non-synonymous rates of nucleotide substitution considering the relative likelihood of nucleotide and codon changes. *Molecular Biology and Evolution* **2**, 150–174.

Li W-H, Tanimura M & Sharp PM (1988). Rates and dates of divergence between AIDS virus nucleotide sequences. *Molecular Biology and Evolution* **5**, 313–330.

Li W-H, Tanimura M & Sharp PM (1987). An evaluation of the molecular clock hypothesis using mammalian DNA sequences. *Journal of Molecular Evolution* **25**, 330–342.

Li W-H, Gouy M, Sharp PM, O'hUigin C & Yang Y-W (1990). Molecular phylogeny of Rodentia, Lagomorpha, Primates, Artiodactyla, and Carnivora and molecular clocks. *Proceedings of the National Academy of Sciences of the USA* **87**, 6703–6707.

Lockhart PJ, Steel MA, Hendy MD & Penny D (1994). Recovering evolutionary trees under a more realistic model of sequence evolution. *Molecular Biology and Evolution* **11**, 605–612.

Logan NA (1994). *Bacterial Systematics*. Blackwell Scientific Publications, Oxford.

Logsdon JM Jr & Palmer JD (1994). Origins of introns—early or late? *Nature* **369**, 526.

Long M & Langley CH (1993). Natural selection and the origin of *jingwei*, a chimeric processed functional gene in *Drosophila*. *Science* **260**, 91–95.

Lynch M & Jarrell PE (1993). A method for calibrating molecular clocks and its application to animal mitochondrial DNA. *Genetics* **135**, 1197–1208.

Maddison DR (1991). The discovery and importance of multiple islands of most parsimonious trees. *Systematic Zoology* **40**, 315–328.

Maddison WP (1989). Reconstructing character evolution on polytomous cladograms. *Cladistics* **5**, 365–377.

Maddison WP & Maddison DR (1992). *MacClade: Analysis of Phylogeny and Character Evolution. Version 3.0*. Sinauer Associates, Sunderland, MA.

Maddison DR, Ruvolo M & Swofford DL (1992). Geographic origins of human mitochondrial DNA—phylogenetic evidence from control region sequences. *Systematic Biology* **41**, 111–124.

Marjoram P & Donnelly P (1994). Pairwise comparisons of mitochondrial DNA sequences in subdivided populations and implications for early human evolution. *Genetics* **136**, 673–683.

Martin AP, Naylor GJP & Palumbi SR (1992). Rates of mitochondrial DNA evolution in sharks are slow compared with mammals. *Nature* **357**, 153–155.

Martin AP & Palumbi SR (1993a). Protein evolution in different cellular environments: cytochrome *b* in sharks and mammals. *Molecular Biology and Evolution* **10**, 873–891.

Martin AP & Palumbi SR (1993b). Body size, metabolic rate, generation time, and the molecular clock. *Proceedings of the National Academy of Sciences of the USA* **90**, 4087–4091.

Martín-Campos JM, Comerón JM, Miyashita N & Aguadé M (1992). Intraspecific and interspecific variation at the *y-ac-sc* region of *Drosophila simulans* and *Drosophila melanogaster. Genetics* **130**, 805–816.

Mattick JS (1994). Introns: evolution and function. *Current Opinion in Genetics and Development* **4**, 823–831.

Maynard Smith J (1968). 'Haldane's dilemma' and the rate of evolution. *Nature* **219**, 1114–1116.

Maynard Smith J (1989). *Evolutionary Genetics*. Oxford University Press, Oxford.

Maynard Smith J & Szathmáry E (1995). *The Major Transitions of Evolution*. W.H. Freeman & Co., Oxford.

Maynard Smith J, Smith NH, O'Rourke M & Spratt BG (1993). How clonal are bacteria? *Proceedings of the National Academy of Sciences of the USA* **90**, 4384–4388.

McClearn GE, Johansson B, Berg S, Pedersen NL, Ahern F, Petrill SA & Plomin R (1997). Substantial genetic influence on cognitive ability in twins 80 or more years old. *Science* **276**, 1560–1563.

McDonald JF (1993). Evolution and consequences of transposable elements. *Current Opinion in Genetics and Development* **3**, 855–864.

McDonald JH & Kreitman M (1991a). Adaptive evolution at the *Adh* locus in *Drosophila. Nature* **351**, 652–654.

McDonald JH & Kreitman M (1991b). Neutral mutation hypothesis test. *Nature* **354**, 116.

McGeoch DJ, Cook S, Dolan A, Jamieson FE & Telford EAR (1995). Molecular phylogeny and evolutionary timescale for the family of mammalian herpesviruses. *Journal of Molecular Biology* **247**, 443–458.

McVean GT & Hurst LD (1997). Evidence for a selectively favourable reduction in the mutation rate of the X chromosome. *Nature* **386**, 388–392.

Mindell DP (ed.) (1997). *Avian Molecular Evolution and Systematics*. Academic Press, San Diego.

Mirkin B, Muchnik I & Smith TF (1995). A biologically consistent model for comparing molecular phylogenies. *Journal of Computational Biology* **2**, 493–507.

Mitchell-Olds T (1995). The molecular basis of quantitative genetic variation in natural populations. *Trends in Ecology and Evolution* **10**, 324–328.

Mooers AØ & Harvey PH (1994). Metabolic rate, generation time, and the rate of molecular evolution in birds. *Molecular Phylogenetics and Evolution* **3**, 344–350.

Mooers AØ, Nee S & Harvey PH (1994). Biological and algorithmic correlates of phenetic tree pattern. In *Phylogenetics and Ecology* (P Eggleton & RI Vane-Wright, eds), pp. 233–251. Academic Press, London.

Moran NA, Munson MA, Baumann P & Ishikawa H (1993). A molecular clock in endosymbiotic bacteria is calibrated using insect hosts. *Proceedings of the Royal Society of London B* **253**, 167–171.

Moran NA, van Dohlen CD & Baumann P (1995). Faster evolutionary rates in endosymbiotic bacteria than in cospeciating insect hosts. *Journal of Molecular Evolution* **41**, 727–731.

Moriyama EN & Gojobori T (1992). Rates of synonymous substitution and base composition of nuclear genes in *Drosophila*. *Genetics* **130**, 855–864.

Moriyama EN & Hartl DL (1993). Codon usage bias and base composition of nuclear genes in *Drosophila*. *Genetics* **134**, 847–858.

Moriyama EN & Powell JR (1997a). Synonymous substitution rates in *Drosophila* mitochondrial versus nuclear genes. *Journal of Molecular Evolution* **45**, 378–391.

Moriyama EN & Powell JR (1997b). Codon usage bias and tRNA abundance in *Drosophila*. *Journal of Molecular Evolution* **45**, 514–523.

Morell V (1996). Life's last domain. *Science* **273**, 1043–1045.

Mouchiroud D, Gautier C & Bernardi G (1988). The compositional distribution of coding sequences and DNA molecules in humans and murids. *Journal of Molecular Evolution* **27**, 311–320.

Mouchiroud D, Gautier C & Bernardi G (1995). Frequencies of synonymous substitutions in mammals are gene-specific and correlated with frequencies of nonsynonymous substitutions. *Journal of Molecular Evolution* **40**, 107–113.

Murti JP, Bumbulis M & Schimenti JC (1994). Gene conversion between unlinked sequences in the germline of mice. *Genetics* **137**, 837–843.

Nakamura Y, Wada K, Wada Y, Doi H, Kanaya S, Gojobori T & Ikemura T (1996). Codon usage tabulated from the international DNA sequence databases. *Nucleic Acids Research* **24**, 214–215.

Nee S, Holmes EC, May RM & Harvey PH (1994a). Extinction rates can be estimated from molecular phylogenies. *Philosophical Transactions of the Royal Society B* **344**, 77–82.

Nee S, May RM & Harvey PH (1994b). The reconstructed evolutionary process. *Philosophical Transactions of the Royal Society, London* **344**, 305–311.

Nee S, Holmes EC, Rambaut A & Harvey PH (1995). Inferring population history from molecular phylogenies. *Philosophical Transactions of the Royal Society, London* **349**, 25–31.

Nee S, Mooers AØ & Harvey PH (1992). Tempo and mode of evolution revealed from molecular phylogenies. *Proceedings of the National Academy of Sciences of the USA* **89**, 8322–8326.

Nei M (1987). *Molecular Evolutionary Genetics*. Columbia University Press, New York.

Nei M (1992). Age of the common ancestor of human mitochondrial DNA. *Molecular Biology and Evolution* **9**, 1176–1178.

Nei M (1995). Genetic support for the out-of-Africa theory of human evolution. *Proceedings of the National Academy of Sciences of the USA* **92**, 6720–6722.

Nei M & Graur D (1984). Extent of protein polymorphism and the neutral mutation theory. *Evolutionary Biology* **3**, 418–426.

Nelson G & Platnick NI (1981). *Systematics and Biogeography: Cladistics and Vicariance*. Columbia University Press, New York.

Nevo E (1978). Genetic diversity in nature—patterns and theory. *Evolutionary Biology* **23**, 217–246.

Nichol ST, Spiropoulou CF, Morzunov S, Rollin PE, Ksiazek TG, Feldmann H, Sanchez A, Childs J, Zaki S & Peters CJ (1993). Genetic identification of a hantavirus associated with an outbreak of acute respiratory illness. *Science* **262**, 914–917.

Nixon KC & Carpenter JM (1996). On simultaneous analysis. *Cladistics* **12**, 221–241.

Noor MA (1995). Speciation driven by natural selection in *Drosophila*. *Nature* **375**, 674–675.

Novacek MJ & Norell MA (1982). Fossils, phylogeny, and taxonomic rates of evolution. *Systematic Zoology* **31**, 366–375.

Nuttall GHF (1904). *Blood Immunity and Blood Relationship*. Cambridge University Press, Cambridge.

O'Brien SJ & Dean M (1997). In search of AIDS-resistance genes. *Scientific American* **277**, 28–35.

Ochman H & Wilson AC (1987). Evolution in bacteria: evidence for a universal substitution rate in cellular genomes. *Journal of Molecular Evolution* **26**, 74–86.

Ohta T (1973). Slightly deleterious mutant substitutions in evolution. *Nature* **246**, 96–97.

Ohta T (1987). Very slightly deleterious mutations and the molecular clock. *Journal of Molecular Evolution* **26**, 1–6.

Ohta T (1992). The nearly neutral theory of molecular evolution. *Annual Review of Ecology and Systematics* **23**, 263–286.

Ohta T (1993a). Amino acid substitution at the *Adh* locus of *Drosophila* is facilitated by small population size. *Proceedings of the National Academy of Sciences of the USA* **90**, 4548–4551.

Ohta T (1993b). An examination of the generation-time effect on molecular evolution. *Proceedings of the National Academy of Sciences of the USA* **90**, 10676–10680.

Ohta T (1995). Synonymous and nonsynonymous substitutions in mammalian genes and the nearly neutral theory. *Journal of Molecular Evolution* **40**, 56–63.

Ohta T & Kimura M (1971). On the constancy of the evolutionary rate in cistrons. *Journal of Molecular Evolution* **1**, 18–25.

Ohta T & Tachida H (1990). Theoretical study of near neutrality. I. Heterozygosity and rate of mutant substitution. *Genetics* **126**, 219–229.

O'hUigin C & Li W-H (1992). The molecular clock ticks regularly in muroid rodents and hamsters. *Journal of Molecular Evolution* **35**, 377–384.

Olsen GJ & Woese CR (1993). Ribosomal RNA: a key to phylogeny. *Federation of American Societies for Experimental Biology* **7**, 113–123.

Olsen GJ, Woese CR & Overbeek R (1994). The winds of (evolutionary) change: breathing new life into microbiology. *Journal of Bacteriology* **176**, 1–6.

Orgel LE & Crick FHC (1980). Selfish DNA: the ultimate parasite. *Nature* **284**, 604–607.

Orr HA (1995). The population genetics of speciation: the evolution of hybrid incompatibilities. *Genetics* **139**, 1805–1813.

Orr HA & Coyne JA (1992). The genetics of adaptation: a reassessment. *American Naturalist* **140**, 725–742.

Ou C-Y, CiesielskI CA, Myers G, Bandea CI, Luo C-C, Korber BTM, Mullins JI, Schochetman G, Berkelman RL, Economou AN, Witte JJ, Furman LJ, Satten GA, MacInnes KA, Curran JW, Jaffe HW, Laboratory Investigation Group & Epidemiologic Investigation Group (1992). Molecular epidemiology of HIV transmission in a dental practice. *Science* **256**, 1165–1171.

Page RDM (1994). Maps between trees and cladistic analysis of historical associations among genes, organisms, and areas. *Systematic Biology* **43**, 58–77.

Pate RDM (1996a). On consensus, confidence, and 'total' evidence. *Cladistics* **12**, 83–92.

Page RDM (1996b) Temporal congruence revisited: comparison of mitochondrial DNA sequence divergence in cospeciating pocket gophers and their chewing lice. *Systematic Biology* **45**, 151–167.

Page RDM & Charleston MA (1997a). From gene to organismal phylogeny: reconciled trees and the gene tree/species tree problem. *Molecular Phylogenetics and Evolution* **7**, 231–240.

Page RDM & Charleston MA (1997b). Reconciled trees and incongruent gene and species trees. In *Mathematical Hierarchies in Biology*, Volume 37 (B Mirkin, FR McMorris, FS Roberts & A Rzhetsky, eds). American Mathematical Society, Providence, RI.

Page RDM, Clayton DH & Paterson AM (1996). Lice and cospeciation: a response to Barker. *International Journal of Parasitology* **26**, 213–218.

Palmer JD (1997). The mitochondrion that time forgot. *Nature* **387**, 454–455.

Palmer JD & Logsdon JM Jr (1991). The recent origins of introns. *Current Opinion in Genetic Development* **1**, 470–477.

Palumbi SR (1989). Rates of molecular evolution and the fraction of nucleotide positions free to vary. *Journal of Molecular Evolution* **29**, 180–187.

Pamilo P & Nei M (1988). Relationships between gene trees and species trees. *Molecular Biology and Evolution* **5**, 568–583.

Paterson AM & Gray RD (1997). Host–parasite cospeciation, host switching, and missing the boat. In *Host–Parasite Evolution: General Principles and Avian Models*, (DH Clayton & J Moore, eds), pp. 236–250. Oxford University Press, Oxford.

Penny D (1982). Towards a basis for classification: the incompleteness of distance measures, incompatibility analysis and phenetic classification. *Journal of Theoretical Biology* **96**, 129–142.

Penny D, Hendy MD & Steel MA (1992). Progress with methods for constructing evolutionary trees. *Trends in Ecology and Evolution* **7**, 73–79.

Penny D, Watson EE & Steel MA (1993). Trees from languages and genes are very similar. *Systematic Biology* **42**, 382–384.

Perutz MF (1983). Species adaptation in a protein molecule. *Molecular Biology and Evolution* **1**, 1–28.

Pesole G, Sbisá E, Preparata G & Saccone C (1992). The evolution of the mitochondrial D-loop region and the origin of modern man. *Molecular Biology and Evolution* **9**, 587–598.

Philippe H, Sörhannus U, Baroin A, Perasso R, Gasse F & Adoutte A (1994). Comparison of molecular and palaeontological data in diatoms suggests a major gap in the fossil record. *Journal of Evolutionary Biology* **7**, 247–264.

Piatigorsky J & Wistow G (1991). The recruitment of crystallins: new functions procede gene duplication. *Science* **252**, 1078–1079.

Poinar G & Poinar R (1995) *The Quest for Life in Amber*. Addison-Wesley, Reading, Ma.

Potts WK & Wakeland EK (1993). Evolution of MHC genetic diversity: a table of incest, pestilence and sexual preference. *Trends in Genetics* **9**, 408–412.

Powell JR & Moriyama EN (1997). Evolution of codon usage bias in *Drosophila*. *Proceedings of the National Academy of Sciences of the USA* **94**, 7784–7790.

Power JP, Lawlor E, Davidson F, Holmes EC, Yap PL & Simmonds P (1995). Molecular epidemiology of an outbreak of infection with hepatitis C virus in recipients of anti-D immunoglobulin. *Lancet* **345**, 1211–1213.

Preston C & Engels WR (1996). *P*-element-induced male recombination and gene conversion in *Drosophila*. *Genetics* **144**, 1611–1622.

Quattro JM, Woods HA & Powers DA (1993). Sequence analysis of teleost retina-specific lactate dehydrogenase C: evolutionary implications for the vertebrate lactate dehydrogenase gene family. *Proceedings of the National Academy of Sciences of the USA* **90**, 242–246.

Rand DM (1994). Thermal habitiat, metabolic rate and the evolution of mitochondrial DNA. *Trends in Ecology and Evolution* **9**, 125–131.

Reynaud C-A, Dahan A & Weill J-C (1987). A gene conversion program during the ontogenesis of chicken B cells. *Trends in Genetics* **3**, 248–251.

Ridley M (1986). *Evolution and Classification: The Reformation of Cladism*. Longman, London.

Ridley M (1996). *Evolution* (2nd edn). Blackwell Science, Cambridge, MA.

Rivera MC & Lake JA (1992). Evidence that eukaryotes and eocyte prokaryotes are immediate relatives. *Science* **257**, 74–76.

Rodríguez F, Oliver JL, Marín A & Medina JR (1990). The general stochastic model of nucleotide substitution. *Journal of Theoretical Biology* **142**, 485–501.

Roger AJ & Doolittle F (1993). Why introns-in-pieces. *Nature* **364**, 289–290.

Ruvolo M (1996). A new approach to studying modern human origins: hypothesis testing with coalescence time distributions. *Molecular Phylogenetics and Evolution* **5**, 202–219.

Ruvolo M (1997). Molecular phylogeny of the hominoids: inferences from multiple independent DNA sequence data sets. *Molecular Biology and Evolution* **14**, 248–265.

Ruwende C, Khoo SC, Snow RW, Yates SNR, Kwiatkowski D, Gupta S, Warn P, Allsopp CEM, Gilbert SC, Peschu N, Newbold CI, Greenwood BM, Marsh K & Hill AVS (1995). Natural selection of hemi- and heterozygotes for G6PD deficiency in Africa by resistance to severe malaria. *Nature* **376**, 246–249.

Saccone C, Pesole G & Preparata G (1989). DNA microenvironments and the molecular clock. *Journal of Molecular Evolution* **29**, 407–411.

Sanderson MJ (1995). Objections to bootstrapping phylogenies: a critique. *Systematic Biology* **44**, 299–320.

Sarich VM & Wilson AC (1967). Immunological time scale for hominoid evolution. *Science* **158**, 1200–1203.

Schliewen UK, Tautz D & Pääbo S (1994). Sympatric speciation suggested by monophyly of crater lake cichlids. *Nature* **368**, 629–632.

Scholtissek C (1987). Molecular aspects of the epidemiology of virus disease. *Experientia* **43**, 1197–1201.

Sharp PM & Li W-H (1989). On the rate of DNA sequence evolution in *Drosophilia. Journal of Molecular Evolution* **28**, 398–402.

Sharp PM & Matassi G (1994). Codon usage and genome evolution. *Current Opinion in Genetic Development* **4**, 851–860.

Sharp PM, Robertson DL, Gao F & Hahn BH (1994). Origins and diversity of human immunodeficiency viruses. *AIDS* **8**, S27–S42.

Sharp PM, Robertson DL & Hahn BH (1995). Cross-species transmission and recombination of 'AIDS' viruses. *Philosophical Transactions of the Royal Society of London B* **349**, 41–47.

Shields DC, Sharp PM, Higgins DG & Wright F (1988). 'Silent' sites in *Drosophila* genes are not neutral: evidence of selection among synonymous codons. *Molecular Biology of Evolution* **5**, 704–717.

Shimmin LC, Chang BH-J & Li W-H (1993). Male-driven evolution of DNA sequences. *Nature* **362**, 745–747.

Shimmin LC, Chang BH-J & Li W-H (1994). Contrasting rates of nucleotide substitution in the X-linked and Y-linked zinc finger genes. *Journal of Molecular Evolution* **39**, 569–578.

Shoemaker DD, Ross KG & Arnold ML (1996). Genetic structure and evolution of a fire ant hybrid zone. *Evolution* **50**, 1958–1976.

Shpaer EG & Mullins JI (1993). Rates of amino acid change in the envelope protein correlate with pathogenicity of primate lentiviruses. *Journal of Molecular Evolution* **37**, 57–65.

Shuster SM & Wade MJ (1991). Equal mating success among male reproductive strategies in a marine isopod. *Nature* **350**, 608–610.

Slatkin M & Maddison WP (1989a). A cladistic measure of gene flow inferred from phylogenies of alleles. *Genetics* **123**, 603–613.

Slatkin M & Maddison WP (1989b). Detecting isolation by distance using phylogenies of genes. *Genetics* **126**, 249–260.

Slatkin M & Hudson RR (1991). Pairwise comparisons of mitochondrial DNA sequences in stable and exponentially growing populations. *Genetics* **129**, 555–562.

Slightom JL, Chang L-YE, Koop BF & Goodman M (1985). Chimpanzee fetal $^{G}\gamma$ and $^{A}\gamma$ globin gene nucleotide sequences provide further evidence of gene conversions in hominine evolution. *Molecular Biology and Evolution* **2**, 370–389.

Slightom JL, Koop BF, Xu P & Goodman M (1988). Rhesus fetal globin genes. *Journal of Biological Chemistry* **263**, 12427–12438.

Smith AB (1994). *Systematics and the Fossil Record: Documenting Evolutionary Patterns*. Blackwell Scientific Publications, Oxford.

Smith TB (1993). Disruptive selection and the genetic basis of bill size polymorphism in the African finch *Pyrenestes*. *Nature* **363**, 618–620.

Sniegowski PD, Gerrish PJ & Lenski RE (1997). Evolution of high mutation rates in experimental populations of *E. coli*. *Nature* **387**, 703–705.

Sober E (1988). *Reconstructing the Past: Parsimony, Evolution, and Inference*. MIT Press, Cambridge, MA.

Sober E (1993). Experimental test of phylogenetic methods. *Systematic Biology* **42**, 85–89.

Spratt BG, Zhang Q-Y, Jones DM, Hutchison A, Brannigan JA & Dowson CG (1989). Recruitment of a penicillin-binding protein gene from *Neisseria flavescens* during the emergence of penicillin resistance in *Neisseria meningitidis*. *Proceedings of the National Academy of Sciences of the USA* **86**, 8988–8992.

Spratt BG, Bowler LD, Zhang Q-Y, Zhou J & Maynard Smith J (1992). Role of interspecies transfer of chromosomal genes in the evolution of penicillin resistance in pathogenic and commensal *Neisseria* species. *Journal of Molecular Evolution* **34**, 115–125.

Steel M (1994). The maximum likelihood tree is not unique. *Systematic Biology* **43**, 560–564.

Stephan W & Mitchell SJ (1992). Reduced levels of DNA polymorphism and fixed between-population differences in the centromeric region of *Drosophila ananassae*. *Genetics* **132**, 1039–1045.

Stewart C-B, Schilling JW & Wilson AC (1987). Adaptive evolution in the stomach lysozymes of foregut fermenters. *Nature* **330**, 401–404.

Stringer CB & Andrews P (1988). Genetic and fossil evidence for the origin of modern humans. *Science* **239**, 1263–1268.

Stryer L (1983). *Biochemistry*. WH Freeman & Company, New York.

Sueoka N (1988). Directional mutation pressure and neutral molecular evolution. *Proceedings of the National Academy of Sciences of the USA* **85**, 2653–2657.

Sueoka N (1992). Directional mutation pressure, selective constraints and genetic equilibria. *Journal of Molecular Evolution* **34**, 95–114.

Sullivan DT (1995). DNA excision repair and transcription: implications for genome evolution. *Current Opinion in Genetics and Development* **5**, 786–791.

Swofford DL (1991). When are phylogeny estimates from molecular and morphological data incongruent? In *Phylogenetic Analysis of DNA Sequences* (MM Miyamoto & J Cracraft, eds), pp. 295–333. Oxford University Press, New York.

Swofford DL, Olsen GJ, Waddell PJ & Hillis DM (1996). Phylogenetic inference. In *Molecular Systematics* (2nd edn) (DM Hillis, C Moritz & BK Mable, eds), pp. 407–514. Sinauer Associates, Sunderland, MA.

Taddei F, Radman M, Maynard Smith J, Toupance B, Gouyon PH & Godelle B (1997). Role of mutator alleles in adaptive evolution. *Nature* **387**, 700–702.

Tajima F (1983). Evolutionary relationship of DNA sequences in finite populations. *Genetics* **105**, 437–460.

Tajima F (1989). Statistical method for testing the neutral mutation hypothesis by DNA polymorphism. *Genetics* **123**, 585–595.

Takahata N (1993). Allelic genealogy and human evolution. *Molecular Biology and Evolution* **10**, 2–22.

Takahata N & Nei N (1990). Allelic genealogy under overdominant and frequency-dependent selection and polymorphism of major histocompatibility complex loci. *Genetics* **124**, 967–978.

Takahata N, Satta Y & Klein J (1992). Polymorphism and balancing selection at major histocompatibility complex loci. *Genetics* **130**, 925–938.

Tamura K (1994). Model selection in the estimation of the number of nucleotide substitutions. *Molecular Biology and Evolution* **11**, 154–157.

Tatusov RL, Koonin EV & Lipman DJ (1997). A genomic perspective on protein families. *Science* **278**, 631–637.

Tautz D & Schlötterer C (1994). Simple sequences. *Current Opinion in Genetics and Development* **4**, 832–837.

Tautz D, Trick M & Dover GA (1986). Cryptic simplicity in DNA is a major source of genetic variation. *Nature* **322**, 652–656.

Taylor MJF, Shen Y & Kreitman ME (1995). A population genetic test of selection at the molecular level. *Science* **270**, 1497–1499.

Templeton AR (1991). Human origins and analysis of mitochondrial DNA sequences. *Science* **255**, 737.

Thomasson HR, Edenberg HJ, Crabb DW, Mai X-L, Jerome RE, Li T-K, Wang S-P, Lin Y-T, Lu R-B & Yin S-J (1991). Alcohol and aldehyde dehydrogenase genotypes and alcoholism in Chinese men. *American Journal of Human Genetics* **48**, 677–681.

Thompson CB (1992). Creation of immunoglobin diversity by intrachromosomal gene conversion. *Trends in Genetics* **8**, 416–422.

Ticher A & Graur D (1989). Nucleic acid composition, codon usage, and the rate of synonymous substitution in protein-coding genes. *Journal of Molecular Evolution* **28**, 286–298.

Tsui L-C (1992). The spectrum of cystic fibrosis mutations. *Trends in Genetics* **8**, 392–398.

Valdes AM, Slatkin M & Freimer NB (1993). Allele frequencies at microsatellite loci: the stepwise mutation model revisited. *Genetics* **133**, 737–749.

Vigilant L, Stoneking M, Harpending H, Hawkes K & Wilson AC (1991). African populations and the evolution of human mitochondrial DNA. *Science* **253**, 1503–1507.

Wakeley J (1996). The excess of transitions among nucleotide substitutions: new methods of estimating transition bias underscore its significance. *Trends in Ecology and Evolution* **11**, 158–163.

Walsh BJ (1987). Sequence-dependent gene conversion: can duplicated genes diverge fast enough to escape conversion? *Genetics* **117**, 543–557.

Ward RH, Frazier BL, Dew-Jager K & Pääbo S (1991). Extensive mitochondrial diversity within a single Amerindian tribe. *Proceedings of the National Academy of Sciences of the USA* **88**, 8720–8724.

Watson JD, Hopkins NH, Roberts JW, Steitz JA & Weiner AM (1987). *Molecular Biology of the Gene* (4th edn). The Benjamin/Cummings Publishing Company Inc., Menlo Park, CA.

Weatherall DJ (1991). *The New Genetics and Clinical Practice* (3rd edn). Oxford University Press, Oxford.

Webster RG, Bean WJ & Gorman OT (1995). Evolution of influenza viruses: rapid evolution and stasis. In *Molecular Basis of Virus Evolution* (A Gibbs, CH Calisher & García-Arenal F, eds), pp. 531–543. Cambridge University Press, Cambridge.

Weill J-C & Reynaud C-A (1996). Rearrangement/hypermutation/gene conversion: when, where and why? *Immunology Today* **17**, 92–97.

Wevrick R & Buchwald M (1993). Mammalian DNA-repair genes. *Current Opinion in Genetics and Development* **3**, 470–474.

Wheeler W (1996). Optimization alignment: the end of multiple sequence alignment in phylogenetics? *Cladistics* **12**, 1–9.

Wilkinson M (1996). Majority-rule reduced consensus trees and their use in bootstrapping. *Molecular Biology and Evolution* **13**, 437–444.

Whittam TS & Nei M (1991). Neutral mutation hypothesis test. *Nature* **354**, 115–116.

Wilson AC, Carlson SS & White TJ (1977). Biochemical evolution. *Annual Review of Biochemistry* **46**, 573–639.

Wolfe KH & Shields DC (1997). Molecular evidence for an ancient duplication of the entire yeast genome. *Nature* **387**, 708–713.

Wolfe KH, Li W-H & Sharp WH (1987). Rates of nucleotide substitution vary greatly among plant mitochondrial, chloroplast and nuclear DNAs. *Proceedings of the National Academy of Sciences of the USA* **84**, 9054–9058.

Wolfe KH, Sharp WH & Li W-H (1989). Mutation rates differ among regions of the mammalian genome. *Nature* **337**, 283–285.

Woodwark M, Skibinski DOF & Ward RD (1992). A study of interlocus allozyme heterozygosity correlations: implications for neutral theory. *Hereditary* **69**, 190–198.

Woodwark M, Skibinski DOF & Ward RD (1993). Analysis of distributions of single-locus heterozygosity as a test of neutral theory. *Genetics Research* **62**, 223–230.

Wray GA, Levinton JS & Shapiro LH (1996). Molecular evidence for deep Precambrian divergences among metazoan phyla. *Science* **274**, 568–573.

Wu C-I & Li W-H (1985). Evidence for higher rates of nucleotide substitution in rodents than in man. *Proceedings of the National Academy of Sciences of the USA* **82**, 1741–1745.

Yang Z (1996). Among-site variation and its impact on phylogenetic analyses. *Trends in Ecology and Evolution* **11**, 367–371.

Yang Z, Goldman N & Friday A (1994). Comparison of models for nucleotide substitution used in maximum likelihood phylogenetic estimation. *Molecular Biology and Evolution* **11**, 316–324.

Yuhki N & O'Brien SJ (1990). DNA variation of the mammalian major histocompatibility complex reflects genomic diversity and population history. *Proceedings of the National Academy of Sciences of the USA* **87**, 836–840.

Zharkikh A (1994). Estimation of evolutionary distances between nucleotide sequences. *Journal of Molecular Evolution* **39**, 315–329.

Zuckerkandl E (1987). On the molecular evolutionary clock. *Journal of Molecular Evolution* **26**, 34–46.

Zukerkandl E & Pauling L (1965). Evolutionary divergence and convergence in proteins. In *Evolving Genes and Proteins* (V Bryson & HJ Vogel, eds), pp. 97–166. Academic Press, New York.

Index

Page numbers in *italics* refer to figures